ENVIRONMENTAL TECHNOLOGY AND ENGINEERING TECHNIQUES

Basic Concepts and Health Interventions

ENVIRONMENTAL TECHNOLOGY AND ENGINEERING TECHNIQUES

Basic Concepts and Health Interventions

Edited by

Moayad N. Khalaf, PhD

Michael Olegovich Smirnov, PhD

Porteen Kannan, PhD

A. K. Haghi, PhD

AAP | APPLE ACADEMIC PRESS

Apple Academic Press Inc.
4164 Lakeshore Road
Burlington ON L7L 1A4
Canada

Apple Academic Press, Inc.
1265 Goldenrod Circle NE
Palm Bay, Florida 32905
USA

Exclusive worldwide distribution by CRC Press, a member of Taylor & Francis Group

International Standard Book Number-13: 978-1-77188-849-3 (Hardcover)
International Standard Book Number-13: 978-0-42932-571-7 (eBook)

Library and Archives Canada Cataloguing in Publication

Title: Environmental technology and engineering techniques : basic concepts and health interventions / edited by Moayad N. Khalaf, PhD, Michael Olegovich Smirnov, PhD, Porteen Kannan, PhD, A.K. Haghi, PhD.
Names: Khalaf, Moayad N., editor. | Smirnov, Michael Olegovich, editor. | Kannan, Porteen, editor. | Haghi, A. K., editor.
Description: Includes bibliographical references and index.
Identifiers: Canadiana (print) 20200216864 | Canadiana (ebook) 20200216899 | ISBN 9781771888493 (hardcover) | ISBN 9780429325717 (ebook)
Subjects: LCSH: Green technology.
Classification: LCC TD145 .E62 2020 | DDC 628—dc23

CIP data on file with US Library of Congress

Apple Academic Press also publishes its books in a variety of electronic formats. Some content that appears in print may not be available in electronic format. For information about Apple Academic Press products, visit our website at **www.appleacademicpress.com** and the CRC Press website at **www.crcpress.com**

About the Editors

Moayad N. Khalaf, PhD

Professor of Polymer Chemistry, Chemistry Department, College of Science, University of Basrah, Basrah, Iraq

Moayad N. Khalaf, PhD, is a Professor of Polymer Chemistry in the Department of Chemistry, College of Science at the University of Basrah in Iraq. Professor Khalaf has more than 27 years of professional experience in the petrochemical industry, working with the state company for petrochemical industries in Iraq. In 2005, he joined the Chemistry Department at the University of Basrah where now he is lecturing on most polymer related subjects. Dr. Khalaf has supervised more than 12 MSc and 4 PhD students. He holds 19 Iraqi patents and has published more than 100 scientific papers in peer-reviewed conference and journals. He also works on modifying the chemical security and safety strategy at the department of chemistry of the University of Basrah through funds from CRDF Global. Dr. Khalaf also received funds totaling $350,000 US to date from the Arab Science and Technology Foundation and the Iraqi Ministry of Higher Education and Scientific research to support his research in the fields of polymer chemistry and water treatment. His research interests include modified polymer for corrosion inhibitor, demulsifier, additive for oil; polymer additives (light stabilizer and antioxidant); lubricants; waste polymers recycling; lingnosulfonate for well drilling; mechanical properties of composite polymer; rheological properties of polymer and composite using nanofiller; water desalination; preparation of flocculent from waste polystyrene; ground water as source for industrial water and agriculture; preparation of polymer for solar cell application; and preparation of conductive polymer.

Professor Khalaf received his BSc in Chemistry Science, MSc in Physical-Organic Chemistry, and PhD in Polymer Chemistry from the University of Basrah in Iraq.

Michael Olegovich Smirnov, PhD

Senior Researcher of "Federal State Pryanishnikov All-Russian Scientific Research Institute of Agrochemistry, Moscow, Russia"

Michael Olegovich Smirnov, PhD, is currently a senior research fellow of Federal State Pryanishnikov All-Russian Scientific Research Institute of

Agrochemistry. He is an expert in plant physiology and agrochemistry. He has published 50 original scientific papers, reviews, and registered patents for inventions in these areas.

Porteen Kannan, PhD

Porteen Kannan, PhD, is an Assistant Professor in the Department of Veterinary Public Health at Madras Veterinary College, Tamil Nadu Veterinary and Animal Sciences University, India. The research activities of Dr. Kannan include food safety and anti-microbial resistance. He performed his postdoctoral studies from the US Department of Agriculture, Maryland, USA, with a specialization on foodborne pathogens. He has published his work in both national and international journals. He is actively involved in mentoring both MVSc and PhD students.

A. K. Haghi, PhD

Editor-in-Chief, International Journal of Chemoinformatics and Chemical Engineering and Polymers Research Journal;
Member, Canadian Research and Development Center of Sciences and Cultures (CRDCSC), Montreal, Quebec, Canada

A. K. Haghi, PhD, is the author and editor of 165 books, as well as 1000 published papers in various journals and conference proceedings. Dr. Haghi has received several grants, consulted for a number of major corporations, and is a frequent speaker to national and international audiences. Since 1983, he served as professor at several universities. He served as Editor-in-Chief of the *International Journal of Chemoinformatics and Chemical Engineering* and *Polymers Research Journal* and is on the editorial boards of many International journals. He is also a member of the Canadian Research and Development Center of Sciences and Cultures (CRDCSC), Montreal, Quebec, Canada. He holds a BSc in urban and environmental engineering from the University of North Carolina (USA), an MSc in mechanical engineering from North Carolina A&T State University (USA), a DEA in applied mechanics, acoustics, and materials from the Université de Technologie de Compiègne (France), and a PhD in engineering sciences from Université de Franche-Comté (France).

Contents

Contributors

Ahmed M. Abdel-Azeem
Department of Botany, Faculty of Science, Suez Canal University, Ismailia 41522, Egypt

Rafail A. Afanas'ev
Pryanishnikov All-Russian Scientific Research Institute of Agrochemistry, d. 31A, Pryanishnikova St., Moscow 127550, Russia

M. Vijaya Bharathi
Cattle and Buffalo Breeding Unit, Post Graduate Research Institute in Animal Science, Kattupakkam, Tamil Nadu Veterinary and Animal Sciences University, Chennai, Tamil Nadu, India

Adali O. Castañeda-Facio
School of Chemistry. Universidad Autónoma de Coahuila. Boulevard Venustiano Carranza and José Cárdenas s/n. Republica Oriente. C.P. 25280, Saltillo, Coahuila, México

Gloria Castellano
Departamento de Ciencias Experimentales y Matemáticas, Facultad de Veterinaria y Ciencias Experimentales, Universidad Católica de Valencia San Vicente Mártir, Guillem de Castro-94, E-46001 València, Spain

José J. Cedillo-Portillo
School of Chemistry. Universidad Autónoma de Coahuila. Boulevard Venustiano Carranza and José Cárdenas s/n. Republica Oriente. C.P. 25280, Saltillo, Coahuila, México

Vasant V. Chabukswar
Department of Chemistry, Nowrosjee Wadia College, 19 Late Prin. V. K. Joag Path, Pune 411001, Maharashtra, India

Karina G. Espinoza-Cavazos
School of Chemistry. Universidad Autónoma de Coahuila. Boulevard Venustiano Carranza and José Cárdenas s/n. Republica Oriente. C.P. 25280, Saltillo, Coahuila, México

A. K. Haghi
Former Editor-in-Chief, International Journal of Chemoinformatics and Chemical Engineering and Polymers Research Journal; Member, Canadian Research and Development Center of Sciences and Cultures (CRDCSC), Canada

Kalpana N. Handore
Modern Education Society's College of Engineering (MESCOE), Pune, Maharashtra, India

Porteen Kannan
Department of Veterinary Public Health and Epidemiology, Madras Veterinary College, Chennai 600007, Tamil Nadu

Waleed F. Khalil
Department of Pharmacology, Faculty of Veterinary Medicine, Suez Canal University, Ismailia 41522, Egypt

Moayad N. Khalaf
Chemistry Department, College of Science, University of Basrah, Basrah, Iraq

Genrietta E. Merzlaya
Pryanishnikov All-Russian Scientific Research Institute of Agrochemistry, d. 31A, Pryanishnikova St., Moscow, 127550, Russia

Santosh Mishra
G. H. Raisoni College of Engineering and Management, Wagholi, Pune, Maharashtra, India

B. S. Pradeep Nag
Department of Veterinary Obstetrics and Gynaecology, National Dairy Research Institute, Karnal, India

V. Naveenkumar
Department of Veterinary Preventive Medicine, Madras Veterinary College, Chennai, Tamil Nadu, India

Sukanchan Palit
Consultant, 43, Judges Bagan, Haridevpur 700082, Kolkata, India

M. Nithya Quintoil
Department of Veterinary Public Health, Rajiv Gandhi Institute of Veterinary Education and Research, Puducherry

P. Raja
Department of Animal Biotechnology, Madras Veterinary College, Tamil Nadu Veterinary and Animal Sciences University, Chennai-600007, Tamil Nadu, India

S. Wilfred Ruban
Department of Livestock Products Technology, Veterinary College, Bengaluru 560024, Karnataka

Aidé Sáenz-Galindo
School of Chemistry. Universidad Autónoma de Coahuila. Boulevard Venustiano Carranza and José Cárdenas s/n. Republica Oriente. C.P. 25280, Saltillo, Coahuila, México

Fatma M. Salem
Department of Plant Protection, Desert Research Center, El-Matarya, Cairo 11753, Egypt

Sherif Y. Saleh
Department of Biochemistry, Faculty of Veterinary Medicine, Suez Canal University, Ismailia, 41522, Egypt

A. Sangeetha
Department of Veterinary Public Health and Epidemiology, Veterinary College and Research Institute, Orathanadu, Thanjavur-614625, India

Sumit B. Sharma
Modern Education Society's College of Engineering (MESCOE), Pune, Maharashtra, India

Hussein A. Shnawa
Polymer Research Center, University of Basrah, Basrah, Iraq

Ekaterina I. Shishatskaya
Siberian Federal University, 79 Svobodnyi Av., Krasnoyarsk 660041, Russia
Institute of Biophysics SB RAS, Federal Research Center "Krasnoyarsk Science Center SB RAS," 50/50 Akademgorodok, Krasnoyarsk 660036, Russia

Michael Olegovich Smirnov
Pryanishnikov All-Russian Scientific Research Institute of Agrochemistry, d. 31A, Pryanishnikova St., Moscow 127550, Russia

Sabu Thomas
Siberian Federal University, 79 Svobodnyi Av., Krasnoyarsk 660041, Russia
International and Interuniversity Centre for Nano Science and Nano Technology, Mahatma Gandhi University, Kottayam, Kerala, India

Francisco Torrens
Institut Universitari de Ciència Molecular, Universitat de València, Edifici d'Instituts de Paterna, P. O. Box 22085, E-46071 València, Spain

Tatiana Gr. Volova
Siberian Federal University, 79 Svobodnyi Av., Krasnoyarsk 660041, Russia
Institute of Biophysics SB RAS, Federal Research Center "Krasnoyarsk Science Center SB RAS," 50/50 Akademgorodok, Krasnoyarsk 660036, Russia

Sherif M. Zaki
Department of Microbiology, Faculty of Science, Ain Shams University, Abbassia-Cairo 11566, Egypt

Abbreviations

AAc	acrylic acid
AAP	accumulation-associated protein
AAV-I	avian adenovirus group I
ADI	acceptable daily intake
AFM	atomic force microscope
AGPT	agar gel precipitation test
AGID	agar gel immunodiffusion
AH	alcoholic hepatopathy
AIDS	acquired immunodeficiency syndrome
AI	artificial intelligence
ALS	amyotrophic lateral sclerosis
ALD	alcoholic liver disease
ALT	alanine amino transferase
AO	antioxidant
ASS	administration and services
ASH	alcoholic steatohepatitis
ATR	attenuated total reflection
AVR	avian reovirus
BHT	butylated hydroxytoluene
BF	bursa of Fabricius
BHA	butylated hydroxyanisole
BHs	black holes
BSE	bovine spongiform encephalopathy
CAR-T	chimeric antigen receptor T-cell
CAM	chorioallantoic membrane
CAV	chicken anemia virus
CIP	cleaning in place
CD	canine distemper
CDV	canine distemper virus
CEA	carcinoembryonic antigen
CF	cystic fibrosis
CFTR	CF transmembrane conductance regulator
CG	clinical genetics
CIE	counter immunoelectrophoresis

CJD	Creutzfeldt Jacob disease
CL	charged lepton
CNS	central nervous system
CP	charge–parity
CTCs	circulating tumor cells
CVD	cardiovascular disease
DHC	dehydrochlorination
DL	deep learning
DNA	deoxyribonucleic acid
EAC	ehrlich ascites carcinoma
EAEC	enteroaggregative *E. coli*
ELISA	enzyme-linked immunosorbent assay
EMR	electromagnetic radiation
EMTs	epithelial to mesenchymal transitions
EOS	essential oils
EPR	electron paramagnetic resonance
EPS	extracellular polymeric substances
EPR	electron paramagnetic resonance
ESBO	epoxidized soybean oil
ESO	epoxidized sunflower
FFs	force fields
FHOs	fractal hybrid orbitals
GALT	gut-associated lymphoid tissues
GHE	greenhouse effect
GLADIS	Global Alcoholic Liver Disease Survey
GMOs	genetically modified organisms
GMP	good manufacturing practice
GRBs	γ-ray bursts
GWs	gravitational waves
HCl	hydrogen chloride gas
HCC	hepatocellular carcinoma
HDPE	high density polyethylene
HEV	hepatitis E virus
HIV	human immunodeficiency virus
HNC	head and neck
HVR	hypervariable
IBD	inflammatory bowel disease
IBD	infectious bursal disease
IBV	infectious bronchitis virus
IBDV	infectious bursal disease virus

ILT	infectious laryngotracheitis virus
IPPI	Iran Polymer and Petrochemical Institute
IP	identity preservation
IO	immuno-oncology
IR	immune response
KCS	keratoconjunctivitis sicca
LAr	liquid argon
LARL	limonoate A-ring lactone
LAT	Latex Agglutination Test
LC	lethal concentration
LDPE	low density polyethylene
LPS	lipopolysaccharide
MB	methylene blue
MCV	maximum constant value
MDGs	Millennium Development Goals
ML	machine learning
mPCR	multiplex polymerase chain reaction
MST	median survival time
MWNTs	multiple walls nanotubes
MYB	myeloblastosis
NDV	Newcastle disease virus
NS	natural selection
P3S	proteins' tertiary structure
PC	primary care
PCR	polymerase chain reaction
PBT	pine bark tannin
PBCA	poly (butyl cyanoacrylate)
PDV	phocine distemper virus
PHAs	polyhydroxyalkanoates
PNAG	poly-N-acetylglucosamine
PP	polypropylene
PPM	precision personalized medicine
prep-HPLC	preparative high-performance liquid chromatography
PTE	periodic table of the element
PVC	polyvinyl chloride
QSARs	quantitative structure–activity relationships
RA	rheumatoid arthritis
RDs	rare diseases
RE	restriction enzymes
RF	risk factor

RFLP	restriction fragment length polymorphism
RLs	research laboratories
RNA	ribonucleic acid
RRDs	respiratory RDs
RT-PCR	reverse-transcriptase PCR
SAAG	serum-ascites albumin gradient
SEM	scanning electron microscopy
SEPTIC	sensing of phage-triggered ion cascade
SH	smoking habit
SLAM	signaling lymphocytic activation molecule
SNO	Sudbury Neutrino Observatory
SLE	systemic lupus erythematosus
SPF	specific pathogen free
SRID	single radial immunodiffusion
SVS	state variety sections
SWNTs	single cylindrical wall nanotubes
TNF	tumor necrosis factor
TRS	teaching–researcher
UV	ultraviolet
VC	Valencia Community
vvIBDV	very virulent IBDV
VNT	Virus Neutralization Test
VS	vincristine sulfate
WHO	World Health Organization

Preface

The crucial interdependence between humans and their environment is explored and illuminated in this revealing overview of the major environmental issues facing society in the twenty-first century.

This book presents a novel picture in current advances in research of theoretical and practical frameworks of environmental problems and solutions taken from the latest empirical research findings.

It provides a modern and concise introduction to environmental technology principles and the dynamic nature of environmental systems.

This new volume focuses on the aspects of new techniques that are particularly valuable for solving environmental problems. The complex environmental issues are presented in simple terms to help readers grasp the basics and solve relevant problems. It provides readers with the fundamentals of environmental science and a toolbox for putting them into practice and is a concise, accessible, and hands-on volume designed for students and professionals working in the chemical and environmental sciences. Timely and comprehensive discussions of applications to real-world environmental concerns are a central focus of this research-oriented book.

It also provides a condensed and integrated tool for students learning and covers key concepts in the rapidly developing field of environmental sustainability.

This book is a must-have reference for researchers and practitioners in environmental science and engineering.

This volume exposes students to a broad range of environmental topics, as well as discussion of relevant regulations and practices.

It places more emphasis on scientific principles, ethics, and safety, and focuses less on engineering design.

Key features of this book:

- It is a concise yet reliable guide that explains key concepts in environmental science and technology.
- It provides the essential background for understanding and solving the most frequent environmental problems.

- It treats fundamental aspects of chemistry and physics as applied to environmentally relevant problems, and gives a state of the art account of the field.
- It includes examples and complex problems based on real data that expose scientists to the real problems and data they will face in their careers.
- It is clearly written for all professionals in environmental science and environmental engineering as well as soil science.
- It provides integrated tools in explaining complex chemical problems of environmental importance.

PART I

Environmental Sustainability

CHAPTER 1

Pollution by Tropospheric Ozone in the Valencia Community

FRANCISCO TORRENS[1*] and GLORIA CASTELLANO[2]

[1]*Institut Universitari de Ciència Molecular, Universitat de València, Edifici d'Instituts de Paterna, P. O. Box 22085, E-46071 València, Spain*

[2]*Departamento de Ciencias Experimentales y Matemáticas, Facultad de Veterinaria y Ciencias Experimentales, Universidad Católica de Valencia San Vicente Mártir, Guillem de Castro-94, E-46001 València, Spain*

[*]*Corresponding author. E-mail: torrens@uv.es*

ABSTRACT

Biofuels generation from edible feedstocks is in competition with food production, increases food price and affects' region food security. Ozone is harmful because it causes grave damages to lung tissue. Ozone forms by other pollutants emissions produced by vehicular traffic, industries, and heating. People demand authorities that quick and understandably inform of O_3 levels *via* Internet and media. No threshold exists below which O_3 has no effect. Air pollution is caused by substances and forms of energy that alter air quality, in such a way that they involve risks, damages, and bothers of any nature to persons and goods. The most important effects of tropospheric O_3 occur on vulnerable populations and, in the Mediterranean Region, during anticyclone. People are exposed to a complex mixing of pollutants that all contributes to a certain health state. People do the following actions vs. tropospheric O_3: (1) beg authorities that inform tropospheric O_3 levels, (2) complain to administrations writing of compulsory plans of air-quality improvement, (3) walk and use bicycle and public transport, and (4) reduce electricity consumption. A conspiracy of silence exists. Sustainability is important in both environmental and social senses. For better understanding among people, the language used by the orthodoxy should be utilized.

1.1 INTRODUCTION

Setting the scene: why O_3 is harmful and pollution by tropospheric ozone O_3 in the Valencia Community. Ozone is harmful because it can cause grave damages to lung tissue. Ozone forms by other pollutants (*precursors*) emissions produced by vehicular traffic, industries, and heating. People can demand authorities that quick and understandably inform of O_3 levels *via* Internet and media. There is no threshold below which O_3 has no effect. Air pollution is caused by substances and forms of energy that alter air quality, in such a way that they involve risks, damages, and bothers of any nature to persons and goods. The most important effects of tropospheric O_3 occur on vulnerable populations and, in the Mediterranean Region, during hot periods. People are exposed to a complex mixing of pollutants that all can contribute to a certain health state. People can do the following actions against tropospheric O_3: (1) beg authorities that inform tropospheric O_3 levels, (2) complain to administrations writing of compulsory plans of air-quality improvement, (3) walk and use bicycle and public transport, (4) reduce electricity consumption. There is a conspiracy of silence. Sustainability is important in both environmental and social (control of population growth) senses. If one wants to be better understood by people, the language used by the orthodoxy should be utilized.

The following hypotheses (H) were proposed on biofuel economic and ethical risks:

H1. (Fidel Castro, 2003). Biofuel generation from edible feedstocks is in competition with food production.
H2. (Fidel Castro, 2003). Generation of biofuels from edible feedstocks increases food price.
H3. (Fidel Castro, 2007). The use of food to produce biofuel will affect region food security.

The role of both left and right wings shaping politics was informed.[1] The trap of diversity was discussed as how neoliberalism fragmented the identity of the *working class*.[2] The societies of the persons without value were analyzed as 4th Industrial Revolution, desubstantiation of capital and generalized devalorization.[3] The working class was reviewed as a possible subject of change in 21st century.[4]

In an earlier publication, converting cellulose, sucrose, and fructose into biofuel, and new biofuels was discussed.[5–7] Seagrass *Posidonia oceanica* was argued as an indicator of seawater quality.[8] Chlorofluorocarbons, ozone levels evaluation, people actions, tropospheric O_3, and air quality were reviewed.[9] In the present report, it was reviewed why O_3 is harmful and the pollution by tropospheric O_3 in the Valencia Community. Tropospheric O_3 permitted proposing questions (Q), answers (A), and hypothesis (H) on air quality.

1.2 WHY IS OZONE HARMFUL?

Ozone is harmful because it can cause grave damages to lung tissue.[10] Every O_3 molecule presents one free O-atom, which is searching for some other molecule with which to combine. Ozone oxidizes the inner protective mucosa of lung and perforates it. Then, it attacks the underlying tissues, which react with the defence mechanism (*inflammation*). Possible consequences are congestions, anomalous secretions, and obstruction of the affected zone, with the consequent loss of respiratory capacity. The victim shows other discomforts (e.g., persistent cough, headache, shedding tears easily), which could become incapacitating. Individual sensitivity vs. O_3 varies, and effects depend on inhaled volume and, so, physical effort and air O_3 concentration, which in the Mediterranean Region, is greater in anticyclonic (dry environment and intense sunshine) days, and vs. height, in which circumstances, old persons, young children, and chronic patients with breathing troubles, to be on the safe side, should abstain from carrying out great physical efforts at free air. One should not confuse O_3 of the air that people breathe and stratospheric O_3 layer that impedes pass of ionizing radiations, which depletion lastly worries people.

1.3 POLLUTION BY OZONE IN THE VALENCIA COMMUNITY

Ecologistas en Acción organized I Day on O_3 Pollution in Valencia Community (VC) proposing Q/A (*cf.* Fig. 1.1).[11]

Q1. How does O_3 form?
A1. Other pollutants (*precursors*) emissions produced by vehicular traffic, industries, and heating.
Q2. What can people do?
A2. Demand authorities that quick/understandably inform of O_3 levels *via* Internet/media.
Q3. What are the problems of the pollution by O_3 at the Mediterranean?
A3. There is no threshold below which O_3 has no effect.

FIGURE 1.1 Day on O_3 pollution in Valencian Community, València, Spain, September 21, 2017.

Ferrero proposed some questions and answers on effects of air pollution on health.[12]

Q4. What is air pollution?[13]

A4. Substances and forms of energy that alter air quality, in such a way that they involve risks, damages, and bothers of any nature to persons and goods.
Q5. Is it an avoidable risk or not?
Q6. How is measured nowadays the impact on health?
Q7. Why is smog harmful?
Q8. How does O_3 form?

She discussed susceptibility factors: (1) age, (2) children/women, (3) vigorous physical activity, (4) previous pathologies.

She provided the following conclusions (C).

C1. The most important effects occur on vulnerable populations and during hot periods.
C2. People are exposed to a complex pollutants mixing that all can contribute to certain health state.

She proposed additional questions and answer.

Q9. Are many cases of asthma and cold caused by O_3?
Q10. Should the beginning of the chain:

Poverty → stress → blood pressure → … → cholesterol → diseases be treated?

A10. Diseases are multifactorial; people think that pollutants are the cause.

The following additional questions and answers were proposed.

Q11. Actions vs. tropospheric O_3, what can people do?
A11. (1) Beg authorities that inform tropospheric O_3 levels, (2) complain to administrations writing of compulsory plans of air-quality improvement, (3) walk/use bicycle/public transport, (4) reduce electricity consumption.

1.4 POLLUTION BY TROPOSFERIC OZONE IN THE VALENCIA COMMUNITY

Ecologistes en Acció organized II Day on Tropospheric O_3 Pollution in VC proposing Q/A.[14]

Q1. How does O_3 form?

A1. Origin is other pollutants (*precursors*) emissions produced by traffic, industries, and heating.

Q2. What can people do?

A2. Demand authorities that quick/understandably inform of tropospheric O_3 levels *via* Internet/media.

Q3. Actions facing tropospheric O_3, what can people do?

Parra raised the following questions on the quality of the air and tropospheric O_3.[15]

Q4. How does tropospheric O_3 form?

Q5. Where does tropospheric O_3 come from?

Q6. What does the law on the quality of the air say?

Sintes proposed question and hypothesis on the effects of air pollution on health.[16]

Q7. Why do people die (*cf.* Fig. 1.2)?[17,18]

H1. The conspiracy of silence.

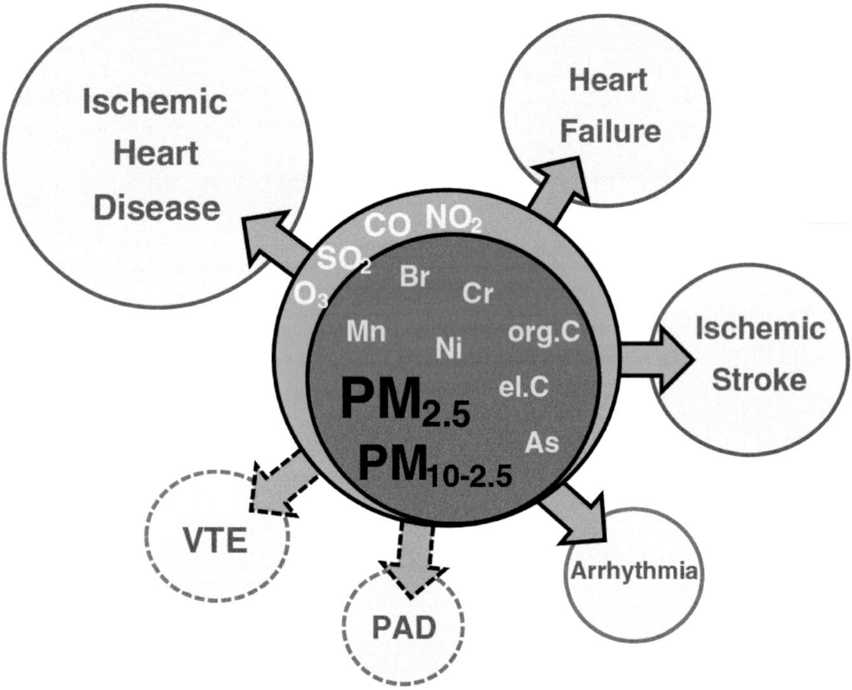

FIGURE 1.2 Air pollution–cardiovascular diseases relations. PAD, peripheral artery disease; VTE, venous thromboembolism.

Source: Reprinted with permission from Ref. [18]. © 2013 Elsevier.

1.5 FINAL REMARKS

From the preceding results and discussion, the following final remarks can be drawn.

1. Biofuels generation from edible feedstocks is in competition with food production, increases food price, and will affect region food security.
2. Ozone is harmful because it can cause grave damages to lung tissue.
3. Ozone forms by other pollutants (*precursors*) emissions produced by vehicular traffic, industries, and heating.
4. People can demand authorities that quick and understandably inform of O_3 levels *via* Internet and media.
5. There is no threshold below which O_3 has no effect.
6. Air pollution is caused by substances and forms of energy that alter air quality, in such a way that they involve risks, damages, and bothers of any nature to persons and goods.
7. The most important effects of tropospheric O_3 occur on vulnerable populations and during hot periods.
8. People are exposed to a complex mixing of pollutants that all can contribute to a certain health state.
9. People can do the following actions against tropospheric O_3: (1) beg authorities that inform tropospheric O_3 levels, (2) complain to administrations writing of compulsory plans of air-quality improvement, (3) walk and use bicycle and public transport, (4) reduce electricity consumption.
10. There is a conspiracy of silence.
11. Sustainability is important in both environmental and social (control of population growth) senses.
12. If one wants to be better understood by people, the language used by the orthodoxy should be utilized.

ACKNOWLEDGMENTS

The authors thank support from Generalitat Valenciana (Project No. PROMETEO/2016/094) and Universidad Católica de Valencia *San Vicente Mártir* (Project No. 2019-217-001).

KEYWORDS

- **troposphere**
- **ozone damage**
- **air pollution**
- **climate change**
- **Spain**
- **forecasting**
- **need assessment**

REFERENCES

1. Ross, S. J. *Hollywood Left and Right: How Movie Stars Shaped American Politics*; Oxford University: New York, NY, 2011.
2. Bernabé D. La Trampa de la Diversidad: Cómo el Neoliberalismo Fragmentó la Identidad de la Clase Trabajadora; Akal: Tres Cantos, Madrid, Spain, 2018.
3. Piqueras, A. Las Sociedades de las Personas sin Valor: Cuarta Revolución Industrial, Des-substanciación del Capital, Desvalorización Generalizada; El Viejo Topo: Vilassar de Dalt, Barcelona, Spain, 2018.
4. Tarín Sanz, A., Rivas Otero, J. M., Eds., La Clase Trabajadora: ¿Sujeto de Cambio en el Siglo XXI?; Siglo XXI: Tres Cantos, Madrid, Spain, 2018.
5. Torrens, F.; Castellano, G. Book of Abstracts, 6th International Symposium on Ozone Applications, Havana, Cuba, June 28–July 1, 2010; The Ozone Research Center of Cuba: Havana, Cuba, 2010; P-1.
6. Torrens, F.; Castellano, G. Book of Abstracts, V International Symposium on Environment, Havana, Cuba, June 28–July 1, 2010; The Ozone Research Center of Cuba: Havana, Cuba, 2010; P-1.
7. Torrens, F.; Castellano, G. Book of Abstracts, II Ozone Therapy International Congress from FIOOT, Havana, Cuba, June 28–July 1, 2010; The Ozone Research Center of Cuba: Havana, Cuba, 2010; P-1.
8. Castellano, G.; Tena, J.; Torrens, F. Classification of Polyphenolic Compounds by Chemical Structural Indicators and its Relation to Antioxidant Properties of *Posidonia oceanica* (L.) Delile. *MATCH Commun. Math. Comput. Chem.* **2012,** *67*, 231–250.
9. Torrens, F.; Castellano, G. Chlorofluorocarbons, Ozone Levels Evaluation, People Actions, Tropospheric O_3, and Air Quality. In *Chemical Technology and Informatics in Chemistry with Applications*; Vakhrushev, A. V., Mukbaniani, O. V., Susanto, H., Eds.; Apple Academic–CRC: Waretown, NJ, 2019; pp 195–203.
10. Brater, J. Lexicon der Rätselhaften Körpervorgänge; Eighborn: Frankfurt am Main, Germany, 2002.
11. Book of Abstracts, Contaminació per Ozó al País Valencià, València, Spain, September 21, 2017; Ecologistas en Acción: València, Spain, 2017.

12. Ferrero, A. Book of Abstracts, Contaminació per Ozó al País Valencià, València, Spain, September 21, 2017; Ecologistas en Acción: València, Spain, 2017; O-3.
13. Boldo Pascua, E. La Contaminación del Aire; La Catarata: Madrid, Spain, 2016.
14. Book of Abstracts, Contaminació per Ozó Troposfèric al País Valencià, València, Spain, May 29, 2018; Ecologistes en Acció: València, Spain, 2018.
15. Parra, M. Book of Abstracts, Contaminació per Ozó Troposfèric al País Valencià, València, Spain, May 29, 2018; Ecologistes en Acció: València, Spain, 2018; O-1.
16. Sintes, M. Book of Abstracts, Contaminació per Ozó Troposfèric al País Valencià, València, Spain, May 29, 2018; Ecologistes en Acció: València, Spain, 2018; O-2.
17. Ballester, F. Contaminación atmosférica, cambio climático y salud. *Rev. Esp. Salud Pública* **2005,** *79*, 159–175.
18. Martinelli, N.; Olivieri, O.; Girelli, D. Air Particulate Matter and Cardiovascular Disease: A Narrative Review. *Eur. J. Intern. Med.* **2013,** *24*, 295–302.

CHAPTER 2

Nanocomposites for Environmental Pollution Control: A Far-Reaching Review

SUKANCHAN PALIT

Consultant, 43, Judges Bagan, Haridevpur 700082, Kolkata, India
E-mail: sukanchan68@gmail.com; sukanchan92@gmail.com

ABSTRACT

The domain of environmental protection and environmental engineering science are moving toward one visionary challenge to another. Environmental remediation is today undergoing drastic and dramatic challenges. Global warming, global climate change, water scarcity, and frequent environmental disasters are urging the scientific community to move toward more innovation and vast scientific intuition. Mankind's immense scientific adjudication, man's vast scientific grit and determination, and the world of scientific and technological validation will all lead a visionary way in the true realization of environmental engineering science. In the similar vein, composite science and material science are the needs of human scientific progress today. Nanotechnology is today merged with almost every branch of science and engineering today. This direction of scientific research is depicted profoundly by the author. Fiber reinforced composites and nanocomposites are the marvels of composite science today. In this chapter, the author also rigorously delineates the success and vision of science in the field of green nanocomposites and fiber reinforced composites. The world of material science is slowly moving toward a new scientific genre and new scientific revelation. Scientific stance, vast scientific imagination, and the vision of engineering science will all be the forerunners toward a newer age in the field of composite science and material science. This well-researched treatise profoundly delineates the recent scientific advancements in the field of nanocomposites, green nanotechnology, composite science, and material science.

2.1 INTRODUCTION

The vast world of chemical engineering science and composite science are today in the midst of scientific introspection and regeneration. In the similar vein, material science and green nanotechnology are in the course of scientific vision and deep scientific understanding and discernment. Green nanotechnology, advanced materials, and green chemistry are the marvels of science and engineering today. Stringent environmental regulations and the ever-growing concerns for global climate change are urging the scientific community to gear forward toward newer vision and new scientific and engineering innovations. In this chapter, the author deeply delineates the success and the vision of nanotechnology, composite science, and the vast world of nanocomposites. The other direction of this well-researched treatise is the mitigation of global environmental pollution. Mankind's immense scientific prowess and scientific sapience, the world of technological validation, and the scientific imagination behind environmental remediation will lead a long and effective way in unraveling the scientific truth of environmental engineering science. In the global scientific scenario, global warming, climate change, and frequent environmental disasters are destroying the vast scientific fabric of vision, might, and determination. This chapter will surely open up newer vistas of scientific regeneration and scientific prowess in the field of nanocomposite applications in environmental protection. Environmental sustainability is the other side of the visionary coin. Sustainable development whether it is social, economic, energy, or environmental are the utmost need of global scientific progress today. This chapter opens up a new era in the field of nanotechnology, composite science, and environmental protection. Human race will thus usher in a new beginning in the field of science and engineering as nanotechnology and environmental protection enters a new age of deep scientific vision and scientific might and forbearance.

2.2 THE AIM AND OBJECTIVE OF THIS STUDY

Humanity today stands firm in the midst of vast scientific and engineering advancements. Space exploration, the advancements in nuclear science, and engineering and the success of science of sustainability will all lead a long and effective way in the true unraveling of engineering science and technology globally today. The main aim and objective of this study is to unfold and uncover the scientific vision in the applications of nanotechnology and composite science in the furtherance of science and the advancement of human society. The author pointedly focuses on the recent scientific research pursuit in the field of nanotechnology and composite science with a deep scientific understanding of environmental remediation. The other areas of scientific endeavor are the advancements in the science of environmental and energy sustainability. Systems engineering and reliability engineering are the pillars of research pursuit in global scenario today. This chapter thus opens up newer thoughts and newer scientific sapience and pragmatism in the field of systems science and environmental remediation. The main thrust area of this chapter is to unfold the scientific intricacies of the applications of nanocomposites in environmental protection. Science, technology, and engineering are today in the path of newer regeneration. Globally climate change and frequent environmental disasters are urging the scientific community to gear forwards toward new innovations and new directions.

2.3 WHAT DO YOU MEAN BY NANOCOMPOSITES?

Nanocomposites are multiphase solid materials where one of the phases has one, two, or three dimensions of less than 100 nm, or structures having nanoscale distances between the different phases that construct the material. The main concept behind nanocomposite is to veritably use building blocks with dimensions in the nanometer range to design, create and envision new materials. In the wider and broadest sense, this definition will include porous media, colloids, gels, and copolymers. The success and vision of material science today rests on the scientific ingenuity of nanocomposites. Material scientists globally are today in the forefront of vision and innovation. This chapter targets the application areas of nanocomposites and environmental remediation.

2.4 THE SCIENTIFIC DOCTRINE OF THE SCIENCE OF ENVIRONMENTAL PROTECTION

The science of environmental protection and remediation are in the avenues of new scientific sagacity and deep scientific provenance. Global research and development initiatives in environmental remediation are still latent and undeveloped as global warming, drinking water contamination, and industrial wastewater treatment are in the avenues of immense disaster. Arsenic and heavy metal contamination of drinking water and groundwater are veritably destroying the global scientific fabric. Human civilization and human scientific progress are thus in a state of immense distress and devastation. The challenges of environmental protection are thus industrial wastewater treatment, drinking water treatment, and groundwater decontamination. In such a situation, chemical process engineering, material science, and polymer science are the marvels of scientific progress today. In this century, nanotechnology is linked with every branch scientific research pursuit. The status of environmental and energy sustainability are equally at a disastrous situation. The target of the scientific domain and the civil society is to gear toward more innovations in traditional and nontraditional environmental engineering techniques. Thus, there will be a greater scientific emancipation in the field of environmental sustainability and environmental remediation. The target of science and engineering globally today will be toward more new technologies and greater emancipation of the science of sustainability. The scientific doctrine of the science of environmental protection will surely usher in a new era in the field of drinking water treatment, industrial wastewater treatment, and groundwater remediation. Developing and developed nations around the world are in the midst of deep scientific comprehension and vast scientific divination. Here comes the immense importance of novel separation techniques such as membrane science and nonconventional environmental engineering tools such as advanced oxidation processes. In this well-researched treatise, the author deeply treads through a weary path in the field of nanotechnology, environmental remediation, and the science of sustainability.

2.5 SIGNIFICANT SCIENTIFIC RESEARCH PURSUIT IN THE FIELD OF NANOCOMPOSITES AND COMPOSITE SCIENCE

Nanocomposites and nanomaterials are revolutionizing the scientific landscape globally today. Material science and nanotechnology are the immediate needs of scientific revolution globally today. Composite science as well as polymer science are changing the research genre in the field of material science, nanomaterials, and engineered nanomaterials. In this section, the author with immense scientific vision comprehends the recent scientific research pursuit in the field of nanocomposites and composite science with a clear view of furtherance of science and engineering.

Camargo et al.[1] discussed with immense scientific lucidity synthesis, structure, properties, and new application opportunities in the domain of nanocomposites. Nanocomposites, a high-performance material exhibit excellent property combinations and unique design parameters. With an estimated annual growth rate of 25% and fastest demand in engineering plastics and elastomers, their potential is extremely striking and are useful in several areas ranging from packaging and biomedical engineering applications.[1] The profundity of science and engineering, the world of scientific challenges and the needs of human society will all lead a long and effective way in unraveling the scientific truth behind material science and nanotechnology.[1] In this well-researched review, the three types of matrix nanocomposites are presented underlining the immense need of these materials, their processing methods, and some research results on structure, properties, and potential applications, the need of use of these materials in space applications and other interesting aspects such as safety and environmental management of the use of these materials.[1] Technological vision, deep scientific provenance, and vast scientific ingenuity stands as major hallmarks toward scientific progress in material science and green nanomaterials.[1] Nanocomposites are the composites in which at least one of the phases shows a dimension in the nanometer range and in the verge of new scientific revolution and understanding. Possible and important uses of natural materials such as clay-based minerals, chrysotile, and lignocellulosic fibers are highlighted in minute details. Nanocomposites are the smart materials of tomorrow. Nanocomposite materials can be classified according to their matrix materials in three different categories such as: (1) ceramic matrix nanocomposites, (2) metal matrix nanocomposites, and (3) polymer matrix nanocomposites.[1] Nanocomposite systems, including those reinforced with carbon nanotubes have been extensively applied and investigated since the 1990s and accordingly there has been a steady research and development forays since then. The authors discussed in lucid details potentials and opportunities in nanocomposites, processing of nanocomposites, structure and properties of nanocomposites, reinforcement nanocomposite systems, and the vast domain of application of nanocomposites.[1] Immensely outstanding potentials of nanocomposites can be exemplified by the research and development initiatives around the world.[1] Technological adroitness, scientific vision, and the challenges of engineering science are all the hallmarks toward a newer era in the field of nanocomposites, composite science, and material science. This well-researched treatise rigorously points toward the scientific success, the vision of material science and composite science, and opens new doors of innovation in the vast domain of nanotechnology.[1]

Bhattacharya[2] deeply comprehended with lucid and cogent insight polymer nanocomposites and made a comparison between carbon nanotubes, graphenes, and clay as nanofillers. Scientific divination, technology, and engineering ardor in the field of composite science and nanotechnology are in the vistas of new scientific regeneration.[2] Nanofilled polymeric matrices have veritably demonstrated outstanding mechanical, electrical, and thermal properties. In this article, the author reviewed and investigated the processing of carbon nanotube, graphene, and clay montmorillonite platelet as potential nanofillers to form nanocomposites.[2] The vast scientific challenges and the scientific intricacies of nanofilled polymeric composites are investigated in minute details. Fillers were previously used to enhance the properties of the polymers and reduce the cost of polymeric products. But with the advancement of science and technology, fillers became an integral part in many engineering applications particularly for increasing the mechanical properties of the polymer.[2] Traditional fillers include talc, glass fibers, carbon black, and calcium carbonate particles in the micrometer range. Nanofillers can be characterized on the basis of their dimensions and size. They are generally one-dimensional which include nanotubes and nanowires, two-dimensional such as nanoclays and graphene and three-dimensional such as spherical and cubical nanoparticles.[2] Carbonaceous nanofillers demonstrate excellent properties due to their mechanical strength and high aspect ratio. Today carbon nanotubes have attracted immense attention because of their unusual structures and properties.[2] Nanomaterials and engineered nanomaterials applications are today in the forefront of scientific regeneration. In the similar manner, polymer processing is today surpassing vast and versatile scientific boundaries. The authors deeply discussed the areas of processing techniques for developing nanocomposites and deeply summarize the properties. The other vital areas of research pursuit are the needs of developing nanocomposites. The scientific vision and the scientific stances of application of nanocomposites are: (1) fillers should have excellent mechanical properties, (2) fillers should have high aspect ratio, (3) fillers should have high surface area, and (4) they should be well dispersed without any agglomeration.[2] The ultimate and effective properties of polymer processing are (1) melt processing, (2) solvent processing, (3) in-situ polymerization, (4) electrospinning, and 4) layer by layer assembly. The entire technological vision of polymer processing of nanocomposites i investigated in lucid details in this chapter.[2] The other areas of this marvelous research endeavor are the properties of nanocomposites which are: (1) reinforcement, (2) electrical conductivity, (3) thermal conductivity, (4) thermal stability, (5) glass transition temperature, (6) barrier and membrane separation properties, and (7) flammability resistance.[2] The author also gives a future outlook and a current progress on the state of nanoparticle filled polymer nanocomposites and their potential applications.[2] The domain of polymer science and material science are in the path of newer scientific rejuvenation. This treatise opens up newer thoughts and newer visionary arenas in the field of polymer nanofillers and its immense polymer processing domain. Smart materials and polymer nanomaterials are the marvels of science and engineering today. The needs of application of nanofillers in polymeric nanomaterials are deeply investigated and envisioned in this chapter.[2]

Puggal et al.[3] discussed and elucidated in a review synthesis, characterization and mechanical properties of polymer nanocomposites. Today, polymer nanocomposites have diverse applications and are being used in packaging, sports equipment, automobile sector and biomedical engineering and sciences.[3] This is due to their excellent property combinations and vast design feasibility. This chapter reviews the various methods for the excellent dispersion of nanofillers, their coating on fibers and the properties improvement.[3] Epoxy being highly cross linked possesses many excellent properties mainly high stiffness and strength, chemical resistance in addition to appropriate performance at elevated temperatures.[3] This treatise deeply focuses on fabrication of nanocomposites, methods of dispersion of nanofillers in matrix phase, characterization tools, properties enhancement, and a deep discussion on future scope of polymer nanocomposites applications. The science and engineering of polymer nanocomposites are today surpassing vast and versatile scientific boundaries.[3] Mankind's immense scientific prowess, the challenges and vision of material science and nanotechnology will surely open newer areas of scientific profundity and scientific provenance in the field of nanocomposite science in decades to come.[3] This treatise targets these areas of polymer science and composite science particularly nanofillers and nanomaterials such as graphene and carbon nanotubes.[3]

Fawaz et al.[4] deeply discussed with vast scientific farsightedness synthesis of polymer nanocomposites. Polymer nanocomposites are hybrid materials which are organic and inorganic with at least one dimension of the filler phase less than 100 nm.[4] Polymer nanocomposites are synthesized via various methods that can be categorized as: (1) melt intercalation, (2) template synthesis, (3) exfoliation adsorption, and (4) in-situ polymerization intercalation.[4] The authors deeply discussed with provenance and scientific sagacity the needs of science and engineering of polymer processing with a clear vision toward furtherance of polymer science, composite science, and material science.[4]

Research and development initiatives in the field of polymer science and composite science are in the vitas of scientific and engineering regeneration. Today there is a veritable gap between science and its applications. This is true for material science and nanotechnology applications. Environmental engineering techniques and material science should be integrated toward the clear vision of proliferation of scientific discernment and scientific understanding. The author in this entire chapter pointedly focuses on the success of material science and composite science in this respect.

2.6 SIGNIFICANT SCIENTIFIC RESEARCH PURSUIT IN THE FIELD OF NANOTECHNOLOGY

Nanotechnology is the marvel and the vignette of research endeavor globally today. Global climate change, the monstrous issue of global warming, depletion of fossil fuel resources and frequent environmental disasters are urging the scientific community to gear toward newer innovations and new futuristic thoughts. The domain of nanotechnology integrated with environmental protection needs to be envisioned and broadened with the march of human civilization. There are visionary scientific endeavors in the field of nanomaterials and engineered nanomaterials today. The world of science and engineering today stands mesmerized as nanomaterials and nanocomposites are in the path of new rejuvenation. In this section, the author deeply pronounces and pointedly focuses on the application of nanotechnology in a broader sense.

The Energy and Resources Institute Report[5] described and discussed with lucid insight in a status treatise nanotechnology developments in India. Science and technology in developing nations such as India stands in the midst of deep scientific vision and scientific upheavals. A focus on capability, governance, and nanotechnology developments in India are discussed in minute details. This report delineates an overview on nanoscience and technology, predominant role of the public sector in nanotechnology in India, nanotechnology and development issues in India, policy support for nanotechnology in India and challenges and opportunities for nanotechnology development.[5] A developing country perspective is described and discussed with nanotechnology research on the forefront.[5] This report provides a status overview with respect to key players, government programs, nanotechnology and development issues in India with reference to agriculture, water and industry. Today, nanotechnology stands in the midst of deep scientific provenance and vision. Nanotechnology highly promises to deliver novel products and processes or increase the performance of the existing ones across diverse sectors and diverse areas of scientific endeavor. They veritably include interventions in wide range of areas such as water, health, energy, agriculture, and environment that can ensure solutions to vast issues in developing countries. Several industry related sectors like pharmaceuticals, electronics, automobiles, textile engineering, chemicals and manufacturing sectors, information and communication technology, as well as biotechnology and biomedical engineering appear envisioned to gain from nanotechnology and nano-engineering applications. Science and engineering of environmental protection and energy engineering are in the process of newer scientific rejuvenation and newer scientific sagacity.[5] Developing countries are yet to garner resources and envision the application domain of nanotechnology. The authors of this treatise vastly target the scientific needs, the deep scientific profundity, and the vision behind nanotechnology applications in developing countries such as India. This report deeply overviews the key developments in diverse sectors like energy, water, agriculture, health, environment etc.[5] Energy storage, production, and conversion within renewable energy framework and integration with nanotechnology and nano-engineering are the other cornerstones of this research endeavor. The other salient features of this treatise policy support to nanotechnology research and development initiatives in the Indian context. The report also targets the key emerging challenges facing the country which are to optimally enhance the capacity building initiatives in nanotechnology applications in diverse sectors of the country's economy.[5] Worldwide public sector research and development in nanotechnology is immensely thriving in several developed and well-developed nations. Here comes the importance of research and development forays in developing countries. This treatise will surely open up new windows of scientific innovation and scientific instinct in the field of nanotechnology applications in developing world.[5]

European Commission Report[6] described with vast scientific farsightedness successful European nanotechnology research. The challenges, the vision, and the intricacies in nanotechnology endeavor globally are in the path of new scientific rejuvenation. Nanotechnology is the new frontier of science and technology in Europe, developing and developed world. The status of research forays in nanotechnology is groundbreaking and surpassing vast and versatile scientific frontiers. Technological motivation, scientific sapience, and scientific adroitness in the field of nanotechnology are today globally in the path of regeneration.[6] Nanomedicine with its diagnostics, drug delivery, and regenerative medicine sub clusters, nanotechnology for environmental engineering and Internet and communication technology applications are the vignettes and marvels of scientific endeavor in European Union today. Today, technological advancements and scientific vision are two opposite sides of the visionary coin of nanotechnology and environmental remediation.[6] Energy production, storage, distribution and energy sustainability are the needs of human civilization and scientific progress today.[6] The first and most important global environmental challenge is water and groundwater remediation which could be veritably used by water supply networks to eliminate metal and hazardous materials traces from drinking water. Another is monitoring industrial pollution in air, water, and soil. The use of nanomaterials and engineered nanomaterials is present in every clusters of human scientific endeavor. This report targets the scientific profundity and the vast scientific vision behind nanomaterials applications in environmental and energy domains. Carbon capture is an important component of European's environmental strategy and is the hallmark of every scientific research pursuit in nanotechnology and nano-engineering.[6] Molecular electronics, nanolithography, extremely thin films, and transistors are increasingly propelling the research and development forays in nanotechnology.[6] Nanomaterials and engineered nanomaterials applications are the marvels of science and engineering today. In the similar manner, smart materials, novel textiles, and advanced functional materials are the success and vision of nanotechnology today. This report vastly points out toward the success of science and the vision of technology and engineering in the field of nanotechnology applications.[6]

The Royal Society Report[7] discussed with vast lucidity and foresight opportunities and uncertainties in nanoscience and nanotechnologies. This report discussed with immense scientific vision nanomaterials, nanometrology, electronics, optoelectronics, and information and communication technologies, bionanotechnology and nanomedicine, nanomanufacturing, and the industrial applications of nanotechnologies, adverse health, safety and environmental effects and social and ethical issues in nanotechnology applications.[7] The science and engineering of nanotechnology, nanomaterials, and engineered nanomaterials are today in the path of newer scientific regeneration. The remit of this study is to summarize the current state of scientific knowledge about the whole domain of nanotechnologies and target the further emancipation of science and technology.[7]

Science and technology in the global scenario are moving at a rapid pace. In the similar vein, nanotechnology and nano-engineering are surpassing vast scientific boundaries. Technological revalidation, scientific motivation, and deep scientific revelation are the needs of present-day human civilization. Today, the world stands in the midst of ever-growing environmental concerns. Thus, the imminent need of novel separation processes, conventional and nonconventional environmental engineering tools. In this treatise, the author pointedly focuses on the scientific success, the deep scientific adroitness and the scientific ingenuity behind application of nanotechnology in diverse areas of engineering and science.

2.7 SIGNIFICANT RESEARCH PURSUIT IN THE FIELD OF NANOMATERIALS AND ENGINEERED MATERIALS

The world of science and engineering of material science today stands challenged as human civilization and human scientific profundity gears forward toward newer vision. Renewable energy is the coinword of today's scientific endeavor. In this sphere, application of nanomaterials and engineered nanomaterials will transform the scientific landscape. In this section, the author deeply stresses on the scientific facts and the vast scientific ingenuity in the field of nanomaterials and the broad domain of nanotechnology.

Arivalagan et al. [8] discussed with cogent insight nanomaterials and its potential applications. Nanomaterials are defined as engineered materials with at least one dimension in the range of 1–100 nm. The present chapter redefines and re-envisions on what are nanomaterials, how is it generated, and how importance nanomaterials it has.[8] Nanotechnology shortened to "nanotech" is the study of manipulating matter on an atomic scale. This treatise defines and discusses synthesis of nanomaterials, properties of nanomaterials, and application domain of nanomaterials. Nanotechnology and nano-engineering are today transforming human life and civilization.[8] The marvel of science and engineering today are nanomaterials and engineered nanomaterials. The areas of green nanomaterials are today applied to developing countries issues such as water, energy, food, shelter, and education. Green engineering and green nanotechnology are changing the face of human scientific endeavor. Biofuels and renewable are the pillars of engineering today.[8] The authors in this chapter concentrate on the "green" aspect of nanotechnology and nano-engineering in the future emancipation of science and technology.[8]

Jaksic et al.[9] discussed with vast scientific profundity functionalization of artificial freestanding composite nanomembranes. Nanomembranes are the next generation nanomaterials applications. Artificial nanomembranes may be defined as synthetic free standing structures with a thickness below 100 nm and a very large aspect ratio. Recently, science and engineering surpassed vast frontiers as nanomembranes whose structures are at the same time ultrathin, with large lateral dimensions and literally robust to stand freely without additional substrate came to the scientific forefront. The most widespread nanomembranes are surely the lipid bilayers which make the protective outer cover of the living cells. In contrast to the lipid bilayer membranes, the man-made structures or artificial synthetic membranes can be fabricated from a wide range of materials, both organic and inorganic.[9] The treatise discussed nanomembrane functionalization with a clear vision toward emancipation of nanomembranes applications.[9] The main cornerstone of this treatise is the classification of nanomembrane functionalization methods. Technology and engineering science of nanomembrane are in the avenues of newer scientific rejuvenation. This chapter targets the intricacies and the vision of nanomembrane science.[9]

Kumar et al.[10] elucidated with immense scientific profundity nanomembranes and its applications. This review primarily focuses on the use of nanomembranes for mass transfer diodes that act analogous to solid state devices based on electron conduction. This well-researched treatise discussed nanomembrane filtration processes, nanomembrane properties, nanomembrane laminations, and the vast domain of applications of nanomembranes.[10]

Science and engineering of nanomaterials and engineered nanomaterials are the vignettes of research endeavor in nanotechnology globally today. The successes of nanocomposites application in environmental remediation are the cornerstones of global research pursuit today. The author in this entire chapter targets the application areas of engineered nanomaterials.

2.8 SIGNIFICANT RESEARCH ENDEAVOR IN THE FIELD OF ENVIRONMENTAL PROTECTION

Environmental protection science needs to be re-envisioned and revamped with the passage of scientific history and time. Technological profundity and the vision of environmental engineering science will all lead a long and effective way in the true realization and the true emancipation of environmental protection globally today.

Katman[11] discussed with immense scientific divination and provenance global climate change, environment, and energy and the global challenges and opportunities to global stability. The success of science and engineering today rests on environmental remediation, water purification, industrial wastewater treatment, and provision of clean drinking water. The struggle for human existence today is aggravating as science and technology moves forward.[11] The authors in this book deeply comprehended on the topics of energy and environmental sustainability in a developed country perspective. Global warming scenario today stands in the midst of deep scientific introspection and technological vision. This treatise rigorously points out toward Turkey's development as regards energy and environmental sustainability applications.[11] A new ethical approach to environmental protection that is ecocentrism is the hallmark of this well-researched treatise. An analysis of Turkish environmental energy policies in a regional framework is another cornerstone of this treatise. Environmental protection and the vast world of environmental engineering science in the global scenario lie in the midst of scientific divination and scientific ingenuity. The authors in this groundbreaking treatise target the scientific needs, the scientific intricacies, and the scientific adroitness of energy and environmental sustainability globally.[11]

Barrow[12] described and discussed with cogent insight environmental management and development. Technological verve, motivation, and vision are today in the midst of a grave disaster as environmental catastrophes globally destroys the scientific firmament.[12] Environmental management is evolving rapidly and is being increasingly applied in developing countries. The other areas are the transboundary and global issues bordering the issues of energy and environmental sustainability. The global issues discussed in this book are resource management issues by sector and environmental tools and policies. Developing nations are in the crucial juncture of environmental disasters and unimaginable drinking water problems.[12] History of human civilization thus is in the crucial juxtaposition of vision and introspection. The author rigorously points toward the sustainability issues linked with development in developing as well as developed nations around the world.[12]

Environmental remediation, industrial pollution control, and water purification according to the civil society should be the cornerstones of human scientific development. Technology and engineering science has practically no answers to the monstrous issues of arsenic and heavy metal groundwater contamination. This chapter enlightens and envisions the needs of sustainability—energy, environmental, social, and economic. The author deeply points toward these monstrous global scientific issues.

2.9 THE STATUS OF GROUNDWATER AND DRINKING WATER TREATMENT IN THE GLOBAL SCENARIO

The status of groundwater and drinking water remediation globally stands in the midst of scientific ingenuity and vast scientific sagacity. The futuristic vision of chemical process engineering and environmental engineering are today challenging the vast scientific firmament. In the similar vision, nanotechnology, composite science, and material science are the needs of human civilization and in the path of newer scientific rejuvenation. The global water crisis is unimaginable and in the same vision groundbreaking. The imminent need of science and technology globally is environmental protection and further scientific emancipation in the field of environmental engineering. In India and Bangladesh, arsenic and heavy metal groundwater contamination are devastating the scientific firmament. The situation is beyond control and immediate global attention is needed. In such a crucial juxtaposition, novel separation processes, advanced oxidation processes, conventional and nonconventional environmental engineering techniques assumes immense importance. Nanotechnology, material science, and composite science are the other scientific avenues of research endeavor.

2.10 GROUNDWATER REMEDIATION AND NANOCOMPOSITES

Groundwater remediation and the application area of nanocomposites are today the imminent needs of human civilization today. Nanomaterials applications in environmental protection are in the similar vein scientific vision of modern science. Man's immense scientific discernment, mankind's vast scientific acuity, and the futuristic vision of engineering and science will all lead an effective way in the true emancipation of nanotechnology and environmental engineering. Nanocomposites and nanomaterials applications should be re-envisioned and revamped with the passage of scientific history and time. Today, composite science and nanotechnology are the opposite sides of the visionary coin. The contribution of nanotechnology in environmental engineering is the imminent need of the hour. The success and vision of science, technology, and engineering of environmental protection will surely open doors of innovation to other areas of research pursuit in material science, nano-engineering, polymer science, and applied sciences.

2.11 NANOTECHNOLOGY, THE SCIENTIFIC VISION, AND THE SCIENTIFIC SAGACITY

The sagacity of science and the ingenuity of engineering science are today changing the vast scientific frontiers. Nanotechnology and its scientific vision are in the midst of forbearance and vast astuteness. Global water scarcity and global climate change are the immediate concerns of human scientific progress today. There is a huge scientific hiatus in the domains of nanomaterials, engineered nanomaterials, nanocomposites, and environmental protection. Here comes the need of a comprehensive treatise in the field of nanocomposites and environmental protection. This treatise will surely open new avenues in research sagacity and research sapience in the field of environmental protection and environmental engineering science.

2.12 ENVIRONMENTAL SUSTAINABILITY, ENVIRONMENTAL PROTECTION, AND THE CHALLENGES OF CIVILIZATION

Environmental sustainability and environmental protection are today in the midst of immense scientific overhauling and deep scientific provenance. Science, technology, and engineering have today few answers to the growing environmental engineering research questions globally. The areas of sustainability are today globally neglected. The challenges of human civilization need to be re-envisioned and revamped with the passage of scientific history and time. The target of research pursuit in nanotechnology should be more scientific emancipation in the field of nanomaterials and engineered nanomaterials applications. The success of human civilization and the success of human scientific progress depends today the vast world of nanotechnology, material science and environmental remediation. The vision of Dr Gro Harlem Brundtland, former Prime Minister of Norway on the science of "sustainability" needs to be redefined and re-envisioned with the course of scientific history and vast scientific ingenuity. Environmental sustainability is the utmost need of the hour as civilization trudges forward toward a new century. Today, environmental sustainability and the vast world of water purification are the two opposite sides of the visionary coin. Provision of basic human needs such as water, electricity, food, shelter, and education are highly neglected in many developed and developing nations around the world. In the similar vein, energy and environmental sustainability are yet to be envisioned in many governments around the world. In this chapter, the author reiterates the success and the need of technology and engineering science in tackling global environmental and energy sustainability. The author rigorously points toward the integration of environmental protection with nanotechnology. Nanotechnology is the absolute need of human civilization and human scientific progress today. Man's immense scientific prowess and sapience, mankind's vast scientific ingenuity and the needs of environmental sustainability will open newer thoughts and newer vistas in research and development globally. Water purification, drinking water treatment, and industrial wastewater treatment are the challenges and the vision of science and engineering today. This well-researched treatise will surely target the domains of energy and environmental sustainability as a further realization of science and engineering globally today.

2.13 THE SCIENCE OF ARSENIC AND HEAVY METAL GROUNDWATER REMEDIATION

The world of science and engineering stands veritably challenged as arsenic and heavy metal groundwater poisoning destroys the vast global scientific fabric. Developing and developed nations around the world are in the throes of the world's largest environmental engineering disaster that is arsenic drinking water contamination. Science and technology in the global scenario needs to be re-envisioned and re-organized as research and development initiatives in developing and developed nations are in the midst of deep scientific intricacies and vast barriers. Today, research and development initiatives in the field of water science and technology should be targeted toward innovation and scientific vision in the field of traditional and nontraditional environmental engineering techniques. Novel separation processes such as membrane science needs to be envisioned and envisaged with the progress of science and technology. Developing nations particularly India and Bangladesh are in the threshold of an unimaginable environmental disaster—the groundwater arsenic contamination. Groundwater remediation stands in the midst of scientific comprehension and vision. The march of human civilization today are replete with innovations as well as disasters. Here comes the need of engineering science and technology. Human mankind's immense scientific profundity, environmental remediation's scientific intricacies, and the needs of sustainability will veritably lead a long and effective way in the true emancipation of environmental science today. The author pointedly focuses on these areas in this treatise.

2.14 THE SAGACITY OF SCIENCE, SCIENTIFIC COMPREHENSION, AND THE MARCH OF MODERN SCIENCE

Water pollution control, drinking water treatment, and industrial wastewater treatment today stands in the midst of deep scientific introspection and immense scientific vision. Research and development initiatives globally are entangled in the midst of immense scientific intricacies. Modern science and water purification science are the challenges and the vision of human civilization today. The author deeply comprehends the needs of energy and environmental sustainability in the march of human civilization today. Modern science and modern engineering science needs to be envisioned as regards its integration with nanotechnology, material science, polymer science, and composite science. The sagacity of science and technology globally thus needs to be re-envisioned and revamped as civilization moves forward. The ever-growing concerns of industrial wastewater treatment and integrated water resource management will surely urge the scientific domain globally to move toward greater emancipation and greater realization of environmental sustainability. Modern science and modern technology are today huge colossus with a vast and definite vision of its own. This treatise widely redefines the success of environmental remediation and environmental sustainability in tackling global environmental changes, global warming, and the vicious domain of global climate change. Scientific sagacity and scientific vision will thus usher in a new avenue in the field of nanotechnology, nanomaterials, and composite science in the similar manner. Global water issues are today moving toward disastrous directions. The civilization has plunged in the deep whirlpool of comprehension and vast scientific introspection. Technology, engineering, and science have practically few answers to the intricacies and barriers in environmental engineering and nanotechnology applications. Here comes the need of scientific and engineering sagacity globally today. In this chapter, the author vastly stresses on the needs of sustainability whether it is environmental or energy in the further emancipation of science and technology.

2.15 FUTURE RECOMMENDATIONS OF THIS STUDY AND FUTURE FLOW OF SCIENTIFIC THOUGHTS

Today, the world of environmental engineering science and nanotechnology are moving forward at a drastic pace and surpassing one engineering boundary over another. The questions of environmental and energy sustainability are still today unanswered as science and technology moves forward. Future recommendations of this study will veritably target the scientific astuteness, the scientific nuances, and the vast scientific ingenuity in the field of nanocomposites and the vast domain of nanotechnology. Technological and engineering challenges in the field of nanotechnology are today vast and ever-growing. The future of science and technology lies in the hands of material scientists and nanotechnologists. So, future flow of scientific thoughts should be in the direction of greater scientific emancipation of nanotechnology.

2.16 CONCLUSION, OUTLOOK, AND SCIENTIFIC PERSPECTIVES

Modern science and the progress of engineering and technology are in the midst of deep scientific forbearance and vast scientific astuteness. Today, scientific and technological challenges of human civilization are vast, versatile and groundbreaking. Technology has few answers to the ever-growing concerns of provision of pure drinking water and the concerns of global environmental sustainability. The application of chemical process engineering, applied chemistry, biotechnology and environmental engineering in human society's progress are of immense importance today. In the similar vein, nanotechnology is today making vast inroads in the avenues of human progress and modern science today. Scientific progress today globally needs to usher in a new era in the field of environmental engineering, human factor engineering, and reliability engineering. The success of engineering science and technology are today changing the face of human society today. The author deeply ponders with vision and might the success of nanotechnology, composite science and material science in the furtherance and challenge of human civilization today. This well-researched treatise vastly opens new thoughts and newer outcomes in the futuristic avenues in environmental remediation and industrial waste water treatment. The vast vision and the challenges of application of nanocomposites in environmental protection are today changing the face of human civilization. The future outlook of nanotechnology and composite science needs to be re-envisioned and revitalized as human civilization and human scientific progress crosses one boundary over another. Material science and composite science are the marvels and pantheons of science and engineering today. This chapter unfolds the deep scientific profundity and the scientific ingenuity in the field of composite science and the vast holistic world of nanotechnology. Nanotechnology capability, the scientific validation of the applications of nanotechnology and the future vision of composite science and material science will all lead a long and visionary way in the true realization of nanocomposite science in decades to come. The author deeply targets the areas of environmental protection and environmental sustainability where there are vast and versatile applications of nanotechnology. The future outlook and the future scientific perspectives will surely target the primary needs of human society such as water, energy, food, shelter, education, and the vast world of sustainability. The success and the vision of global science lie in the hands of sustainability. This chapter targets these intricate issues.

KEYWORDS

- **composites**
- **nanotechnology**
- **vision**
- **review**
- **environment**
- **sustainability**

IMPORTANT WEBSITES FOR REFERENCE

https://www.azonano.com/article.aspx?ArticleID=1832
https://en.wikipedia.org/wiki/Nanocomposite
www.understandingnano.com/nanocomposites-applications.html
https://www.nature.com › subjects
www.scielo.br/scielo.php?script=sci_arttext&pid=S1516-14392009000100002
https://www.tandfonline.com/loi/ynan20
https://www.msm.cam.ac.uk/research/research.../composite-and-nanocomposite-materi...
https://www.emeraldinsight.com/doi/10.1108/01445151111117683
https://www.sciencedirect.com/science/article/pii/S1748013217305649
www.jnpn.org
https://www.hindawi.com/journals/ijc/2018/4749501/
https://www.elsevier.com/books/nanomaterials...nanocomposites/.../978-0-12-814615-.
https://warwick.ac.uk/fac/sci/wmg/research/iinm/
https://www.sigmaaldrich.com/technical-documents/articles/.../lightweight-metal.html
https://www.empa.ch/web/s204/nanoparticles-and-nanocomposites

REFERENCES

1. Camargo, P. H. C.; Satyanarayana, K. G.; Wypych, F. Nanocomposites: Synthesis, Structure, Properties and New Application Opportunities. *Mater. Res.* **2009,** *12* (1), 1–39.
2. Bhattacharya, M. Polymer Nanocomposites—a Comparison Between Carbon Nanotubes, Graphene, and Clay as Nanofillers. *Materials* **2016,** *9*, 262. DOI:10.3390/ma9040262

3. Puggal, S.; Dhall, N.; Singh, N.; Litt, M. S. A Review on Polymer Nanocomposites: Synthesis, Characterization and Mechanical Properties. *Indian J. Sci. Technol.* **2016,** *9* (4), 1–6.
4. Fawaz, J.; Mittal, V. Synthesis of Polymer Nanocomposites: Review of Various Techniques. In *Synthesis Techniques for Polymer Nanocomposites*, 1st Ed.; Mittal, V., Ed.; Wiley-VCH Verlag GmBH & CO: Germany, 2015.
5. The Energy and Resources Institute Report. Nanotechnology Developments in India—a Status Report, (The International Development Research Centre, Canada), Part of the Project: Capability, Governance and Nanotechnology Developments: a Focus on India, 2009.
6. European Commission Report. Successful European Nanotechnology Research, Outstanding science and Technology to Match the Needs of the Future Society, Edited by the European Commission Directorate General for Research and Innovation, Directorate Industrial Technologies, 2011.
7. The Royal Society Report. Nanoscience and Nanotechnologies: Opportunities and Uncertainities. The Royal Society and The Royal Academy of Engineering, United Kingdom, 2004.
8. Arivalagan, K.; Ravichandran, S.; Rangasamy, K.; Karthikeyan, E. Nanomaterials and its Potential Applications. *Int. J. Chem. Tech. Res.* **2011,** *3*, (2), 534–538.
9. Jaksic, Z.; Matovic, J. Functionalization of Artificial Freestanding Composite Nanomembranes. *Materials* **2010,** *3*, 165–200. DOI: 10.3390/ma3010165
10. Kumar, A.; Gayakwad, A.; Nagale, B. D. A Review: Nanomembrane and Application. *Int. J. Innov. Res. Sci. Engineering Technol.* **2014,** *3*, (1), 8373–8381.
11. Katman, F. Global Climate Change, Environment and Energy: Global Challenges and Opportunities to Global Stability; Cambridge Scholars Publishing: United Kingdom, 2014.
12. Barrow, C. J. Environmental Management and Development; Routledge (Taylor and Francis Group): United Kingdom, 2005.

CHAPTER 3

Environmental Performance of Bio-Based Polymer Additives: Thermal Stabilizers and Antioxidants

HUSSEIN A. SHNAWA[1*] and MOAYAD N. KHALAF[2]

[1]*Polymer Research Center, University of Basrah, Basrah, Iraq*

[2]*Chemistry Department, College of Science, University of Basrah, Basrah, Iraq*

[*]*Corresponding author. E-mail: hussanqi@yahoo.com*

ABSTRACT

Polymers thermal stabilizer and antioxidants are essential additives to protect polymeric materials against the thermal degradation both during processing and applications. In recent years, bio-based polymers and polymers additives derived from renewable resources have become increasingly important as sustainable and eco-efficient products. It can replace totally or partially the products based on petrochemical-derived stocks. This chapter deals with the bio-based materials in the field of polymers additives. Most polymers in general, and especially polyolefins, are highly sensitive to thermal degradation and thermal oxidation degradation processes during processing and even when exposed to oxidant atmospheres during its life. Hence, end-users using polymers formulations require the addition of stabilizer to preserve their physical and mechanical properties for long periods. In this chapter, some of the main subjects such as the recent advances regarding and researches in the field of natural polymers additives from renewable resources are considered with a particular interest on the impact of thermal stabilizer and antioxidants from renewable resources.

3.1 GENERAL INTRODUCTION TO THERMAL OXIDATION DEGRADATION

It is well known that the use of polymeric material extends to cover most branches of our life but this global use is not without limitations. One of the limiting factors in the applications of plastics and resins at high temperatures or in outdoor uses are their tendency to not only transfer behaviors from rigid or stiffness to soft or melt states but also to loss of requested polymer properties. The main reason for this failure is attributed to the thermal degradation or to thermal oxidation degradation. Indeed unless correctly inhibited, significant thermal degradation and/or oxidation can start at conditions lower than that at processing and lower than those at which mechanical failure is likely to occur.[1]

Certainly, the thermo-oxidative degradation affects thermoplastic materials properties such as brittleness color, tensile strength and elongation, impact strength, bending strength, intrinsic viscosity, and even loss in weight. The term *thermal oxidation degradation* (ageing) means the influence on polymers by heat energy in presence of oxygen (air). And then the next results, the polymers will lose all efficient physical and mechanical properties.[2]

The changes in polymer properties are usually undesirable as they lead to a worsening of all polymer performances. It is important to study polymer degradation and stabilization in order to:

- minimize the thermal degradation reactions during the thermal process and molding by determination of the appropriate processing conditions and introducing the suitable stabilizers types. Without these additives, the polymers' processes will become more difficult.
- estimate and consequently to increase the life of polymeric objects subjected to the natural or hard usage conditions; to ensure enhanced and controlled processing and service life for polymer products.
- evaluate and select the request lifetime of short- or long-term products from polymers by using the correct concentration levels of the stabilizers.

In polymers, the thermal oxidation and thermal degradation are auto-complex process and are caused by the free radicals mechanisms through three main steps: (i) at the beginning, initiation (alkyl free radicals formation by induced of heat or light, and even by shear), (ii) propagation (not long after, peroxy radicals or other species of free radicals formation, chains scissions, and crosslinking considered as secondary degradation reactions), and, finally, (iii) termination reactions (free radical coupling or quenching).

Most polymers are susceptible to thermal degradation by free radicals via breaking or cleavage of their polymeric chains or by crosslinking bridges between chains. For this reason, these reactions lead to changes in all polymer properties.[3,4,5] Furthermore, most types of oxidation degradation follow a similar basic pattern, especially the oxidation of additions polymers (polyolefin). The conventional mechanism for thermal oxidation degradation is that of a free radicals autoxidation-based process, which involves the major three steps of thermal degradation. The main steps of thermal- or photo-degradation process in most of polymer systems are similar to that which is followed in polymerization reactions in term of their sequences and numbers. In brief, these steps are:

1. **Initiation stage:**
   On the whole, this step can occur either at the thermal conditions of processing and molding or by the conditions of uses. From the mechanical point of view, the initiation of thermal oxidation degradation (and photo oxidation degradation) starts from the loss of a hydrogen atom (or side group) from the polymer backbones (as shown as RH in Fig. 3.1) as a result of energy input from heat or light. This creates a highly reactive and unstable polymer site with unpaired electron "free radical" (R·) with releasing of a free hydrogen atom with an unpaired electron (H). All these chemical species are highly active sources of oxidative process in polyolefin and can inter by high speed reactions with oxygen or other molecules.
2. **Propagation stage:**
   The propagation stage can involve a wide variety of high-speed reactions and one of these is the free radical (R) reactions with an oxygen (O_2) molecule to form a peroxy radical (ROO), which can then remove a hydrogen atom from another polymer chain to form a hydro peroxide (ROOH) and so regenerate the free radical (R). The hydroperoxide can then split into two new free radicals, (RO) + (OH), which will continue to propagate the reaction to other polymer molecules. The degradation or oxidation reactions in the propagation stage, if they start, they are less probability to stop in the absence of stabilizers or antioxidants.
3. **Termination stage:**
   To this end, after change in the chemical structure and loss of most of native polymer properties, termination stage is achieved by coupling or "mopping up" the free radicals to create the final reaction products or relativity less-active molecules. This can occur naturally by combining

free radicals or it can be assisted by addition stabilizers into the plastic. Reactions 3.1 and 3.2 are examples to termination stage. Again, all these reactions are undesirable reactions and cause changes in polymer properties even at low levels.

$2R^{\cdot} \rightarrow R\text{-}R$ (crosslink) ....(3.1)

$ROO^{\cdot} + R^{\cdot} \rightarrow ROOR$ (more stable products) ....(3.2)

The main degradation steps, starting and final products, as well as the intermediate compounds of oxidation degradation cycles are illustrated by the schematic diagram given in Figure 3.1.[5]

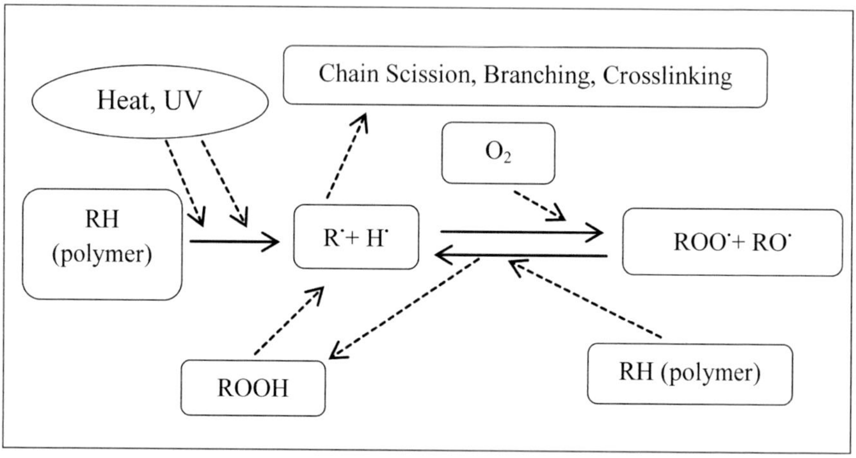

FIGURE 3.1 Polyolefin thermal oxidation-degradation process and its products.

In Figure 3.1, RH represents a polymer chain, R• represents an alkyl radicals; RO• represents an alkoxy radical; ROO• represents a peroxy radical; ROOH represents a hydroperoxide molecule.

Without a doubt, the thermal degradation as well as the thermal oxidation degradation reactions and their speeds are directly influenced by original polymer structure properties. For example, the crystallinity, the rate of oxidation, is inversely proportional to degree of crystallinity because these structures do not allow oxygen to diffuse into the polymer system. Moreover, the crystal regions in polymer matrix show more resistance to degradation and oxidation. In the same way, tacticity, stereo regular polymers side groups per main single chain, are leads to more crystallinity and to more resistance to oxidation. Molecular weight also has direct effect but, in solid polymers, both the rate of initiation and the rate of termination stages are reduced. In solution, oxidation rate decreases with increase in molecular weight.[4,6,7]

Furthermore, the effects of chemical structure clearly appears in irregular structures such as internal double bonds and in highly branching structures, which leads to low crystals regions in polymer matrix. Accordingly, the branching structures cause higher rate of oxidation due to the presence of more number of tertiary H atoms and slow termination of tertiary peroxy radicals. For these reasons, the high density polyethylene (HDPE) polymer is more resistance to oxidation degradation than the low density polyethylene (LDPE).

Presence of double bonds, on the other hand, results in more allylic positions labile or active groups, which are inherently more reactive and hence increases oxidation rate Above all, film thickness also has an indirect effect. Generally the rate of oxidation reaction decreases with increasing in thickness of polymer sample where the degradation or oxidation reactions become slow due to the diffusion of oxidation elements is controlled by rate of diffusion.. Impurities and additives, many ingredients, for example, catalyst residues, processing aids, metallic impurities, etc., have been considered to be the main reasons for increasing oxidation rate.[4,6,7]

In summary, throughout the compounding, molding, and extrusion process, polymeric products are exposed to the effect of heat, shear, and oxygen. And also during end uses, almost all commercial polymers are susceptible to conditions that tend to initiate oxidation and degradation processes in their polymer chains and change their chemical structures following this, their properties will change even before the finished product has been packed for shipment. It follows, then, to maintain a polymer's original molecular weight, appearance, physical properties, and mechanical properties throughout its planned processing and design lifetimes, thermal or heat stabilizers and antioxidants are essential additives in thermoplastic compounds.

3.2 BIO-BASED ANTIOXIDANTS AND THERMAL STABILIZERS

In the past few decades, concerns about the environment, climate changes, and limited fossil resources has led to intensive researches of alternatives resources to fossil-based polymers and polymer additives. Bio-based products represent a readily available and alternative resource for many fossil-based products. For example, the addition of antioxidants is the most used method to inhibit or to prevent the thermal oxidation degradation process that may occur in many polymeric materials during its processing and its service life. However, the safety of synthetic antioxidant is not full undoubtedly. For these reasons, the evaluation and potential use of natural antioxidants for polyolefin stabilization are now receiving special attention and has became the main goal of many research groups and authors.

Concurrently, more studies for new polymers antioxidants and other additives has been encouraged in the last years because of some doubts about the effects of the reaction products of synthetic phenolic antioxidants on human health.[8–10]

Practically, synthetic antioxidants such as butylated hydroxytoluene (BHT) and butylated hydroxyanisole (BHA) are used for stabilization of the polymers, some food products, as well as other hydrocarbons in industrial practice. But about several years ago, some warnings emerged regarding the effects of phenolic antioxidants and their reaction products on human health and on the environmental systems. Recently, polymers synthetic antioxidants and some heavy metal thermal stabilizers have been reported as nonsafe materials for human health. During the storage periods or during the service life, the antioxidants can be extracted or migrated to the food when used for stabilization the polymeric food containers. Generally, most of these warnings have not been solved yet and many attempts has been intensified on the use of some alternative natural antioxidants (from plants) to prevent the side effects of industrial antioxidants, especially for food packaging applications.[11,12]

In the same manner, the production and uses of thermoplastic polymers such as polyvinyl chloride (PVC), low and high density polyethylene (LDPE, HDPE), and polypropylene (PP) in the world continues to develop and grow. These materials are successfully used in many economical activities. By contrast, the successful and broad uses of plastic materials in many applications, for example, in the automotive industry, the electronics sector, and the packaging and manufacturing of consumer goods, are substantially attributable to the incorporation of additives into virgin (and recycled) resins.[13]

Above all, the properties of polymer are not only influenced by the matrix of polymers but also by the additives types and their composition, which is specifically tailored for the respective applications. Usually, up to ten different additives with concentration ranging from some 100 ppm to several weight percent are used in order to adjust properties like viscosity, brittleness, as well as light and thermal stability.[13,14]

Therefore, replacement of petroleum-based raw materials, especially polymers and polymer additives, by green and renewable resources, such as materials and polymers from plant resources, constitutes a major contemporary challenge in terms of both economic and environmental aspects (environmental friendly and biodegradable). Generally, natural polymer additives, which impart low toxicity, total, or partial biodegradability, are economically and technically viable to substitute the conventional petrochemical-based products.

The fundamental benefits from uses and applications of bio-based additives or bio-based polymers as alternative materials for petroleum-derived products or polymers blend and composites are, among others: prevention of an imbalance of supply and demand of products manufactured from nonrenewable fossil raw materials, sustainable waste management, carbon emission reduction, biodegradability of materials or facilitated recycling process, low toxicity, and low costs production.[15]

Without a doubt, throughout the compounding, molding, extrusion, and end uses almost all commercial polymers are susceptible to conditions that tend to initiate oxidation and degradation processes in their polymer chains. Following that, changes occur in their chemical structures and the outcome losing their properties even before the finished product has been packed for shipment. To maintain a polymers' original molecular weight, appearance, physical properties, and mechanical properties throughout its planned processing and design lifetimes, thermal or heat stabilizers and antioxidants are essential additives in all thermoplastic formulations.[14,15]

Incorporation of small amounts of chemical substances that are able to suppress and/or delay the thermal oxidation and degradation processes is considered a critical step for polymers manufacturing. This method is known as polymer stabilization by additives, and the chemical substances are named thermal stabilizers. Generally, most of thermoplastics can be protected from thermal degradation and thermal oxidation by incorporating thermal stabilizers and antioxidants into them. Stabilizers are used to keep the polymer chains and the original molecular structure intact and therefore polymers properties such as strength, stiffness, and toughness can be retained over a longer period.

Antioxidant is any substance which when found into polymers at small amounts (less than 2 wt% per polymer weight), significantly leads to protection of the polymers and plastics (and other hydrocarbon materials) against thermal and/or photo-oxidative degradation processes. These materials at least, must delay the oxidation process that occur in thermal or photo conditions or during natural ageing. Some types of these additives come from natural resources such as vitamins and many naturally occurring plants products (phytochemicals), and other types produced chemically by industrial processes. Antioxidants are able to neutralize reactive molecules (free radicals) and reduce oxidative damage by different mechanisms. They attract intense scientific and economic interest in polymer and petrochemical industries, human health, and many food industries.[16,17]

Synthetic antioxidants are used for stabilization of polymers in industrial practice and as a consequence, more and more attention is paid to the potential use of nontoxic natural antioxidants for stabilization polymers in food packaging applications. Several compounds with different chemical structures may be considered as natural stabilizers in polymers and specifically in polyethylene and polypropylene. Plant phenolic molecules are heterogeneous and large groups of secondary plant metabolites. Antioxidant from plants resources include simple phenolics, phenolic acids, anthocyanins, flavonoids, and tannins, all these compound act key roles in stability of stability of different systems against oxidation, disease resistance, pigmentation, and the harmful effects of environment. Many natural phenolic compounds have been reported to act as high performance antioxidants.[18]

Several natural compounds with different chemical structures may be considered as natural stabilizers in polymers and specifically in polyolefin. Vitamin E, that is, alfa-tocopherol has gained a great interest within the polymer and polymer additives industries due to its their beneficial properties. Its effect as an antioxidant was studied in detail, among others, by Al- Malaika et al.[19–22] and it was shown to be a very efficient stabilizer.

Similarly, lignin is the second-most abundant renewable polymer and comes directly after cellulose. The chemical structure of lignin, as shown in Figure 3.2, contains both aliphatic and aromatic functional groups. The aromatic part consists of benzene ring and sterically hindered phenolic hydroxyl groups; all these functional groups introduce lignin with antioxidant activity. From the scientific and industrial point of view, the resources of lignin can be obtained from the waste or by-product of paper industries or biofuel productions, but it could also be isolated by many chemical processes. This phenolic bio-polymer can be used to produce low-molecular weight phenols and has potential to be used as antioxidant or polymer additive.[23,24]

1. R1=OMe, R2=H: Coniferyl alcohol/guaiacyl

1. R1=R2=OMe: Sinapyl alcohol/syringyl

3. R1=R2=H: p-Coumaryl alcohol

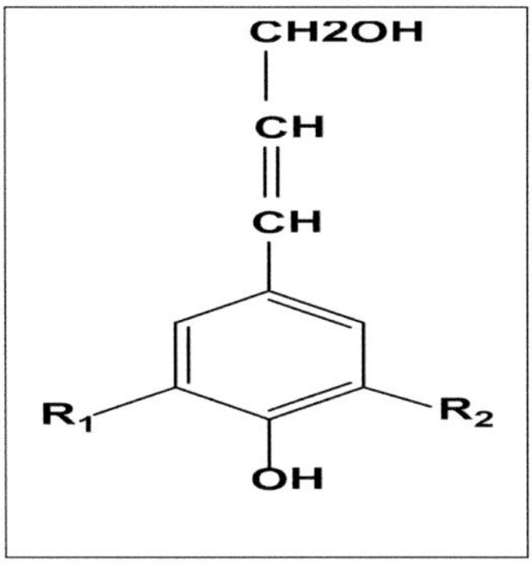

FIGURE 3.2 The three repeated units of lignin.

Lignin as macromolecule or the fine chemicals obtained by chemical treatment (pyrolysis) of this polymer is also considered as prospective antioxidants. Hindered phenolic groups of lignin can act as stabilizers in reactions induced by oxygen reactive species and by heat condition and then slowdown of ageing in composites and biological systems.

In the work of Alexander Arshanitsa et al.,[25] the antioxidant activity of the lignin fractions were tested by their influence on thermo-oxidative destruction of prepared model polyurethane, PU films. The data of TGA method under oxidative conditions clearly testified. In this work, fractionation of lignin with solvents of different polarity is a prospective tool for obtaining of more homogeneous lignin products, which reveal prominent antioxidant activity. The results obtained clearly show that antioxidant activity of fractionated technical lignin is one more advantage for lignin application in PU production.

Another natural product with antioxidant properties is quercetin, that is [2-(3,4-dihydroxyphenyl)-3,5,7-trihydroxy-4H-chromen-4-one] (Figure 3.3), which is produced by plants and can be found in some parts such as seeds, leafs, red grape, red onion, and so on. Quercetin acts as plant pigment and it belongs to a class of plant secondary metabolites known as flavonoids. Such as other plants polyphenols, this natural product have gained large interest in the field of polymers additives, food, and pharmaceuticals due to its antioxidant activity and other properties. Quercetin is capable of scavenging reactive oxygen species and its antioxidant potential is attributed to this free radical scavenging activity. Some investigations were aimed to use this compound as antioxidant for polymer systems. Quercetin has been successfully used as an antioxidant in low-density polyethylene.[26,27]

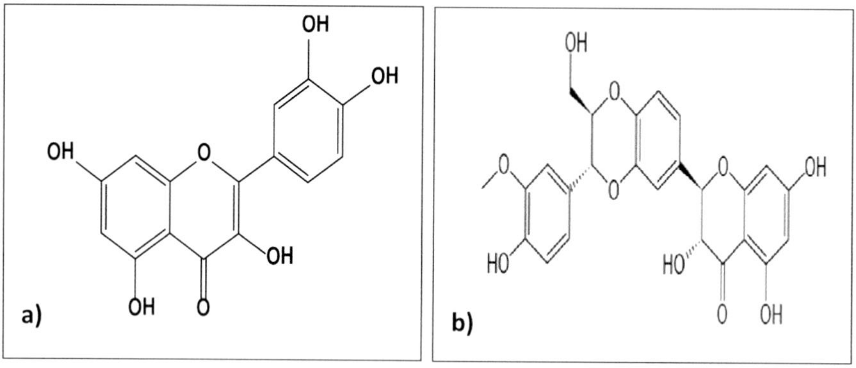

FIGURE 3.3 Chemical structures of quercetin (a) and silibinin (b).

Flavonoids have gained recent interest due to their activity in biological systems. Flavonoids are characterized as containing two or more aromatic rings, each one bearing one or more phenolic hydroxyl groups, and connected by a carbon bridge. The capacity of flavonoids to act as antioxidants depends on their multi-hydroxyl groups' chemical structure. The position of hydroxyl groups and other features in the chemical structure of flavonoids are important for their antioxidant and free radical scavenging activities.

In the study of M. D. Samper et al.,[28] several flavonoids (quercetin, silibinin, etc.; Figure 3.3) and other flavonoids types were investigated as potential sources of antioxidants and UV light stabilizers for PP. Mixtures of PP with each one of the flavonoids were obtained by means of extrusion and their behavior under UV light radiation and thermo-oxidative behavior was analyzed. Thermal stabilization had been assessed in an oxidizing atmosphere by means of differential scanning calorimetry (DSC) both in isothermal and in dynamic conditions. In addition, the effectiveness of these phenolic compounds as thermal stabilizers at high temperature had been quantified with the use of thermogravimetric analysis (TGA).

Simultaneously, stabilization effect of these phenolic products against UV radiation of polypropylene had been estimated by study the morphology changes of the exposed surfaces by scanning electron microscope (SEM);

also, surface chemical changes had been followed by infrared spectroscopy. Global results of this study show that the quercetin and silibinin provide the best results in stabilizing both against oxidation degradation and against photo degradation.

Addition of small amounts of these materials contributes to the development of more environmental-friendly formulations for this type of polymers, and thus allowing full or partial replacement of petrochemical antioxidants. Furthermore, it has been shown that within the group of flavonoids, the silibinin and quercetin provide similar activity to petroleum-based antioxidants widely used in the stabilization of polyolefins.

In the same manner, the potential use of quercetin, a flavonoid-type natural antioxidant, as a stabilizer in polyethylene was explored in the work of Dóra Tátraaljai et al.[29] Its efficiency was compared to that of Irganox 1010, a hindered phenolic antioxidant used routinely in industrial practice. The results of this study showed that quercetin is a very efficient antioxidant. It prevents the formation of long chain branches.

Once again, the efficiency of quercetin is considerably better than that of Irganox 1010, the hindered phenolic antioxidant used as reference stabilizer. The difference in efficiency might be explained with the dissimilar number of active -OH groups on the two molecules, but the stabilization mechanism of quercetin may be also different from that of Irganox 1010.

Curcumin, shown in Figure 3.4, is an oil-soluble natural pigment, practically insoluble in water at acidic and neutral pH, and soluble in alkali. Many publications[30–32] indicate that the strong antioxidant activity of curcumin originates mainly from the phenolic OH groups.

In his previous study, Dóra Tátraaljai et al.,[33] the effect of curcumin, 1,7-bis(4-hydroxy-3-methoxyphenyl)-1,6- heptadiene-3,5-dione, on the melt stability of polyethylene during processing was investigated. The final results reveal that the melt-stabilizing efficiency of curcumin is superior to that of the synthetic antioxidant investigated (Irganox 1010) and is further enhanced by the addition of the phosphonite secondary antioxidant. The changes in the characteristics of the polymer indicate that besides the phenolic OH groups, the linear linkage between the two methoxy phenyl rings of curcumin also participates in the stabilization reactions. The results of their study reveal that the melt stabilizing efficiency and thermo-oxidative stabilization of curcumin in polyethylene was superior to that of the synthetic antioxidant used as reference.

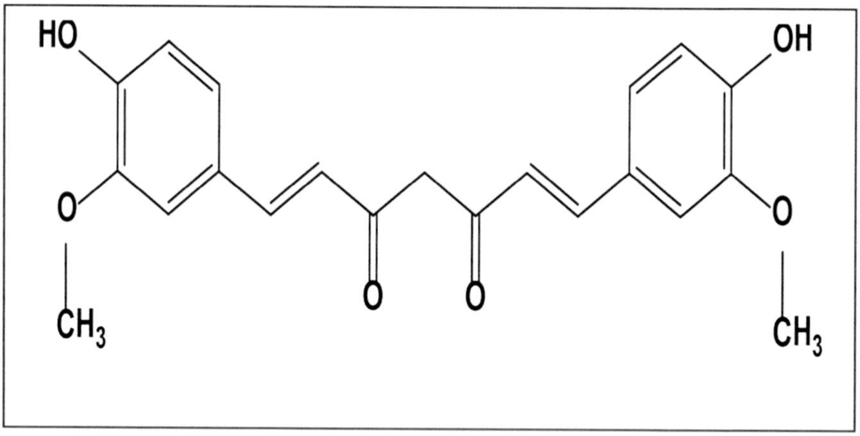

FIGURE 3.4 Chemical structures of curcumin.

On the other hand, in recent years, bio-based polymers, monomers, composites, and additives derived from renewable resources have become increasingly important as sustainable and eco-efficient products. To exemplify, polyurethanes are one of the most important and versatile materials with applications ranging from flexible foams in upholstered furniture to rigid foams as insulators and automotive interiors, which can be made from natural raw material such as plant oil derivatives. So the partially bio-based polyurethane can replace the products based on petrochemical-derived stocks.[34]

Once again, the lignocellulosic materials including agricultural or forestry wastes contain natural polymers (e.g., cellulose, lignin, hemicellulose, and tannins) with more than two hydroxyl groups per molecule. For example, they can be used as additives for thermoplastics and as polyol for PUs preparation and they have pushed scientific researchers to prepare and evaluate various materials analogous to polymeric composites materials, foams, adhesives, etc.[35,36]

Furthermore, corn and soybean, also, are primarily used as feedstock to produce bio-based polyols in North America. The demand for green and bio-polyols in the market is driven for end-user industries such as automotive, packaging, and furniture. Actually, almost 70.0% vehicles manufactured by Ford Motor Company in North America contained seat components that use green and bio-polyols. Cargill Inc. (U.S.), The DOW Chemical Company (U.S.), Johnson Controls Inc. (U.S.), etc., are among the largest manufacturers of green and bio-polyols in North America[37]

As another example, epoxy cross-linked polymers or its pre-polymers are one of the most widely used and versatile compounds in thermosetting resins family. Bio-based epoxy pre-polymers can be successfully developed from vegetable oils or other materials, such as soybean oil, linseed oil, or palm oil, and they hold great potential for renewable resource-based and low-cost materials and can be used for PVC stabilization and plasticization. On the other hand, the raw material used in the production of phenolic resins (mainly phenol and formaldehyde) are obtained in a large scale from nonrenewable sources. Thus, the substitution of these reagents by equivalent ones obtained from nonfossil sources is an interesting alternative, providing both economic and environmental benefits.[38,39]

In fact, for the simultaneous development of petroleum-origin or bio-derived plastic systems, there is a need for parallel developments of sustainable, functional plastic additives. Therefore, many investigations have been done on the development of bio-based (green) polymers, bio-based polymer additives, or polymer modifiers derived from renewable resources such as sugars, polysaccharides, vegetable oils, lignin, tannins, plant phenols, pine resin derivatives, furans, and other natural products or by-products.

As a matter of fact, polymers (synthetic and bio-based polymers) industries are impossible without additives. Additives in plastics provide the means to solve processing problems, property performance limitations, and restricted environmental stability are overcome by additives. An additive is a substance that is incorporated into plastics to achieve a technical effect in processing and in the finished products, and it is intended to be an essential part of the finished articles. Some examples of additives are antioxidants, antistatic agents, fillers, impact modifiers, lubricants, plasticizers, thermal stabilizers, thickeners, and UV absorbers.[40,41]

Indeed, polymers additives are needed not only to make resins processable, but also to improve the properties of the products during use. As the scope of plastics has increased, so has the range of additives: for better mechanical properties, resistance to heat, light and weathering, flame resistance, electrical conductivity, etc. Plastic additives contribute a range of functionality and process ability to plastics.[40] Researches evaluating bio-derived plastic additives has ranged from fermentation products, forest-waste industries, agrowaste polymers, and modified vegetable oils as PVC thermal stabilizers and PVC/NBR alloys modifier[42] to those based on polyphenols, such as lignin, tannins, and virgin olive pomace or other bio-products as antioxidants.[43,44]

Beta-Carotene, as example, is one natural product with antioxidant activity and appears as color pigments in some plants and fruits, as carrot and pumpkins. It is a nonpolar highly conjugated lipophilic hydrocarbon compound; beta-carotene is considered as a chain-breaking antioxidant, as it exhibits high reactivity toward carbon-free radicals, single oxygen, peroxy, alkoxy, and NO_2 radicals. According to Ozhogina and Kasaikina,[45] beta-carotene acts as antioxidant by the formation of free radicals in its conjugated system. Therefore, it is considered as a natural antioxidant that works by chain-breaking mechanism.

Abdel-Razik[46] found that the beta-carotene increases the thermo-oxidative stability of ABS copolymer and he explained its effect by the formation of biradicals, which can react rapidly with oxygen molecules inhibiting the formation of alkoxy and peroxy radicals from the polymer. Preliminary experiments run in the laboratory of Tátraaljai Dora et al.[47,48] revealed that the beta-carotene improves the processing stability of polyethylene. The positive effect is further enhanced by the addition of a phosphonite secondary antioxidant; the efficiency of this additive combination proved to be similar to that of the synthetic phenolic antioxidant package frequently used in industrial practice.

The effect of beta-carotene on the behavior of polyethylene stabilized with alfa-tocopherol and a phosphonite antioxidant was studied under processing and storage conditions by works of Tátraaljai Dóra.[47,48] The Conclusions of this studies about the effect of beta-carotene in polyethylene stabilized with a combination of alfa-tocopherol and phosphonite reveals that the additive with its eleven conjugated double bonds colors the polymer significantly but does not reduce the stabilizing efficiency of antioxidants either during processing or in the course of storage at ambient temperature.

Plants-derived phenolic compounds such as alfa-tocopherol, thymol, carvacrol, tannic acid, and carnosic acid show acceptable antioxidant activity in plastics and rubbers, to provide specific melting conditions or to thermally stabilize polyolefin. Stabilization of PE, PP, and polyvinyl chloride (PVC) with natural antioxidants or thermal stabilizers for packaging materials or other applications is a new challenge both from the scientific and the industrial point of view. Replacing synthetic phenolic antioxidants by natural types could stop worries caused by the possible harmful effects of their reaction products on human health.[49,50]

Poly(vinyl chloride), PVC, is a common commodity plastic, and its production is the third largest, after polyethylene and polypropylene. PVC is one of the widely used and versatile synthetic polymers, which increasingly is being used in various industries such as building and constructions, packaging, electrical, automotive, furniture/office equipment, clothing, and footwear. Hence, PVC is one of the most important polymers. It is applied in almost every aspect of our lives, because it is easily modified and inexpensive. As an example of PVC applications, PVC is the most widely used plastic resin in medical devices. Approximately 25% of all plastic medical products are made of PVC, according to most market estimates. The main reason is the resin's low cost, ease of processing, and the ability to tailor its properties to a wide range of applications.[51]

PVC can be manufactured in a range of flexibilities and rigidities. Also, it can be used in a wide range of temperatures, and it retains its strength and durability at low temperatures. PVC formulations exhibit excellent strength and toughness and it exhibits very good water, fire, and chemical resistance and stability and it is also biocompatible for some applications. PVC is relatively lower cost and has high-performance value and it has safety and cost advantages for a wide variety of medical applications, especially for single-use disposable devices.[51,52]

In an opposing view, PVC is characterized by poor heat and UV resistance due to the presence of some abnormal structures that originate from synthesis. Addition of thermal stabilizers is, therefore, required to prevent thermal degradation during process. The poor thermal stability of PVC leads to an extensive discoloration of the polymer and deterioration of its physical and mechanical properties. Because all these changes are accompanied by deterioration on some of the useful properties of the polymer, it has become the practice to process PVC in the presence of heat stabilizers.[53]

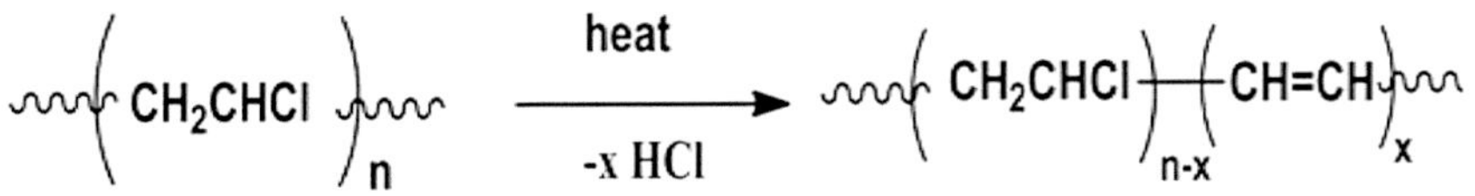

SCHEME 3.1 Thermal dehydrochlorination (DHC) reaction of PVC.

Moreover, PVC has much different selection rules than other polyolefins due to its decomposition mechanisms that include elimination reaction to produce HCl and polyenes, as shown in Scheme 3.1. After that, the polyenes react with oxygen by free radical-based oxidation reactions (Figure 3.5). Because of all these reactions occur at processing or at relativity lower temperatures, the thermal stabilizers consider as essential additives to protect this polymer during processing and using ages. Commonly used thermal stabilizers for PVC belong to one of four categories: lead salts, metal soaps, epoxy resins, organo tin compounds, and some auxiliary stabilizing materials. In the United States, tin stabilizers dominate the market and represent state of the art. Europe has traditionally used lead-based stabilizers for PVC. However, they will completely switch away from lead in the next several years. The most common replacement are solid calcium/zinc-based stabilizers.[53,54]

Commercially, PVC is rarely used as a virgin resin because it is thermally unstable at the temperatures required for processing and even at environmental conditions when used for outdoor applications and also at sterilization temperatures when used into medical tool products. So, it relies on the inclusion of stabilizers and other additives. As a result of the low thermal stability of PVC, it has limited material properties and requires a range of additives, which are determined by the target application. These additives are mainly plasticizers, stabilizers, antioxidants, antistatic agents, antifogging agents, impact modifiers, lubricants, and flame retardants. Among them, the addition of thermal stabilizers, as well as co-stabilizers, produce this polymer to become the most important commodity plastics and it has been produced for more than 60 years on an industrial scale.[55,56]

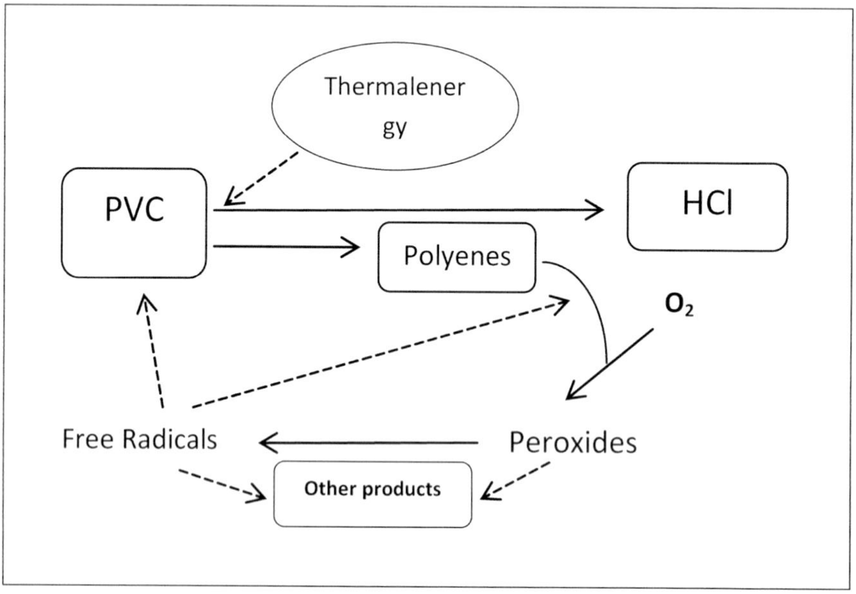

FIGURE 3.5 Thermal degradation process of PVC and its products.

As strong distinction, the thermal degradation of PVC occurs *via* thermal dehydrochlorination (DHC) leading to the release of hydrogen chloride gas (HCl) and the formation of conjugated double bonds in the polymer chains. This results in severe coloration and loss of physical and mechanical properties. Heat stabilizers are typically used in commercial grade PVC, to protect it against not only the high temperatures the resin might confront it during processing, but also the high heat it may encounter in outdoor uses, storage, or autoclaving. Barium/zinc-based additives are very effective heat stabilizers for PVC but are restricted for medical applications in some countries. Alternatives like calcium/zinc formulations are often used to stabilize medical-grade PVC.[56]

Heat stabilizers trap the hydrogen chloride that is generated when PVC decomposes at relativly high temperatures. This prevents discoloration and degradation. Rigid PVC may contain up to 15% by weight of thermal stabilizers. Another additive, Tinuvin®, is used to provide stability to this polymer from exposure to UV light.

Natural or bio-based thermal stabilizer for PVC include metal soap of plant oils and its epoxides; more recently, many papers focused on using the epoxides of unsaturated natural products as secondary thermal stabilizer additives for PVC. Epoxy compounds are well known as typical nonmetallic stabilizers for PVC. They are generally regarded as secondary stabilizers used to enhance the effectiveness of metal soaps. They act as acceptors for the liberated hydrogen chloride and retardants for the appearance of discoloration. The effects of epoxidized sunflower (ESO) on the thermal degradation and stabilization of PVC in the presence of metal carboxylates (Ba/Cd and Ca/Zn stearates) have been investigated. ESO showed excellent properties as secondary stabilizer for PVC.[57]

Epoxidized soybean oil (ESBO), for example, is a vegetable oil widely used as plasticizer and/or stabilizer for PVC in food contact materials. The study carried out by C. Bueno-Ferrer et al.[58] aimed to evaluate the behavior of ESBO as the only plasticizer and stabilizer for PVC at different concentration levels by means of a structural and thermal study. Formulations based on PVC with different amounts of ESBO (from 30 to 50) wt% were fully characterized showing good compatibility and a clear increase in thermal stability. An evaluation of the use of ESBO for PVC stabilization in commercial lids was carried out by using thermogravimetric analysis (TGA). Results of this study showed a significant increase in thermal degradation temperatures of PVC. It was reported that the ESBO stabilization mechanism involves a reaction between the epoxide ring and hydrogen chloride generated during PVC degradation. This reaction prevents PVC from further dehydrochlorination, preserving its color and limiting loss of PVC properties.

Epoxidized sunflower oil (ESO) was used as an organic thermal co-stabilizer for rigid poly(vinyl chloride) (PVC) in the presence and absence of mixture tricalcium dicitrate and mercury (II) acetate as thermal stabilization in the studied in the work of M. T. Taghizadeh.[59] It was found that the additives retard the thermo-oxidative degradation of PVC and reduce the extent of polymer chain scission associated with the thermal degradation of poly(vinyl chloride). The researcher in his conclusion reported that the epoxidized sunflower oil shows excellent properties as a secondary stabilizer for PVC when used in combination with the synergetic metal soaps (Hg/Ca).

Therefore, because of their susceptibility to auto-oxidation, commercial polymers such as PP, PE, and PVC will require stabilizers (antioxidant or thermal stabilizers) to be useful at commercial scale. In the lucrative plastic additives market, sustainable options remain less developed, particularly those additives being both renewable and biodegradable. Several crop wastes contain chemicals and raw materials and many agri-food wastes are considered as biosource for high value-added additives.[60]

Waste material from industrial processing of some crops, nowadays a cost for agri-food companies, contains chemicals and raw material that until now have been exploited only partially. Antioxidants-rich fractions were extracted from hazelnuts, chestnuts, and especially grape seeds. The chemical components of these extracts are tannins and polyphenolic compounds, which represent the major family of natural antioxidants. They are chemical analogues of synthetic stabilizers widely used in plastic industry as well.[60,61,62]

Natural additives obtained from wastes or byproducts of agro-food industry deserve particular attention. Nowadays, huge amounts of biowastes are produced by agri-food processing industry, and their disposal has a significant environmental impact. To improve eco-sustainable and cost-effective waste management, several attempts have been made to convert byproducts into bio-fuels, composites, and feedstock. Moreover, large amounts of byproducts generated during crop growing, processing, and storage contain highly helpful, biologically active substances, whose extraction provides high added-value compounds. More emphatically, most of these by-products are sources of antioxidant, antibacterial, anti-inflammatory compounds, such as polyphenols, stilbenes, flavonoids, etc.[60,61]

Polyphenolic materials, in the same manner, such as flavonoids and condensed tannins extracted from plant materials (from forestry or viticulture by-products) are available in high quantities, being historically used in the leather tanning and adhesives industries. Condensed tannins, shown in Figure 3.6, primarily, are polyphenolic oligomers based on a flavanyl repeat unit with the hydroxylation pattern providing potential for activity antioxidant and chemical modifications activity. Polyphenolic materials extracted from plant materials are available in high quantities. In plants, these compounds offer a range of protective roles, including astringency, pathogen, and UV inhibition.[62–65]

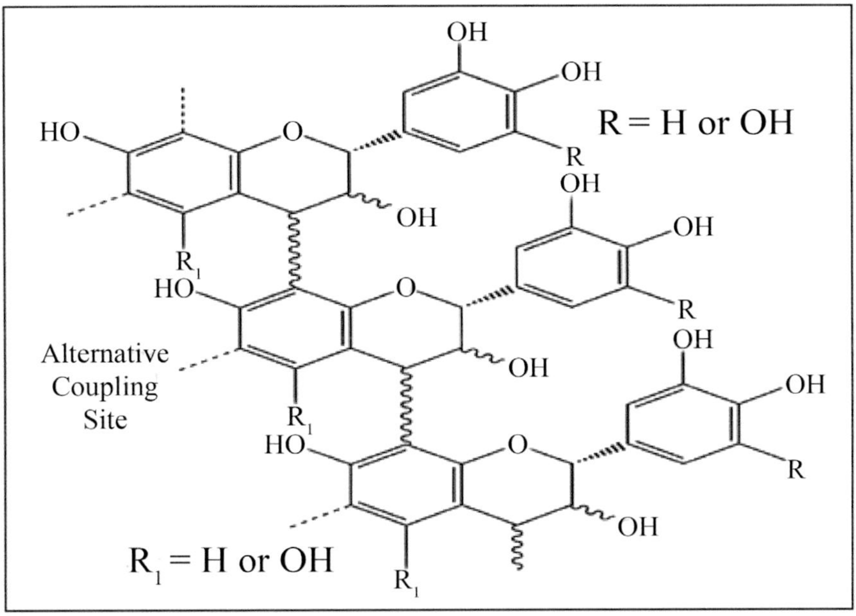

FIGURE 3.6 The chemical structure of condensed tannins.

Tannins have increasing attention due to their potent antioxidant properties and to their marked effects in the prevention of various oxidative stress-associated diseases such as cancer. In the last few years, the identification and development of phenolic compounds or extracts from different plants has become a major area of health and medical-related researches.[66]

Plant phenolics have been considered powerful antioxidants and proved to be more powerful antioxidants than vitamin C and E and carodenoids. Given their relationship to phenolic acids and flavonoids, in terms of phenolic structures with multihydroxyl groups and suitable positions, they have been reported to possess many useful properties, including antioxidant and anti-inflammatory properties. They exert their antioxidant activity by scavenging free radicals, chelating trace metals ions.

Liangliang Zhang and his coworkers[67] extracted phenolic compounds from the bark of *A. mangium* and their antioxidant activities were investigated. The author used MALDI-TOF MS analyses to characterize the structures of condensed tannins isolated from *A. mangium*. The results of these analysis shows that the condensed tannin from *A. mangium* were shown to be heterogeneous oligomers mixtures consisting of procyanidin and prodelphinidin structural units with polymerization degrees up to 9.

In this study, a significant linear relationship between antioxidant potency, antiradical activity, and the content of phenolic compounds of bark extracts was observed. In the conclusion, the researchers suggest that the *A. mangium* bark contains considerable quantities of antioxidant phenols. These extracts may be considered as new sources of natural antioxidants for great importance for the industry, since the extracts of these byproducts are finding increasing applications as active substances for cosmetic and pharmaceutical compositions.

There are extensive studies aimed to evaluate antioxidant activity of tannins extracted from medicinal plants or from agricultural byproducts, such as green tea waste, chestnut and persimmon hulls, wood manufactures, and paper industries.[68–70] Most of these publications attributed the anti-oxidation activity of plant phenolics to different mechanisms. Scavenging of free radical, which initiates peroxidation, is one of the most-known mechanisms by which the natural phenolic antioxidants inhibit oxidation reactions, and then breaks the cycle of generation of new radicals. Formation of chelating complex with metal ions is the second example of mechanism, which was suggested to explain the ability of these materials to act as antioxidants.[18,71]

Most of researches relates with tannins as antioxidants and demonstrated that the tannins exhibited strong activity and may provide alternative and supplements to conventional anti-oxidant feed additives. Extractions of phenols as anti-oxidants from eucalyptus bark were done by G. Vazquez and his coworkers[72] They demonstrated the potential of eucalyptus bark as source for natural antioxidants products and they found that the eucalyptus bark extracts with lower molecular weight showed high antioxidant activity.

The results of Shu-Dong Wei and et al.[73]study which were obtained byMALDI-TOF MS and RP-HPLC analysis point out that the catechin/epicatechin were the basic units occurring in *M. pauhoi* leaf. The authors point to that the condensed tannins isolated from *M. pauhoi* leaves are more effective as free radicals scavenging than the ascorbic acid and butylated hydroxyanisole (BHA), and may be considered as a new source of natural antioxidants for food and nutraceutical products. The results suggest that the tannins are potentially very important biological antioxidants.

In Warren J. Grigsby[74] study, both native tannins and chemically modified tannins have been evaluated as functional additives in synthetic and biodegradable plastics, such as polylatic acid. Based on his preceding works, tannin materials with long ester chain lengths exhibited melt behaviors suitable for processing of plastics. Pine bark tannin (PBT) is used and preferentially esterified with hexanoate or other ester chains to improve tannin hydrophobicity and miscibility within the plastics. According to this research, tannin esters compounding into PLA can lower the PLA glass transition by 5–6 °C. Overall, the researcher illustrated that the general results suggest scope for the use of tannin esters as plastic additives.

In his next study, Warren J. Grigsby[75] esterified and native tannins were evaluated as stabilizer additives in polypropylene and an aliphatic polyester. These two polymers have been also used as parallel plastic systems to assess the plasticization functional attributes provided by the tannin-based additives. It is found that tannin esters have inhibited the UV degradation effects in both polypropylene and aliphatic polyester, and has extended the oxidative induction time of polypropylene. The general results of this study suggest that the tannins and it's derivative have potential activity to replace synthetic additives (and also they act as color material) and improve plastic sustainability credentials.

These properties, for instance, their availability and relative ease to chemically modifications, make tannins attractive as polymer additives or modifiers. They could be modified by acetylation, chemical hydrolysis, condensation, polymerization, etc. They may also be copolymerized with iso-cyanates, formaldehyde, amine-or phenolic resins to elaborate thermosetting binders for increasing the compatibility with polyolefin. Finally, increasing interest in the lowering or replacement of synthetic antioxidants has led the researches to focus into natural antioxidants, especially of plant origin materials.[76–79]

3.3 CONCLUSION

By balance between the presence and amounts of antioxidants or thermal stabilizers in polymeric systems, it can be protecting these materials from the thermal degradation or thermal oxidation degradation. Bio-based materials in the field of polymer additives are considered as important and promising additives due to their unique structural features, chemical modification ability, abundant availability, and low cost. A large number of efficient chemicals and products have been used and developed dependent on the highly phenolic hydroxyl groups, aromatic rings, and other chemical groups of these natural raw material especially flavonoids-based structures. All these properties proved potential activity for thermal stabilization of polymeric materials. Accordingly, the interest is focused more and more on the potential use of natural antioxidants as example, for the stabilization of polymers during processing, as well as during applications.

ACKNOWLEDGMENTS

This chapter was part of a PhD thesis that was completed in the Iran Polymer and Petrochemical institute (IPPI). The authors acknowledge Dr. M. Nekomenesh, Director of IPPI, and Dr. G. Naderi, the head of the international office of this institute, for their huge support and assistance during the period of study.

KEYWORDS

- **polymers additives**
- **polymers degradation**
- **natural antioxidants**
- **bio-based polymer additives**
- **bio-based thermal stabilizers**

REFERENCES

1. Allen, N. S.; Edge, M. *Fundamentals of Polymer Degradation and Stabilization*, 2nd ed.; Elsevier: London, 1992.
2. Mariani, P.; Carianni, G.; La Mantia, F. P. The Antioxidant Effect in Controlling Thermal Degradation of a Low Density Polyethylene Blown Film. *Polym. Degrad. Stab.* **2004,** *85* (3), 1003–1307.
3. Grassie, N.; Scott, G. *Polymer Degradation and Stabilization*; Cambridge University Press: Cambridge, 1985.
4. Hawkins, W. Lincoln. Polymer Degradation and Stabilization. In *Polymers/Properties and Applications*, Vol. 8; Cantow, H.-J., Harwood, H. J., Kennedy, J. P., Ledwith, A., Meibner, J., Okamura, S., Henrici-Olive, G., Olive, S., Eds.; Springer Verlag: Berlin, Heidelberg, 1984.
5. Schnabel, W. *Polymer Degradation: Principles and Practical Applications*; Macmillan: New York, 1981.
6. Peterson, J. D.; Vyazovkin, S.; Wight, C. A. Kinetics of the Thermal and Thermo-Oxidative Degradation of Polystyrene, Polyethylene and Polypropylene. *Macromol. Chem. Phys.* **2001,** *202* (6), 775–784.
7. Kelen, T. *Polymer Degradation*; Van Nostrand: New York, 1983.
8. Chen, G.-Q.; Patel, M. K. Plastics Derived from Biological Sources: Present and Future: A Technical and Environmental Review. *Chem. Rev.* **2012,** *112* (4), 2082–2099.
9. Holmbom, B.; Stefan, W.; Jarl, H.; Suvi, P.; Linda, N.; Patrik, E.; Rainer, S. Knots in Trees: A Rich Source of Bioactive Polyphenols, Chapter 22. In *Materials, Chemicals,*

and Energy from Forest Biomass; Dimitris, S., Argyropoulos, Eds.; ACS Symposium Series, Vol. 954; American Chemical Society: Washington, DC, 2007.
10. Gupta, V. K.; Sharma, S. K. Plants as Natural Antioxidants. *Nat. Prod. Rad.* **2006,** *5* (4), 326–334.
11. Zhang, L.; Liu, S. Investigation of Organic Compounds Migration from Polymeric Pipes into Drinking Water under Long Retention Times. *Procedia Eng.* 2014, *70*, 1753–1761. 12th International Conference on Computing and Control for the Water Industry, CCWI2013; www.sciencedirect.com (accessed Dec 10, 2014).
12. Siripatrawan, U.; Harte, B. R. Physical Properties and Antioxidant Activity of an Active Film from Chitosan Incorporated with Green Tea Extract. *Food Hydrocoll.* **2010,** *24* (8), 770–775.
13. Bart, J. C. J. *Additives in Polymers: Industrial Analysis and Applications*; John Wiley & Sons, Ltd.: England, 2005.
14. Tolinski, M. *Additives for Polyolefins*, First edition; Elsevier Inc.: UK, 2009.
15. Nagarajan, V.; Mohanty, A. K.; Misra, M. Sustainable Green Composites: Value Addition to Agricultural Residues and Perennial Grasses. *ACS Sustain. Chem. Eng.* **2013;** *1* (3), 325–333.
16. Al-Malaika, S. In *Plastic Additives, An A-Z Reference*; Pritchard, Geoffrey, Ed.; Chapman and Hall: London, 1998; p 56.
17. Zweifel, H. *Stabilization of Polymeric Materials*; Springer: Berlin, 1998.
18. Brewer, M. S. Natural Antioxidants: Sources, Compounds, Mechanisms of Action, and Potential Applications. *Comp. Rev. Food Sci. Food Safety* **2011,** *10* (4), 221–247.
19. Al-Malaika, S.; Ashley, H.; Issenhuth, S. The Antioxidant Role of α-tocopherol in Polymers. I. The Nature of Transformation Products of a Tocopherol Formed During Melt Processing of LDPE. *J. Polym. Sci.* **1994,** *32* (16), 3099–3013.
20. Al-Malaika, S.; Goodwin, C.; Issenhuth, S.; Burdick, D. The Antioxidant Role of α-tocopherol in Polymers. II. Melt Stabilizing Effect in Polypropylene. *Polym. Degrad. Stab.* **1999,** *64* (1), 145–156.
21. Al-Malaika, S.; Issenhuth, S. The Antioxidant Role of Alpha-Tocopherol in Polymers. III. Nature of Transformation Products During Polyolefins Extrusion. *Polym. Degrad. Stab.* **1999,** *65* (1), 143–151.
22. Al-Malaika, S.; Issenhuth, S.; Burdick, D. The Antioxidant Role of Vitamin E in Polymers. V. Separation of Stereoisomers and Characterization of Other Oxidation Products of Dl-Alpha-Tocopherol Formed in Polyolefins During Melt Processing. *Polym. Degrad. Stab.* **2001,** *73* (3), 491–503.
23. Alexya, P.; Košíkováb, B.; Podstránska, G. The Effect of Blending Lignin with Polyethylene and Polypropylene on Physical Properties. *Polymer* **2000,** *41* (13), 4901–4908.
24. Pouteau, C.; Dole, P.; Cathala, B.; Averous, L.; Boquillon, N. Antioxidant Properties of Lignin in Polypropylene. *Polym. Degrad. Stab.* **2003,** *81* (1), 9–18.
25. Arshanitsa, A.; Ponomarenko, J.; Dizhbite, T.; Andersone, A.; Gosselink, R. J. A.; Putten, J. van der; Lauberts, M.; Telysheva, G. Fractionation of Technical Lignins as a Tool for Improvement of Their Antioxidant Properties. *J. Anal. Appl. Pyroly.* **2013,** *103* (2013), 78–85.
26. Koontz, J. L.; Marcy, J. E.; O'Keefe, S. F.; Duncan, S. E.; Long, T. E.; Moffitt, R. D. Polymer Processing and Characterization of LLDPE Films Loaded with Alpha-Tocopherol, Quercetin, and Their Cyclodextrin Inclusion Complexes. *J. Appl. Polym. Sci.* **2010,** *117* (4), 2299–2309.

27. Dopico-Garcia, M. S.; Castro-Lopez, M. M.; Lopez-Vilarino, J. M.; Gonzalez-Rodriguez, M. V.; Valentao, P.; Andrade, P. B.; Garcia-Garabal, S.; Abad, M. J. *J. Appl. Polym. Sci.* **2011,** *119* (6), 3553–3359.
28. Samper, M. D.; Fages, E.; Fenollar, O.; Boronat, T.; Balart, R. The Potential of Flavonoids as Natural Antioxidants and UV Light Stabilizers for Polypropylene. *J. Appl. Polym. Sci.* **2013,** *129* (4), 1707–1716.
29. Tátraaljai, D.; Földes, E.; Pukánszky, B. Efficient Melt Stabilization of Polyethylene with Quercetin, a Flavonoid Type Natural Antioxidant. *Polym. Degrad. Stab.* **2014,** *102* (1), 41–48.
30. Maheshwari, R. K.; Singh, A. K.; Gaddipati, J.; Srimal, R. C. Multiple Biological Activities of Curcumin: A Short Review. *Life Sci.* **2006,** *78* (18), 2081–2087.
31. Priyadarsini, K. I.; Maity, D. K.; Naik, G. H.; Kumar, M. S.; Unnikrishnan, M. K.; Satav, J. G.; et al. Role of Phenolic O–H and Methylene Hydrogen on the Free Radical Reactions and Antioxidant Activity of Curcumin. *Free Radical Bio. Med.* **2003,** *35* (5), 475–484.
32. Kim, M. K.; Jeong, W.; Kang, J.; Chong, Y. Significant Enhancement in Radical-Scavenging Activity of Curcuminoids Conferred by Acetoxy Substituent at the Central Methylene Carbon. *Bioorg. Med. Chem.* **2011,** *19* (12), 3793–3800.
33. Tátraaljai, D.; Kirschweng, B.; Kovács, J.; Földes, E.; Pukánszky, B. Processing Stabilization of PE with a Natural Antioxidant, Curcumin. *Eur. Polymer J.* **2013,** *49* (6), 1196–1203.
34. Mohanty, A. K.; Misra, M.; Drzal, L. T. *Natural Fibers, Biopolymers, and Biocomposites*, 1st ed; CRC Press-Taylor & Francis Group: Boca Raton, Florida, 2005.
35. Alma, M. H.; Basturk, M. A. New Polyurethane Type Rigid Foams from Liquified Wood Powders. *J. Mater. Sci. Lett.* **2003,** *22* (17), 1225–1228.
36. Ten, E.; Vermerris, W. Functionalized Polymers from Lignocellulosic Biomass: State of the Art Review. *Polymers* **2013,** *5* (2), 600–642.
37. Mr. Rohan. The Green & Bio Polyol Market. Report by marketsandmarkets; http://www.marketsandmarkets.com/Market-Reports/green-and-bio-polyols-market-1175.html (Publishing Date: January 2014, Report Code: CH 2042) (accessed Dec 10, 2014).
38. Chrysanthos, M. Novel Biobased Epoxy Networks Derived From Renewable Resources: Structure-Property Relationships. Ph.D. Thesis, University of Lyon, 2012. http://theses.insa-lyon.fr/publication/2012ISAL0051/these.pdf (accessed Dec 10, 2014).
39. Semsarzadeh, M. A.; Mehrabzadeh, M.; Arabshahi, S. S. Mechanical and Thermal Properties of the Plasticized PVC-ESBO. *Iranian Polymer J.* **2005,** *14* (9), 769–773.
40. Kumar, A.; Gupta, R. K. *Fundamentals of Polymer Engineering*, Second Edition Revised and Expanded; Marcel Dekker, Inc.: New York, NY, 2003.
41. McKeen, L. W. Plastics Used in Medical Devices (Chapter 3). In *Handbook of Polymer Applications in Medicine and Medical Devices*; Ebnesajjad, S.; Modjarrad, K., Eds.; Elsevier Inc.: Oxford, 2014.
42. Mousa, A.; Heinrich, G.; Kretzschmar, B.; Wagenknecht, U.; Das, A. Utilization of Agrowaste Polymers in PVC/NBR Alloys: Tensile, Thermal, and Morphological Properties. *Int. J. Chem. Eng.* **2012;** http://dx.doi.org/10.1155/2012/121496.
43. Mousa, A.; Heinrich, G.; Gohs, U.; Hassler, R.; Wagenknecht, U. Application of Renewable Agro-Waste-based Olive Pomace on the Mechanical and Thermal Performance of Toughened PVC. *Polymer-Plastics Technol. Eng.* **2009,** *48* (10), 1030–1040.

44. Cerruti, P.; Malinconico, M.; Rychly, J.; Matisova-Rychla, L.; Carfagna, C. Effect of Natural Antioxidants on the Stability of Polypropylene Films. *Polym. Degrad. Stab.* **2009,** *94* (11), 2095–3100.
45. Ozhogina, O. A.; Kasaikina, O. T. Beta-Carotene as an Interceptor of Free Radicals. *Free Radic. Biol. Med.* **1995,** *19* (5), 575–581.
46. Abdel-Razik, E. A. Aspects of Degradation and Stability of ABS Copolymers. I. Effect of Beta-carotene as Antioxidant. *J. Polym. Sci. Part A Polym. Chem.* **1989,** *27*, 343–355.
47. Tátraaljai, D.; Kovács, J.; Pataki, P.; Földes, E.; Pukánszky, B. The Effect of Natural Additives on the Processing Stability of Polyethylene. In Proceedings CD of MoDeSt 2010 conference, Athens; Papaspyrides, C. D., Vouyiouka, S. N., Eds.; Greece. Publ.: MoDeSt Hellenic Organic Committee; September 5-9, 2010, ISBN 978-960-99362-0-0; September 5e9, 2010. Poster Presentations, Topic I, PI. 24.
48. Tátraaljai, D.; Major, L.; Földes, E.; Pukánszky, B. Study of the Effect of Natural Antioxidants in Polyethylene: Performance of Beta-carotene. *Polym. Degrad. Stab.* **2014,** *102*, 33–40.
49. Brocca, D.; Arvin, E.; Mosbaek, H. Identification of Organic Compounds Migrating from Polyethylene Pipelines into Drinking Water. *Water Res.* **2002,** *36* (15), 3675–3680.
50. Siró, I.; Fenyvesi, E.; Szente, L.; de Meulenaer, B.; Devlieghere, F.; Orgoványi, J.; Sényi, J.; Barta, J. Release of Alpha-Tocopherol from Antioxidative Low-Density Polyethylene Film into Fatty Food Simulant: Influence of Complexation in Beta-cyclodextrin. *Food Addit. Contam.* **2006,** *23* (8), 845–853.
51. Tahira, B. E.; Khan, M. I.; Saeed, R.; Akhwan, S. A Review: Thermal Degradation and Stabilization of Poly (Vinyl Chloride). *Int. J. Res.* **2014,** *1* (6), 732–750.
52. Starnes, W. H. Structural and Mechanistic Aspects of the Thermal Degradation of Poly (Vinyl Chloride). *Prog. Polym. Sci.* **2002,** *27* (10), 2133–2170.
53. Owen, E. D. *Degradation and Stabilisation of PVC*; Elsevier Applied Science Publishers LTD.: London & NY, 1984.
54. Nass, L. I.; Heiberger, C. A., Eds. Resin Manufacture and Properties, Volume 1. In *Encyclopedia of PVC*; 2nd Edition, Revised and Expanded; Marcell Dekker, INC.: NY and Basel, 1986.
55. Jakupca, M. Polymer Stabilizers: Current Challenges and Future Trends; www.doverchem.com/LinkClick.aspx?fileticket (accessed Dec 10, 2014).
56. Smith, V.; Magalhaes, S.; Schneider, S. The Role of PVC Additives in the Potential Formation of NAPLs; Report from AMEC to NDARWMD, The Nuclear Decommissioning Authority's Radioactive Waste Management Directorate, Nuclear Decommissioning Authority 2013. All rights reserved. AMEC/PPE/2834/001, Jan 22, 2013; www.nda.gov.uk. (accessed Dec 13, 2014).
57. Benaniba, M. T.; Belhaneche-Bensemra, N.; Gelbard, G. Stabilization of PVC by Epoxidized Sunflower Oil in the Presence of Zinc and Calcium Stearates. *Polym. Degrad. Stab.* **2003,** *82*, 245–249.
58. Bueno-Ferrer, C.; Garrigós, M. C.; Jiménez, A. Characterization and Thermal Stability of Poly(Vinyl Chloride) Plasticized with Epoxidized Soybean Oil for Food Packaging. *Polym. Degrad. Stab.* **2010,** *95* (11), 2207–2212.
59. Taghizadeh, M. T.; Nalbandi, N.; Bahadori, A. Stabilizing Effect of Epoxidized Sunflower Oil as a Secondary Stabilizer for Ca/Hg Stabilized PVC. *eXPRESS Polymer Lett.* **2008,** *2* (1), 65–76.

60. Cerruti, P.; Malinconico, M.; Rychly, J.; Matisova-Rychla, L.; Carfagna, C. Effect of Natural Antioxidants on the Stability of Polypropylene Films. *Polym. Degrad. Stab.* **2009,** *94* (11), 2095–3100.
61. Peltzer, M. A.; Wagner, J. R.; Migallon, A. J. Stabilization of Polymers with Natural Antioxidants (Chapter 2). In *Polymer and Biopolymer Analysis and Characterization*; Zaikov, G. E., Gimenez, A., Eds.; Nova Publishers: New York, 2007.
62. Peltzer, M. A.; Wagner, J. R.; Migallon, A. J. Stabilization of Polymers with Natural Antioxidants (Chapter 2). In *Polymer and Biopolymer Analysis and Characterization*; Zaikov, G. E.; Gimenez, A., Eds.; Nova Publishers: New York, 2007.
63. Zhang, S.-J.; Lin, Y.-M.; Zhou, H.-C.; Wei, S.-D.; Lin, G.-H.; Ye, G.-F. Antioxidant Tannins from Stem Bark and Fine Root of Casuarina Equisetifolia. *Molecules* **2010,** *15* (8), 5658–5670.
64. Zhang, L. L.; Lin, Y. M. HPLC, NMR and MALDI-TOF MS Analysis of Condensed Tannins from Lithocarpus Glaber Leaves with Potent Free Radical Scavenging Activity. *Molecules* **2008,** *13* (12), 2986–2997.
65. Gandini, A. Monomers and Macromonomers from Renewable Resources. In *Biocatalysis in Polymer Chemistry*; Loos, Katja, Ed.; WILEYVCH Verlag GmbH & Co. KGaA: Weinheim, Germeny, 2010; pp: 1–34. http://www.wiley-vch.de/books/sample/3527326189_c01.pdf. (accessed Dec 22, 2014).
66. Zalacain, A.; Carmona, M.; Lorenzo, C.; Blazquez, I.; Alonso, G. L. Antiradical Efficiency of Different Vegetable Tannin Extracts. *J. Am. Leather Chem. Assoc.* **2002,** *97* (4), 137–142.
67. Zhang, L.; Chen, J.; Wang, Y.; Wu, D.; Xu, M. Phenolic Extracts from Acacia Mangium Bark and Their Antioxidant Activities. *Molecules* **2010,** *15* (5), 3567–3577.
68. Sung, S. H.; Kim, K. H.; Jeon, B. T.; Cheong, S. H.; Park, J. H.; Kim, D. H.; Kweon, H. J.; Moon, S. H. Antibacterial and Antioxidant Activities of Tannins Extracted from Agricultural By-Products. *J. Med. Plants Res.* **2012,** *6* (15), 3072–3079.
69. Anelise, S.; Formagio, N.; Carla, R. F.; Volobuff, M. S.; Claudia, A. L.; Cardoso, M. d. C.; Vieira, Z. V. P. Evaluation of Antioxidant Activity, Total Flavonoids, Tannins and Phenolic Compounds in Psychotria Leaf Extracts. *Antioxidants* **2014,** *3* (4), 745–757.
70. Rosales-Castro, M.; González-Laredo, R. F.; Bae, Y.-S.; Kim, J. K.; Morre, J.; Karchesy, J. J. Characterization and Antioxidant Properties of the Condensed Tannins from Alaska Cedar Inner Bark. *Rec. Nat. Prod.* **2014,** *8* (3), 217–227.
71. Nawar, W. F. Lipids. In *Food Chemistry*; Fennema, O., Ed.; 3rd Ed; Marcel Dekker, Inc.: New York, 1996.
72. Vazquez, G.; Santos, J.; Freire, M. S.; Antorrena, G.; Gonzalez-Alvarez, J. Extraction of Antioxidants from Eucalyptus (Eucalyptus Globulus) Bark. *Wood Sci. Technol.* **2012,** *46* (1-3), 443–457.
73. Wei, S.-D.; Chen, R.-Y.; Liao, M.-M.; Tu, N.-W.; Zhou, H.-C.; Lin, Y.-M. Antioxidant Condensed Tannins from Machilus Pauhoi Leaves. *J. Med. Plants Res.* **2011,** *5* (5), 796–804.
74. Grigsby, W. J.; Bridson, J. H.; Lomas, C.; Elliot, J.-A. Esterification of Condensed Tannins and Their Impact on the Properties of Poly(Lactic Acid). *Polymers* **2013,** *5* (2), 344–360.
75. Grigsby, W. J.; Bridson, J. H.; Lomas, C.; Frey, H. Evaluating Modified Tannin Esters as Functional Additives in Polypropylene and Biodegradable Aliphatic Polyester. *Macromol. Mater. Eng.* **2014,** *299* (10), 1251–1258.
76. Yuso, M. de A.; Lagel, M. C.; Pizzi, A.; Fierro, V.; Celzard, A. Structure and Properties of Rigid Foams Derived from Quebracho Tannin. *Mat. Design* **2014,** *63*, 208–212.

77. Basso, M. C.; Pizzi, A.; Lacoste, C.; Delmotte, L.; Al-Marzouki, F. M.; Abdalla, S.; Celzard, A. MALDI-TOF and 13C NMR Analysis of Tannin–Furanic–Polyurethane Foams Adapted for Industrial Continuous Lines Application. *Polymers* **2014,** *6* (12), 2985–3004.
78. Hemingway, R. W.; Karchesy, J. J.; Branham, S. J. *The Chemistry and Significance of Condensed Tannins*. Plenum Press: New York, NY, USA, 1989.
79. Riedl, K. M.; Hagerman, A. E. Tannin-protein Complexes as Radical Scavengers and Radical Sinks. *J. Agric. Food Chem.* **2001,** *49* (10), 4917–4923.

CHAPTER 4

Recent Trends on Smart Bioresponsive Polymeric Materials

KALPANA N. HANDORE[1*], SUMIT B. SHARMA[1], SANTOSH MISHRA[2], and VASANT V. CHABUKSWAR[3]

[1,4]*Department of Chemistry, Nowrosjee Wadia College, 19 Late Prin. V. K. Joag Path, Pune 411001, Maharashtra, India*

[2]*Modern Education Society's College of Engineering (MESCOE), Pune, Maharashtra, India*

[3]*G. H. Raisoni College of Engineering and Management, Wagholi, Pune, Maharashtra, India*

[*]*Corresponding author. E-mail: kalpanahandore@gmail.com*

ABSTRACT

Life is polymeric in its essence as we all are associated with polymers. The important components of living cells proteins, carbohydrates, and nucleic acid are polymeric. Nature uses polymers as constructive elements and as a part of the complicated cell machinery of living things.

Smart polymers are high-performance polymer that change according to the environment also known as stimuli-responsive polymer. Macromolecules are sensitive to number of factors including temperature, wavelength of light, humidity, intensity, chemicals, electrical and magnetic fields which respond in various ways, like altering the color or transparency, becoming conductible or permeable to water or changing shape (shape memory polymers). Small changes in the environment are sufficient to induce large changes in polymer properties. They are used for the production of hydrogels, biodegradable packaging and great extent in biomedical engineering. Humidity sensitive polymer used in self-adaptive wound dressing automatically regulates the moisture balance in and around the wound. These smart polymers are potentially very useful for a variety of applications including some related to biotechnology and biomedicines. The study is focused on the entire features of smart polymers, characterization techniques, and their relevant applications in various field.

4.1 INTRODUCTION

In recent years, smart polymers which are also known as bioresponsive polymers have gained considerable attention as a unique class of polymers. The term "smart polymers" contains a wide spectrum of different compounds with unique potential for various applications. The characteristic features that actually make these polymers "smart" is their ability to respond to very slight changes in the surrounding environment. Based on the structure–property relationship, synthetic and modified natural polymers have been obtained with well controlled composition and architecture. Due to their well-controlled properties, stimuli-responsive polymers can be potentially utilized in many fields such as water treatment, enhanced oil recovery, catalysis, and separation processes,[1–6] etc.

Smart polymers have also been useful as drug delivery systems[7,8] and in the biotechnological field for enzyme immobilization.[9–11] More recently, coating medical devices with stimuli-responsive polymers may significantly widen their range of applications. Devices able to attract or repel certain cells of the body (e.g., for prevention of inflammatory responses or behaving as tissue scaffolds) or devices that should perform as membranes or artificial muscles take advantage of these smart materials.[12–15] A surface grafted with stimuli-responsive polymers has been highlighted for its interfacial and wetting applications. For example, a smart surface with switchable wettability could be potentially used in tissue engineering and antimicrobial coatings.[15–20] In addition, a polymer material responsive to temperature or light could also be potentially utilized as sensor and actuator. Currently, stimuli-responsive polymers have been mainly emphasized for their potential applications in nanotechnology, nanoscience, and nanomedicine.[21–23]

4.2 RECENT DEVELOPMENTS

Physical properties of polymeric materials can be modified by controlling various parameters such as chain conformation, color, configuration, and solubility.

Swelling and deswelling of hydrogels, shape memory of thin films, and self-assembly of polymers in solution to produce aggregates with respect to the external stimuli are a few examples. In the following sections, broad properties of stimuli-responsive polymers are discussed.

As shown in Figure 4.1, the typical stimuli can be logically categorized into three groups: physical (light, electric, magnetic, ultrasound, mechanical, temperature, etc.), chemical (pH, ionic strength, solvent, electrochemical, specific substances, etc.), and biological (enzymes, antibodies, receptors, saccharide, proteins, etc.).[24–27]

A physical interaction or chemical reaction among functional moieties in the smart materials can occur after reception of these physical, chemical, and biochemical signals.

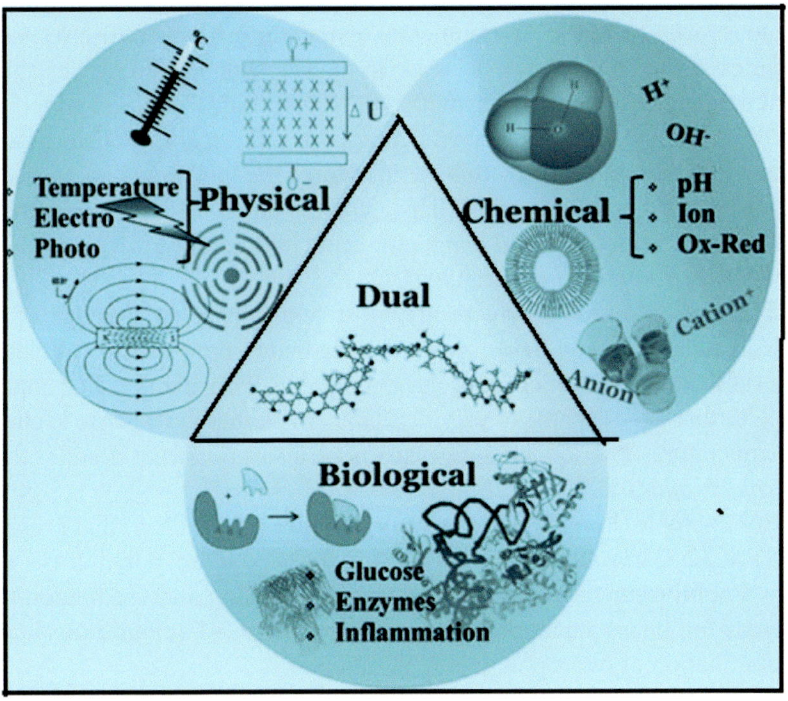

FIGURE 4.1 Classification of smart polymers on the type of stimulus.

4.2.1 CLASSIFICATION OF SMART POLYMERS BASED ON STIMULI THEY ARE RESPONDING

1. Temperature-responsive polymers
2. pH-responsive polymers
3. Photo-responsive polymers
4. Electrically/magnetically responsive polymer

5. Multiresponsive polymer
6. Smart polymer hydrogels

4.2.1.1 TEMPERATURE-RESPONSIVE POLYMERS

Polymers which are sensitive to temperature changes are the most studied as they have potential applications in the biomedical field. Temperature has attracted a great deal of attention because this stimulus can be easily applied and monitored. Temperature-responsive polymers change their properties due to a variation in the environmental temperature. Temperature-responsive polymers undergo an abrupt decrease in physicochemical properties above a certain temperature which is named as the lower critical solution temperature (LCST).[28] Polymer chains behave hydrophilic and remain swollen (in water) below LCST, while above this temperature, the polymer chains become increasingly hydrophobic and collapse. On the other hand, polymers that are hydrophobic below a critical temperature and hydrophilic above it present an upper critical solution temperature (UCST).[29]

Thermoresponsive polymers are most widely studied because of their projected uses in many fields such as biomedicine (drug delivery) functional materials, surface modifiers, and chromatography. Polymer gels first appeared in 1970s that were capable of varying shape or volumes in response to changes in temperature. The temperature responsive also underwent a color change *(thermochromic)*.[30–35]

For example, Poly(N- Isopropylacrylamide) (PNIPAAm) that presents a LCST at 32°C in water solution. Below that temperature, polymer is soluble as hydrophilic interactions due to hydrogen bonding and are predominant, whereas the phase separation occurs above the LCST (cloud point) due to predomination of hydrophobic interaction.

4.2.1.2 pH-RESPONSIVE POLYMERS

pH-responsive polymers are polyelectrolytes which contain structure weak acidic or weak basic groups that either accept or release protons in response to changes in environmental pH. The acidic or basic groups on polyelectrolytes undergo ionization just like acidic or basic groups of monoacids or monobases. However, complete ionization on polyelectrolytes is more difficult due to electrostatic effects. Various natural polymers such as albumin and gelatin also show pH-responsive swelling behavior. The advantage of natural pH-responsive polymers as compared to their synthetic counterparts is their ability to degrade within the body over time, which is ideal for implanted materials or circulating drug delivery systems.

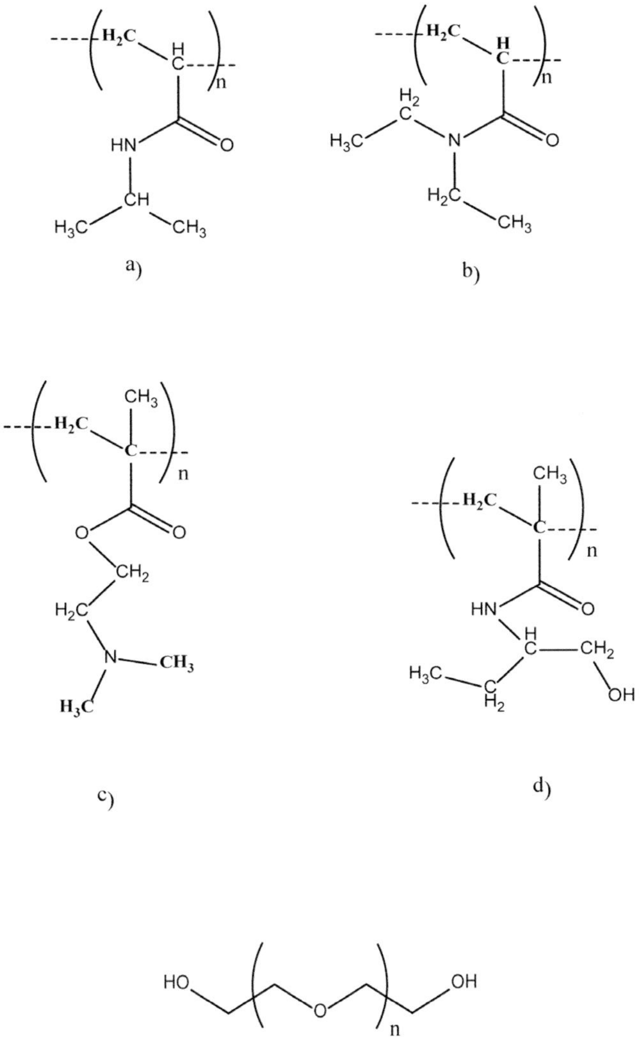

PEG (molecular weight < 200000 gm/mole)

FIGURE 4.2 Polymers showing an LCST: (a) PNIPAAm, LCST 32°C, (b) PDEAAm, LCST 26–35°C), (c) PDMAEMA, LCST 50°C, and(d) Poly-(L)-(hydroxymethyl) propylmethacrylamide) LCST 30°C.

4.2.1.2.1 *Polymers with Acidic Groups*

Polyacids or polyanions are pH-sensitive polymers that possess ionizable acidic groups in their structures carboxylic or sulphonic. Carboxylic group accepts the proton at low pH value and releases proton at high pH value. When pH increases the polymer swells due to electrostatic repulsion of negatively charged groups. The pH in which acids become ionized depends on polymers pKa (depends on the polymers composition). The most commonly studied ionic polymers for pH-responsive behavior include poly(acrylamide) (PAAm), poly (acrylic acid) (PAA), poly(methacrylic acid) (PMAA), poly(diethylaminoethyl methacrylate) (PDEAEMA), and poly(dimethylaminoethyl methacrylate) (PDMAEMA). But polymers containing phosphoric acid derivatives have also been reported.[30,31]

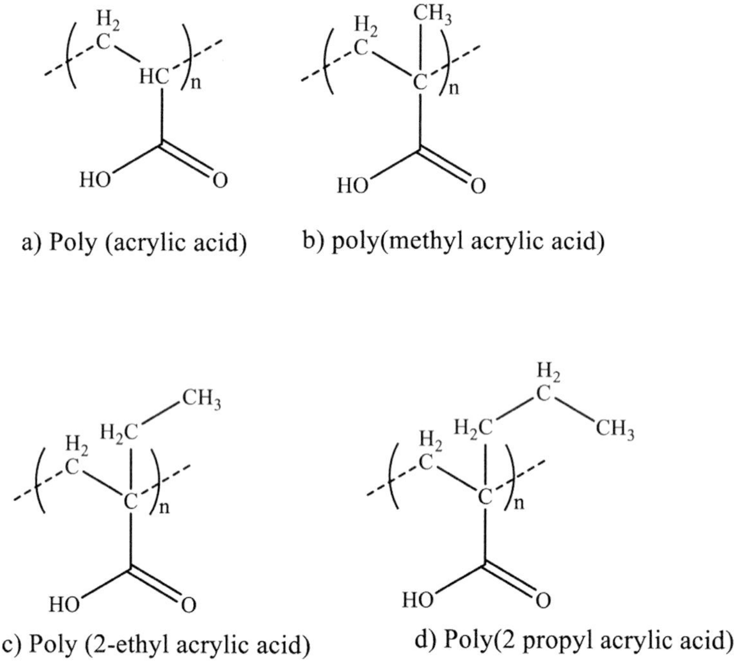

FIGURE 4.3 Chemical structures of pH-sensitive polyacids Poly(acrylic acid) (PAA), poly(methacrylic acid) (PMAA), poly(2-ethyl acrylic acid) PEAA, poly(2-propyl acrylic acid) PPAA.

Another kind of polyacidic polymer is the polysulphonamides (derivative of p-amino benzenesulphonamide).

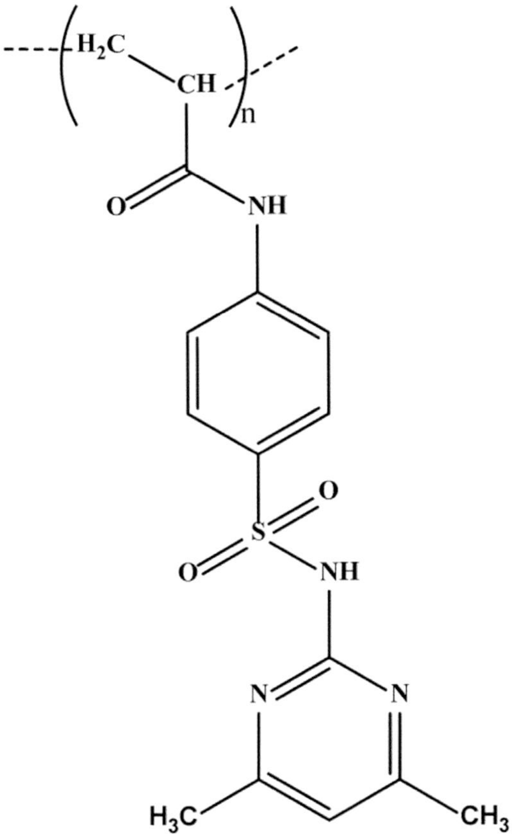

FIGURE 4.4 Chemical structure of 4- amino-N-[4'6-di methyl-2-pyrimidinyl] benzenesulphonamine (sulfomethazine) containing polymer.

4.2.1.2.2 Polymers with Basic Functional Groups

Polybasic or polycations are protonated at high pH value and positively ionized at neutral or low pH they go through a phase transition at pH 5 due to deprotonation of pyridine groups.

4.2.1.3 PHOTO-RESPONSIVE POLYMERS

Polymers which are sensitive to visible light are known as light sensitive polymers. Light can be considered as clean stimulus that allows remote control without physical contact or mechanical apparatus. It enables to change the geometry and dipole moment of photo switching molecule causing variations in their structure. These changes affect on the final properties of polymer, that is, charge, color, binding, alignment, wettability, and permeability. Photoresponsive polymeric systems have been widely studied in many areas such as photo-switches, photo-optical media, photo-mechanical systems, micropatterning, and non-linear optical media. These systems are important because light is nondestructive, can be localized as well as remotely activated. Thus, delivering energy to systems results in photo-triggered sensing, actuating, and transporting.

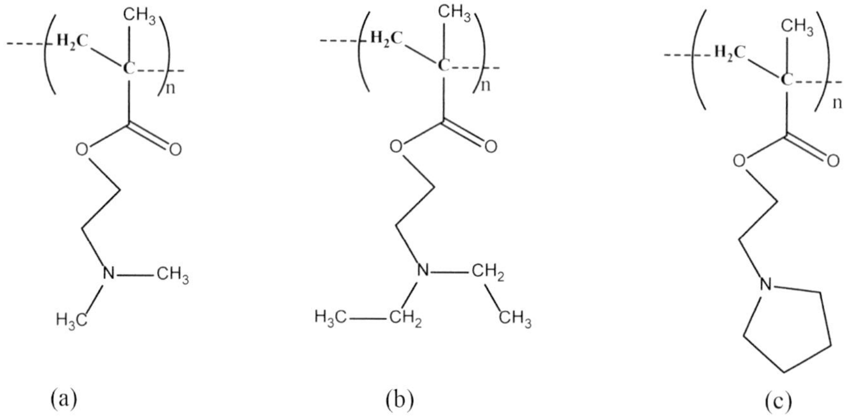

FIGURE 4.5 Chemical structures of pH-sensitive Polybases (a) PDMAEMA, (b) PDEAEMA, and (c) PEPyM.

A photoresponsive polymer exhibits conformational changes in response to light. Such polymers can be constructed by incorporating chromophores that can translate light energy into a change in conformation. Through the introduction of photochromic functionalities, physical and chemical property changes following irradiation have been exploited. Photo-stimulated conformational changes at the microscopic level are amplified to shape changes (contraction or expansion) at the macroscopic level, effectively converting energy from a light beam into mechanical energy.

Photochromism is defined as a reversible transformation of chemical species between two states having observable light absorptions in different regions induced in one or both directions by electromagnetic radiation. During photoisomerization, changes are not restricted to absorption spectra, but various physicochemical properties are altered, such as the refractive index, dielectric constant, oxidation and reduction potential, geometrical structure and luminescence. In photoresponsive functionalities, the response may be associated with chemical processes involving either isomerization or cleavage of chemical bonds. The main chemical processes involved in photochromism are *cis-trans* isomerization (azo compounds, stilbenes), pericyclic reactions (spiropyrans, diarylethenes, and anthracenes), dissociation reactions (triphenylmethaneleuco derivatives), electron transfer (viologens), intramolecular hydrogen transfer (salicylates, triazoles and oxazoles), and intramolecular group transfer (polycyclic quinones).

Following reaction shows some photoresponsive functionalities. In each process, the original molecule changes chemical structure as well as chemical and physical properties. In photoresponsive functionalities, the response may be associated with chemical processes pressure inside and outside the gel drives swelling. Hence, these kinds of polymers swell better in deionized water than in normal water.

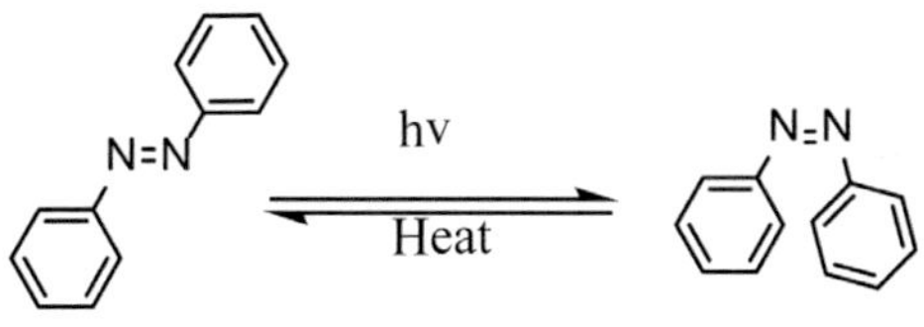

Isomerization of Azocompound

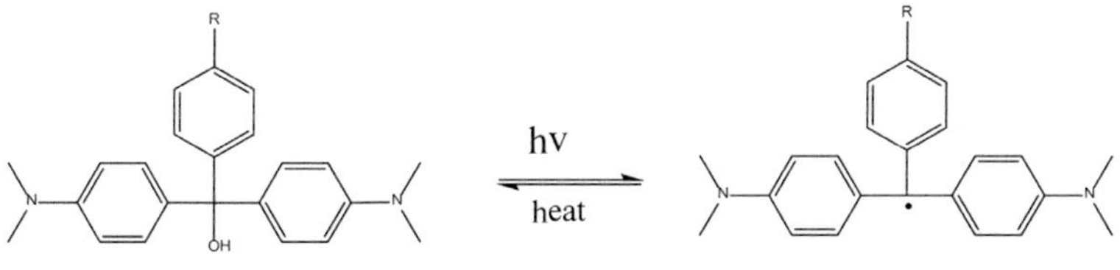

Ionization

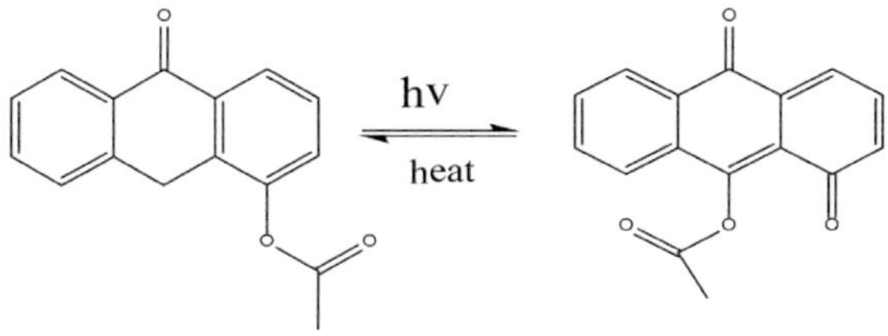

FIGURE 4.6 Photochromic compound.

FIGURE 4.7 Photochromic lens.

Photochromic lenses are a special type of lens that starts out clear and turns dark in the bright sunlight. They offer a convenient way to protect eyes outdoors from UV rays and bright light.

4.2.1.4 *ELECTRIC/MAGNETIC-RESPONSIVE POLYMERS*

Electric and magnetic-responsive systems have attracted considerable attention because of their easy controllability and industrial importance. Electrically responsive polymers are typically conducting polymers that can transform their shape (shrink or swell) when subjected to an electric field.[13] Electrical stimulus can precisely control the response of polymers via the magnitude of the current, duration of an electrical pulse, and intervals between the pulses. Electrically responsive polymers include polythiophene, polyaniline, and polypyrole.

Polymers that respond to changes in magnetic fields are called magnetic-responsive polymer. These polymer systems are usually made of elastomers or gels with small magnetic particles; magnetic particles such as iron (III) oxide particles, ferromagnetic particles, nickel powders. Some polymers that have been explored for developing the magnetic-responsive materials are: poly(N-isopropylacrylamide) (PNIPAAm). The interesting smart electromagneto-responsive characteristics of quick response and fine tuning make the appropriate candidates in broad areas, such as medical therapies, artificial muscle stimulators, actuators, micro-fluidic control and viscosity reduction of crude oil.

Magnetic drug delivery systems possess advantages such as visualization of drug delivery vehicles, ability to control and guide the movement of drug carriers through magnetic fields and thermal heating which has been used to control drug release or produce tissue ablation. Magnetic drug carriers like magnetite, cobalt, ferrite, and carbonyl iron are mainly used and they are biocompatible, non-toxic, and non-immunogenic. Magnetic nanoparticles have also been encapsulated within liposomes. Polyelectrolyte-coated liposomes were highly stable as they showed no significant membrane disruption or leakage of encapsulated contents in the presence of detergent Triton TX-100.

Polymers capable of executing mechanical work under the influence of an electric field, such as expansion, contraction, elongation, and bending are called electroresponsive polymers. Electric fields have advantages as external stimuli. One can conveniently and precisely control the magnitude of the current, as well as the duration of and the intervals between pulses. Polyelectrolyte gels, conductive polymers, ionic-polymer metal composites, and dielectric elastomers are also included in this category. Characteristic properties of polyelectrolyte gels have considerable potential for applications, such as soft and wet scaffolds of cells, soft actuators, and replacement of biological tissues. These materials also attracted a great deal of attention in medicine and mechanical engineering and have projected uses in the delivery of precise doses of drugs, as contact lenses, artificial muscles, mehanical grippers, and as sensors.

Among many electroresponsive polymers, electroresponsive hydrogels have become tempting because of their use in controlled drug delivery as well as their biocompatibility. Response depends on the shape of the gel and its position relative to the electric field. When a hydrogel lies perpendicular to the electrodes, or the gels are touching the electrodes, deswelling or contraction is noticeable. Bending can be observed when the gel lies parallel and are not touching the electrodes. The bending of hydrogels has been studied mostly for production of mechanical devices such as artificial muscles, valves, and switches. Deswelling or contractile behavior of hydrogels has been mainly studied in controlled drug delivery. In 1982, Tanaka observed an electroresponsive contraction and a phase separation of a partially hydrolyzed polyacrylamide gel in contact with platinum electrodes. Tanaka et al.[10,66] recognized deswelling that is due to an electrophoretic pressure gradient. A similar effect for water swollen poly(2-acrylamido-2-methyl-1-propanesulfonic acid) gel was found by Osada and Hasebe in 1990.[10] These workers observed up to 30% loss of absorbed water from a gel in the presence of an electric field.

4.2.1.5 MULTI-STIMULI-RESPONSIVE POLYMERS

Polymers can also exhibit responsive behavior to muliple stimuli- and dual stimuli-responsive systems which have been studied. Combinations of light and temperature, pH and temperature, light and electric field have been reported.[11] There are reports on triple stimuli-responsive polymers that respond to light, heat, and pH. Multi-stimuli-responsive polymeric materials can be obtained from the incorporation of different functional groups responding to different stimuli.

For example, free radical copolymerization of *N*-isopropyl acrylamide and a polymerizable spiropyan derivative with a *N,N*-methylene bis acrylamide cross-linker produces a thermo- and light- responsive polymer. The PNIPAAm backbone will respond to the temperature whereas the spiropyran will respond to light.

4.2.1.6 SMART POLYMER HYDROGELS

Hydrogels are three-dimensional high-molecular weight networks that swell in water or in biological fluids without dissolving. They are organized as three-dimensional networks of polymeric chains, physically and chemically bounded and partially solvated by water molecules.

Hydrogels are cross-linked polymer networks with polymer backbones that are hydrophilic due to presence of alcohol, carboxylic acid, and amide groups.[3,4] Figure 8a and 8b shows representation of a hydrogel polymer particle.

Cross-linking typically connects polymer chains by joining two macromolecules through a smaller molecule. In manufacturing hydrogels polymers, the most common cross-linker is a hydrophilic molecule with two or more polymerizable double bonds.[30] Cross-links between the polymer chains form a three-dimensional network, preventing the polymer from swelling to infinity (prevent dissolution). The degree of cross-linking affects the level of swelling. Higher the cross-linking density, the lower the swelling capacity and the stronger the polymer. Hydrogels have great interest in particular in biomedical applications, thanks to their three-dimensional structure, high water content, good biocompatibility, and mechanical properties, which can be synthesized with required properties depending on the chemical structure, composition and conformation of starting materials, density of linking of polymer chains, hydrophobicity and hydrophilicity for a particular application. Hydrogels can be prepared from either chemical or physical cross-linking. Hydrogels prepared from chemical cross-linking form permanent junction-type networks. On the other hand, hydrogels can be prepared from physical cross-linking, allowing the formation of transient junction-type networks, such as polymer chain entanglements or physical interactions such as hydrogen bonds, ionic or hydrophobic interactions.[30,31]

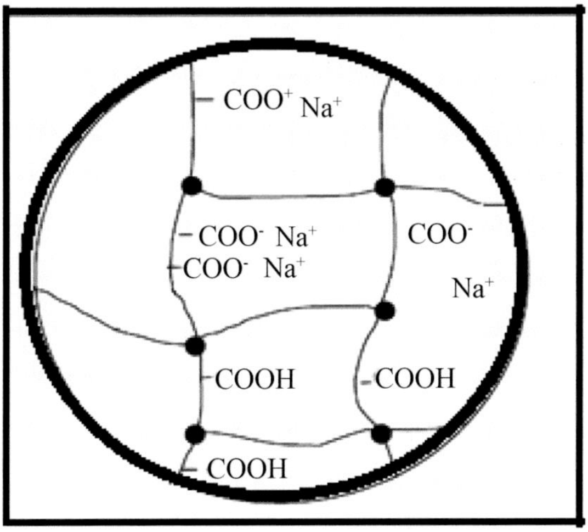

FIGURE 4.8a Polymer network of sodium polyacrylate.

Smart hydrogels exhibit dramatic changes in their swelling behavior, network structure, permeability, and mechanical strength in response to external stimuli, including pH, ionic strength, temperature and electrical or magnetic fields. Stimuli-responsive polymers play an important role in the development of novel smart hydrogels. The most important systems from a biomedical point of view are those sensitive to temperature and pH. Poly(N-isopropylacrylamide) (PNIPAAm) is one of the most extensively studied temperature-responsive polymer for the preparation of smarthydrogels. However, other kind of smart polymers have been used for the preparation of smart hydrogels such as poly(N-vinylcaprolactam) (PVCL), poly(dimethylaminoethylmethacrylate), poly(acrylic acid)(PAAc), and poly(methacrylic acid).

FIGURE 4.8b Polymer hydrogel.

4.3 CHARACTERIZATION OF SMART POLYMERS

Currently, there are many methods available regarding the characterization of smart polymers which rely upon the responsive properties of the polymer.

4.3.1 SWELLING MEASUREMENTS

The equilibrium of swollen materials in distilled water is determined by swelling the dried grafted polymeric materials for different periods of time and weighing them on an analytical balance until constant maximum weight is obtained.[43–49] The degree of swelling of the films (%) is calculated according by the formula:

$$\text{Swelling \%} = (W_s - W_d/W_d) \times 100$$

where W_s and W_d are weights of the swollen and initial catheter, respectively.

4.3.2 THERMOSENSITIVE POLYMERS

Miscibility in high-molecular weight polymer is governed mostly by intermolecular interactions. Miscibility for most miscible polymer blends is caused by specific (i.e., attractive) interactions, which leads to LCST behavior. The LCST of various smart polymers are determined by swelling them in water for the equilibrium time at different temperatures and plotting the swelling percentage as a function of temperature. The inflexion point gives the LCST.[50–53]

4.3.3 pH CRITICAL POINT

The critical pH point is determined in the same way from a plot of swelling percentage in different pH solutions with a buffer of citric acid and sodium phosphate. In pH-sensitive smart materials, the responsive mechanism lies within the side chain groups, branches, and cross-links of a polymer's chemical structure. In polymer networks that contain weak acid or base groups, absorption and adsorption of water can occur simultaneously. The movement of water into the polymer network results in ionization of the acid and base pendant groups. This phenomenon is controlled by the solution's pH, ionic composition, and ionic strength.[54–56]

Acrylic acid (AAc) is one of the most popular monomers that have been grafted onto different polymeric matrices and its polymer or copolymers with pH-sensitive response have a capability to undergo further chemical reaction to produce new functional groups.[57,58]

4.3.4 SURFACE PLASMON RESONANCE SPECTROSCOPY (SPR)

A surface-sensitive characterization based on the evanescent field of the surface plasmon. The latter is an electromagnetic wave traveling along the interface between a metal and a dielectric. Its electric field decays exponentially into both materials over a distance of a few hundred nanometers and the wave has a finite propagation length due to damping processes in the metal. The resulting data are a direct measure of the local average refractive index of the dielectric close to the surface, and with Fresnel calculations either the thickness or the refractive index of thin films at an interface can be determined. Furthermore, a time-dependent measurement mode enables the detection of changes in the local average dielectric constant due to the adsorption of molecules onto the surface or changes in film properties due to an external trigger.[59] Surface plasmons are the quanta of charge-density waves of free electrons in a metal propagating along the interface of a metal and a dielectric medium such as buffer or air. The electromagnetic field of these surface waves reaches its highest intensity at the metal surface and decreases exponentially into the adjacent phase. Therefore, it is influenced by the optical properties of this phase. The strong dependence of the surface plasmons on the refractive index of the dielectric medium can be used for sensor development purposes.[66–68]

4.3.5 FTIR SPECTROSCOPIC METHOD FOR THE DETERMINATION OF THE LCST

This is a new and simple method for determining the LCST in both linear and cross-linked polymers by Fourier-transform infrared (FTIR) spectroscopy.[69,70] This technique has been shown to be useful for probing transitions such as a lipid bilayer transition, the self-association of aqueous surfactants, and studying changes in hydrogen bonding in polymers. In addition, changes in the infrared spectra provide information on the conformation and bonding state of the functional groups involved in the transition. In this method, the FTIR-ATR spectra of PNIPAAm change dramatically in the vicinity of the coil-globule transition temperature. Although the polymer will aggregate and precipitate at the bottom of ATR crystal above LCST, this exactly indicates the occurrence of phase separation, actually the polymer will also aggregate to some extent above LCST using the IR transmittance measurements which was considered.[71–76] It was possible to measure the dependence of the transition parameters on the heating rate. The DSC measurements, by heating at different rates, provide results closely approximating equilibrium. The transition temperature, enthalpy, and entropy of this thermosensitive behavior as well as the transition LCST can be determined.

The LCST of the linear polymers (0.5%, w/v) was determined using an aqueous sample (10 mg by weight). The samples were run in a sealed aluminum crucible under a nitrogen purge at 1°C/min unless otherwise stated. Samples were analyzed in the range from 0 to 60°C. The transition temperature has been defined in previous studies as either the temperature of onset or the peak temperature. In the current work, the phase transition temperature was defined as the maximum of the endothermic transition peak.[77]

4.3.6 CONTACT ANGLE

This method consists of the measurement of advancing and receding contact angles by a dynamic method[76] or by the sessile drop method using the optical contact angle.[79] The measurements of the contact angles for various samples are carried out at room temperature, 1 min after water drops are deposited onto the surface of polymeric films, using direct microscopic measurement of the contact angles. The contact angle is an important parameter in surface science. It is a common measure of the hydrophobicity of a solid surface.

4.3.7 MICROSCOPY

Surface morphology investigation and force-curve measurement with atomic force microscope (AFM) was used for determination of stiffness; microscopic infrared imaging is commonly used to found the distribution of specific compounds on the surface. Specially, the use of a confocal laser microscope enables surface morphology measurement. Surface morphology induced by the external stimulation such as pH or temperature change can be evaluated by the scanning electronic microscopy (SEM), this technique is possible to visualize the morphology surface by means of a micrograph. The composition along the surface and at different thickness of the micrometer order of magnitude can be recorded by Raman spectroscopy. Force-curve measurement by AFM will be performed using a special home-made tip with a colloidal nanosphere attached to the cantilever, which is essential to measure an extremely soft surface. These techniques help to investigate the relationship between the surface stiffness and the composition and other condition of the copolymers, and change of surface stiffness induced by the external stimulation such as pH or temperature change. Using microscopic infrared imaging, the change of spatial distribution of specific chemical species induced by the external stimulation could be investigated. Combination of these results reveal the actual surface behavior as the response to the applied external stimuli (pH and/or temperature change), and it will enable us to optimize the design of the materials.[77]

4.4 APPLICATIONS

Hydrogels and superabsorbent polymers have applications in biomedical fields, pharmaceutical fields, and as personal hygiene products because of their high water content and the consequent biocompatibility. Soft contact lenses made of hydrogels are popular because of their comfort, disposability, and relatively low cost. Tissue engineering, molecular imprinting, wound dressing materials, immune isolation, drug delivery are among many medical applications of hydrogels. Immuno-isolation is the process of protecting implanted material such as biopolymers, cells, or drug release carriers from an immune reaction. Microencapsulation of these implanted cells with hydrogel-based capsules is common to protect them from the immune system. Alginate, chitosan, agarose, and polyethylene glycol have been used for encapsulation. These hydrogels are non-interfering with cellular functions of the inside cells, encapsulation can be done under physiological conditions, and provide microenvironment for the trapped cell survival for longer time.[31–33]

4.4.1 BIOCATALYST

Recently, smart biocatalyst where enzymes are conjugated to stimuli-response polymer have gained significant attention. Based on presence or absence of external stimuli the polymer attached to enzymes changes its conformation to protect the enzyme from external environment and regulate enzyme activity, thus acting as molecular switch.

Smart polymers can be used to design reversible soluble or insoluble biocatalyst. Reversible biocatalysts catalyze on enzyme reaction in their soluble state and thus can be used in reactions with insoluble or poorly soluble substrates. Reversible soluble biocatalysts are formed by the phase separation of smart polymers in aqueous solutions following a small chance in the external conditions, when the enzyme molecule is bound covalently to polymer. As the reaction is complete, the conditions are changed to cause the catalyst to precipitate so that it can be separated from the product and be reused. Stimuli that are used to reuse include pH, temperature, ionic strength, and addition of chemical species like calcium.

Thermo or temperature-responsive biocatalysts are the conjugated product of temperature response polymer and the enzyme. Thermoresponsive polymer shows reversible phase transition at a particular temperature.This temperature is known as LCST, below this temperature polymers exist in a solution state and start precipitating or becoming turbid above this temperature, also known as cloud point.

For example, trypsin immobilized on a pH-responsive copolymer of methylmethacrylate and methacrylic acid is used for repeated hydrolysis of casein. Similarly simplex cells are immobilized inside beads of the thermoresponsive polymer gel as a biocatalyst. A biocatalyst sensitive to magnetic field is produced by immobilizing invertase and γ-Fe_2O_3 in poly(N-isopropylacrylamide-co-acrylamide) gel. The heat generated by exposure of γ-Fe_2O_3 to a magnetic field causes the gel to collapse, which is followed by a sharp decrease in the rate of sucrose hydrolysis.

Industrial biocatalyst polymers such as poly-N-isopropyl acryl amide (PNiAAm), polyvinylcaprolactum (PNVCl), poly(*N,N*-diethyl acrylamide) (PDEAM), poly(*N*-ethylmethacrylamide) (PNEMAM), poly(methyl vinyl ether) (PMVE), and poly(2-ethoxyethyl vinyl ether (PEOVE) are typical examples of thermoresponsive polymers. Among them, PNiPAAm has been extensively used in the development of smart industrial biocatalysts. Structurally, the *N*-isopropyl moiety in the polymer is responsible for its smart behavior in response to the temperature changes, which interacts with the water molecules at temperatures below the LCST, but dissociates above the LCST. The LCST of PNiPAAm is approximately 32.5°C.

4.4.2 SENSORS

Gel has ability to change volume in response to a specific stimulus and is utilized in many sensing designs that are currently under development. A variety of gels with various chemical structures have been proposed, which can collectively achieve response to a broad range of stimuli. Temperature sensors are in high demand for industrial applications and to identify sites of infection, inflammation, or other pathology in healthcare. Many polymer systems were therefore investigated for applications in gel-based temperature sensors. Most of these systems contain polymers that exhibit LCST, that is, shrink when heated, or UCST, that is, swell when heated.

4.4.2.1 USED AS GLUCOSE SENSORS

In designing glucose biosensors, polymers with an extended pi-electron system such as thiophene, pyrrole, and thiazole are widely being explored. This is because of their conductivity, good stability, biocompatibility and reproducibility.

One of the most popular applications of pH-sensitive polymers is the fabrication of insulin delivery systems for the treatment of diabetic patients. Delivering insulin is different from delivering other drugs, since insulin has to be delivered in an exact amount at the exact time of need. Many devices have been developed for this purpose and all of them have a glucose sensor built into the system. In a glucose-rich environment, such as the bloodstream after a meal, the oxidation of glucose to gluconic acid catalyzed by glucose oxidase (GluOx) can lower the pH to approximately 5.8. Glucose oxidase is mostly used in glucose sensing.

4.4.3 BIOMIMETIC ACTUATORS

There have been attempt to mimic the efficient conversion of chemical energy into mechanical energy in living organisms. A cross-linked gel of Poly(vinyl alcohol) chains entangled with the polyacrylic acid chains has good mechanical properties and shows rapid electric field association bending deformations: a gel rod of 1 mm diameter bends semi-circularly within 1 s on the application of electric field. Polymer gels capable of mechanical response to electric field have also been developed using the cooperative binding of the positively charged surfactant molecule to the polyanionic polymer poly (2 acrylamido-2-methyl-1-propane sulfonic acid). Copolymer gels consist of N-Isopropylacrylamide and acrylic acid would be useful for constructing biochemomechanical systems. A pH-induced change in the–COOH ionization of acrylic acid alters the repulsive forces, the attractive force is produced by hydrophobic interactions arising from the dehydration of N-isopropylacrylamide moieties. The biomimetic actuators could be used in future soft machines.

4.4.4 BIOSEPARATIONS

Conjugate systems have been used in physical affinity separation and immunoassays. In affinity precipitation of biomolecule, the bioconjugate is synthesized by coupling a ligand to a water-soluble smart polymer. The ligand polymer conjugate selectively binds the target protein from the crude extract and the protein–polymer complex is precipitated from the solution by the changes in the environment like pH, temperature, ionic strength, or addition of some reagents. Finally the desired protein is dissociated from the polymer and the later can be recovered from the reuse for another cycle. Various ligands like protease inhibitors, antibiotics, nucleotides, metal chelates, and carbohydrates have been used in affinity precipitation.

4.4.5 REDUCES RADIOACTIVE WASTE

Polymer-bound smart catalysts are useful in waste minimization, catalyst recovery, and catalyst reuse. Polymeric smart coatings have been developed that are capable of both detecting and removing hazardous nuclear contamination. Such applications of smart materials involve catalyst chemistry and sensor chemistry.

Recently, scientists in Germany and India are reporting the development of a new polymer that reduces the amount of radioactive waste produced during routine operation of nuclear reactor. In the study the researchers created an absorbent material that unlike unconventional ion exchange resins has the unique ability of disregarding iron bases ions.[42]

4.4.6 BIOTECHNOLOGY AND MEDICINE

Smart polymers may be physically mixed with or chemically conjugated to biomolecules to yield a large family of polymer biomolecule to yield a large family of polymer–biomolecule system that can respond to biological as well as to physical and chemical stimuli. Biomolecules that can be polymer conjugated include proteins and oligopeptides, sugars, polysaccharides, single- and double-stranded oligonucleotides, DNA plasmids, simple lipids, phospholipids, and synthetic drug molecule. These polymer–biomolecule complexes are referred as affinity smart biomaterials or intelligent bioconjugates. Also such polymers have been used in developing smart surfaces and smart hydrogels that can respond to external stimuli. Such polymeric biomaterials have shown a range of different applications in the field of biotechnology and medicine. The researchers have used these polymers for biomedical applications to downstream processing and biocatalyst. The latest thrilling breakthrough was achieved by the group of Stayton and Hoffman, at the University of Washington, USA. The researchers developed a clever way to use smart polymers that provide size selective switches to turn proteins on and off.

4.4.7 SMART POLYMER IN PROTEIN PURIFICATION

Smart polymers are used for the concentration of protein solutions and for the isolation as well as pur*i*fication of biomolecules. Recombinant thermostable lactate dehydrogenase from the thermophile *Bacillusstearo thermophilus* was purified by affinity partitioning in an aqueous two-phase polymer system formed by dextran and a copolymer of N-vinyl caprolactam and 1-vinyl imidazole. The enzyme partitioned preferentially into the copolymer phase in presence of Cu ions. The enzyme lactate dehydrogenase from porcine muscle has better access to the ligands and binds to the column. With the decrease in temperature the polymer molecules undergo transition to a more expanded coil conformation. Finally, the bound enzyme is replaced by the expanded polymer chains. This system was used for lactate dehydrogenase purification.[36]

4.4.8 PROTEIN FOLDING

In order to attain the native structure and function of proteins, the refolding process is a major challenge in currently ongoing biochemical research. Using smart polymer reduces the hydrophobicity of surfactant which facilitates or hinders the conformational transition of unfolded protein, depending upon the magnitude of unfolded protein. Refolding of bovine carbonic anhydrase was examined in presence of PPO–Ph–PEG at various temperatures. The refolding yield of carbonic anhydrase was strongly enhanced and aggregate formation of PPO–Ph–PEG at specific temperature of 50–55°C. Eudragit S-100, a pH-sensitive smart polymer is supposed to increase the rate of refolding and refolding percentage of denatured protein. This was found to assist refolding of α-chymotrypsin, which is known to bind to the polymerrather than non-specifically.[33–35]

4.4.9 AUTONOMOUS FLOW CONTROL IN MICROFLUIDICS

The concept of "lab in a chip" has evolved out of efforts to miniaturize analytical instruments. By using photolithography on a chip, one can create microchannels and work with very small volumes. Smart materials show considerable promise in designing microactuators for autonomous flow control inside these microfluidic channels. Saitoh et al.[42] have explored the use of glass capillaries coated with Poly-N-Isopropylacrylamide for creating an on/off valve for the liquid flow. Below LCST the PNIPAm coated capillary allowed the flow of water, above LCST the flow was blocked as the coating was now hydrophobic. Beebe et al., on the other hand, used a pH-sensitive methacrylate to control the flow inside the microchannels. The hydrogel-based microfluid valve opened or closed depending upon the pH of the solution. The design has the potential of being self-regulating/autonomous since the valve can be controlled by feedback of H^+ produced or consumed in the reaction. Undoubtedly we will see many other innovative designs for such applications in coming years.

4.5 CONCLUSION

In this review we have focused on intelligent polymeric materials, their properties characterization techniques, and various applications in different areas. We have strived to illustrate the versatility and potential of these materials. Applications of polymers in biomedical field are highly significant and more effective safe, less toxic, and comparatively simple. Mechanism of drug delivery and biosepration are so efficient and reliable. Smart polymers play wide role in biomedical applications such as oxygenation of blood and also cardiopulmonary bypass surgery.

4.6 FUTURE IMPORTANCE

Smart materials are the focus of a great deal of how buildings made from smart materials could withstand earthquakes, how airplanes could fly without conventional hydraulic systems or electrical motors to move their control surfaces and how a materials revolution could occur in the electronics industry. Some visionaries have even gone so far as to predict materials that "learn," meaning that their behavior would change and self-correct over time. In the future, new smart polymers with transition temperatures and pH levels in the range at which certain biomolecules are most stable (4–5°C and pH 5–8) will be developed and smart and genetically produced biomaterials and drug delivery systems have made great progress in recent years.

KEYWORDS

- **smart polymers**
- **stimuli responsive materials**
- **hydrogels**
- **biodegradable packaging**
- **drug delivery**

REFERENCES

1. Ganesh, V. A.; Baji, A.; Ramakrishna, S. Smart Functional Polymers—A New Route Towards Creating a Sustainable Environment. *RSC Advances* **2014,** *4* (533), 52–64.
2. Crucho, C. I. C. Stimuli-Responsive Polymeric Nanoparticles for Nanomedicine. Chem *Med. Chem.* **2015,** *10*, 24–38.
3. Koetting, M. C.; Peters, J. T.; Steichen, S. D.; Peppas, N. A. Stimulus-Responsive Hydrogels: Theory, Modern Advances and Applications. *Mater. Sci. Eng.: R.* **2015,** *93,* 1–49.

4. Sun, X. F.; Gan, Z.; Jing, Z. X.; Wang, H. H.; Wang, D.; Jin, Y. A. Adsorption of Methylene Blue on Hemicellulose-Based Stimuli Responsive Porous Hydrogel. *J. Appl. Polym*er *Sci.* **2015,** *132*, 41606.
5. Cong, H. P.; Qiu, J. H.; Yu, S. H. Thermoresponsive Poly (N-isopropylacrylamide)/Graphene/Au Nanocomposite Hydrogel for Water Treatment by a Laser-Assisted Approach. Small **2015,** *11*, 1165–1170.
6. Cao, P. F.; Mangadlao, J. D.; Advincula, R. C. Stimuli-Responsive Polymers and Their Potential Applications in Oil–Gas Industry. *Polym. Rev.* **2015,** *55*, 706–733.
7. Zhang, J. L.; Zhang, M. X.; Tang, K. J.; Verpoort, F.; Sun, T. L. Polymer-Based Stimuli-Responsive Recyclable Catalytic Systems for Organic Synthesis. *Small* **2014,** *10*, 32–46.
8. Wang, Y. F.; Lai, C. L.; Hu, H. W.; Liu, Y.; Fei, B.; Xin, J. H. Temperature-Responsive Nanofibers for Controllable Oil/Water Separation. *RSC Advances* **2015,** *5*, 51078–51085.
9. Guney, O.; Serin, E. Stimuli-Responsive Molecularly Imprinted Hybrid Polymer Gel as A Potential System for Controlled Release. *J. Appl. Polym. Sci.* **2016,** *133*, 42913.
10. Thambi, T.; Park, J. H.; Lee, D. S. Stimuli-Responsive Polymers for Cancer Therapy. Biomater. Sci. **2016,** *4* (1), 55–69.
11. Yang, M. Y.; Tan, L.; Wu, H. X.; Liu, C. J.; Zhuo, R. X. Dual-Stimuli-Responsive Polymer-Coated Mesoporous Silica Nanoparticles Used for Controlled Drug Delivery. *J. Appl. Polym. Sci.* **2015,** *132*, 42395.
12. Chauhan, G. S. Evaluation of Nanogels as Supports for Enzyme Immobilization. *Polym. Int.* **2014,** *63*, 1889–1894.
13. Rzaev, Z. M. O.; Dincer, S.; Piskin, E. Functional Copolymers of N-isopropylacrylamide for Bioengineering Applications. *Progr. Polym. Sci.* 2007, *32*, 534–595.
14. Melendez-Ortiz, H. I.; Alvarez-Lorenzo, C.; Concheiro, A.; Bucio, E. Grafting of N-vinyl Caprolactam and Methacrylic acid Onto Silicone Rubber Films for Drug-Eluting Products. *J. Appl. Polym. Sci.* **2015,** *132*, 41855.
15. Melendez-Ortiz, H. I.; Diaz-Rodriguez, P.; Alvarez-Lorenzo, C.; Concheiro, A.; Bucio. Binary Graft Modification of Polypropylenefor Anti-Inflammatory Drug-Device Combo Products. *J. Pharm. Sci.* **2014,** *103*, 1269–1277.
16. Choi, S.; Tripathi, A.; Singh, D. Smart Nanomaterials for Biomedics. *J. Biomed. Nanotechnol.* **2014,** *10*, 3162–3188.
17. Ionov, L. Polymeric Actuators. *Langmuir* **2015,** *31*, 5015–5024.
18. Mano, J. F. Stimuli-Responsive Polymeric Systems for Biomedical Applications. *Adv. Eng. Mater.* **2008,** *10*, 515–527.
19. Balint, R.; Cassidy, N. J.; Cartmell, S. H. Conductive Polymers: Towards a Smart Biomaterial for Tissue Engineering. *Acta Biomaterialia.* **2014,** *10*, 2341–2353.
20. Alvarez-Lorenzo, C.; Bucio, E.; Burillo, G.; Concheiro, A. Medical Devices Modified at the Surface by Gamma-Ray Grafting for Drug Loading and Delivery. *Exp. Opin. Drug Deliv.* **2010,** *7*, 173–185.
21. Musyanovych, A.; Landfester, K. Polymer Micro- and Nanocapsules as Biological Carriers with Multifunctional Properties. *Macromol. Biosci.* **2014,** 14, 458–477.
22. Yarin, A. L. Stimuli-Responsive Polymers in Nanotechnology: Deposition and Possible Effect on Drug Release. *Math. Model. Natural Phenomena* **2008,** *3*, 1–15.
23. Ganta, S.; Devalapally, H.; Shahiwala, A.; Amiji, M. A Review of Stimuli-Responsive Nanocarriers for Drug and Gene Delivery. *J. Control. Release* **2008,** *126*, 187–204.
24. Schmaljohann, D. Thermo- and pH-Responsive Polymers in Drug Delivery. *Adv. Drug Deliv. Rev.* **2006,** *58*, 1655–1670.

25. Gil, E. S.;Hudson, S. M. *Prog. Polym. Sci.* **2004,** *29*, 1173–1222.
26. Delcea, M.; Möhwald, H.; Skirtach, A. G. *Adv. Drug Deliv. Rev.* **2011,** *63*, 730–747.
27. Liechty, W. B.; Kryscio, D. R.; Slaughter, B. V.; Peppas, N. A. *Annu. Rev. ChemBiomol. Eng.* **2010;** 1:149–173.
28. Seo, J. W.; Shin, U. S. Cationic Effect of an Ionic Copolymer with a Temperature-Responsive Characteristic on the LCST Value: Abroad LCST spectrum of 35 to 46 A Degrees. *Macromol. Res.* **2015,** *23*,1073–1078.
29. Guo, P. P.; Li, H. L.; Ren, W. C.; Zhu, J. L.; Xiao, F.; Xu, S. M.; Wang, J. D. Unusual Thermo Responsive Behaviors of Poly(NIPAM-co-AM)/PEG/PTA Composite Hyd *Mater. Lett.* **2015,** *143*, 24–26.
30. Nakamae, K.; Miyata, T.; Hoffman, A. S. Swelling Behavior of Hydrogels Containing Phosphate Groups. *Macromol. Chem.* **2003,** *193*, 983–990.
31. Miyata, T.; Nakamae, K.; Hoffman, A. S.; Kanzaki, Y. Stimuli-Sensitivities of Hydrogels Containing Phosphate Groups. *Macromol. Chem. Phys.* **2003,** *195*, 1110–1120.
32. Shashkina, Yu. A.; Zaroslov, Yu. D.; Smirnov, V. A.; Philippova, O. E.; Khokhlov, A. R.; Pryakhina, T. A.; Churochkina, N. A. *Polymer* **2003,** *44* (8), 2289–2293.
33. Hoffman, A. S. Bioconjugation of Intelligent Polymers and Recognition Proteins for Use in Diagnostics and Affinity Separation. *Clin. Chem.* **2000,** *46*, 1478–1486.
34. Kukoi, R.; Morita, S.; Ota, H.; Umakoshi, H. Protein Refolding Using Stimuli-Responsive Polymer-Modified Aqueous Two-Phase Systems. *J. Chromatogr. B. Bionmed. Sci. Appl.* **2000,** *743*, 215–223.
35. Chen, Y. J.; Huang, L. W.; Chin, H. C.; Lin, S. C. Temperature-Responsive Polymer-Assisted Protein Refolding. *Enzyme Microb. Technol.* **2003,** *32*, 120–130.
36. Matttiasson, B.; Dainyak, M. B.; Yu Galaev, I. Smart Polymers and Protein Purification. *Polymer-Plastic Technol. Eng.* **1998,** *37* (3), 303–308.
37. Verma, I.; Sonia, M. Gene Therapy-Promise, Problems and Prospects. *Nature* **1997,** *389*, 239–242.
38. Godbey, W. T.; Mikos, A. G. Recent Progress in Gene Delivery Using Non-Viral Transfer Complexes. *J. Control. Release* **2001,** *72*, 115–125.
39. Stayton, P. S.; Hoffman, A. S.; Murty. N. et al. Molecular Engineering of Proteins and Polymers for Targeting and Intercellular Delivery of Therapeutics. *J. Control. Release* **2000,** *65*, 203–220.
40. Stayton, P. S.; Sayed El; Murthy N., et al. Smart' Delivery Systems for Bimolecular Therapeutics. *Orthodontics Craniofacial Res.* **2005,** *8*, 219–225.
41. Stayon, P. S.; Sayed El; Hoffman, A. S. Smart Polymeric Carriers for Enhanced Intercellular Delivery of Therapeutic Macromolecules. *Expert Opin. Biol. Ther.* **2005,** *5*, 23–32.
42. Bhaskarapillai, A.; Sevilimedu, N. V.; Sellergren, B. Synthesis and Characterization of Imprinted Polymers for Radioactive Waste Reduction. *Ind. Eng. Chem. Res.* **2009,** *48* (8), 3730.
43. Messersmith, P. B.; Znidarsich, F. Synthesis and LCST Behavior of Thermally Responsive Poly(N-isopropylacrylamide)/layered Silicate Nanocomposites, in Nanophase and Nanocomposite Materials II. *Mater. Res. Soc. Symposium Proc.* **1997,** *457*, 507–512.
44. Wu, J. H.; Lin, J. M.; Zhou, M.; Wei, C. R. Synthesis and Properties of Starch-Graft-Polyacrylamide/Clay Superabsorbent Composite. *Macromol. Rapid Comm.* **2000,** *21*, 1032–1034.
45. Haraguchi, K.; Takehisa, T.; Fan, S. Effects of Clay Content on the Properties of Nanocomposite Hydrogels Composed of Poly (Nisopropylacrylamide) and Clay. *Macromolecules* **2002,** *35*, 10162–10171.

46. Lin, J. M.; Wu, J. H.; Yang, Z. F.; Pu, M. L. Synthesis and Properties of Poly(acrylic acid)/mica Superabsorbent nanocomposite. *Macromol. Rapid Comm.* **2001,** *22*, 422–424.
47. Haraguchi, K.; Takehisa, T. Nanocomposite Hydrogels: A Unique Organic–Inorganic Network Structure with Extraordinary Mechanical, Optical, and Swelling/De-Swelling Properties. *Adv. Mater.* **2002,** *14*, 1120–1124.
48. Zhang, W. A.; Luo, W.; Fang, Y. E. Synthesis and Properties of a Novel Hydrogel Nanocomposites. *Adv. Mater* **2005,** *59*, 2876–2880.
49. Burillo, G.; Bucio, E.; Arenas, E.; Lopez, G. P. Temperature and pH Sensitive Swelling Behavior of Binary DMAEMA/4VP Grafts on Polypropylene Films. *Macromol. Mater. Eng.* **2007,** *292*, 214–219.
50. Tsukasa, S.; Kazutaka, K.; Takaki, S.; Tomoo, S. UCST and LCST Behavior in Polymer Blends Containing Poly(methylmethacrylate-stat-styrene). *Polymer* **1998,** *39*, 773–780.
51. Ougizawa, T.; Inoue, T.; Kammer, H. UCST and LCST Behavior in Polymer Blends. *Macromolecules* **1985,** *18*, 2089–2092.
52. Ueda, H.; Karasz, F. E. Upper Critical Solution Temperatures in Blends of Chlorinated Polyethylenes. *Macromolecules* **1985,** *18*, 2719–2722.
53. Sato, T.; Endo, M.; Shiomi, T.; Imai, K. UCST Behavior for High-Molecular Weight Polymer Blends and Estimation of Segmental Parameters from Their Miscibility. *Polymer* **1996,** *37*, 2131–2136.
54. Lowman, A. M.; Morishita, M.; Kajita, M.; Nagai, T.; Peppas, N. A. Oral Delivery of Insulin Using pH-Responsive Complexation Gels. *J. Pharm. Sci.* **1999,** *88,* 933–937.
55. Wood, K. M.; Stone, G.; Peppas, N. A. Lectin Functionalized Complexation Hydrogels for Oral Protein Delivery. *J. Control. Release* **2006,** *116*, 66–68.
56. Wood, K. M.; Stone, G. M.; Peppas, N. A. Wheat Germ Agglutinin Functionalized Complexation Hydrogels for Oral Insulin Delivery. *Biomacromolecules* **2008,** *9*, 1293–1298.
57. Bucio, E.; Burillo, G. Radiation Grafting of ph and Thermo Sensitive N-Isopropylacrylamide and Acrylic Acid Onto PTFE Films by Two-Steps Process. *Radiation Phys. Chem.* **2007,** *76*, 1724–1727.
58. Huang, M.; Jin, X.; Li, Y.; Fang, Y. Muddied Guar Gum Matrix Tablet for Controlled Release of Diltiazem Hydrochloride. *Reactive Funct. Polym.* **2006,** *66*, 1041–1046.
59. Harmon, M. E.; Thomas, A. M.; Wolfgang, K.; Curtis, W. F. A Surface Plasmon Resonance Study of Volume Phase Transitions in Nisopropylacrylamidegel Films. *Macromolecules* **2002,** *35*, 5999–6004.
60. Kunz, U.; Katerkamp, A.; Renneberg, R.; Spener, F.; Cammann, K. Sensing Fatty Acid B Protein with Planar and Fiber-Optical 3urface Plasmon Resonance Spectroscopy Devices. *Sensors Actuators B: Chem.* **1996,** *32*, 149–155.
61. Godin, M.; Laroche, O.; Tabard-Cossa, V.; Beaulieu, L. Y.; Grütter, P.; Williams, P. J. *Rev. Sci. Instruments* **2003,** *74*, 4902–4907.
62. Knoll, W. Interfaces and Thin Films as Seen by Bound Electromagnetic Waves. *Ann. Rev. Phys. Chem.*. 1998; 49 569-638.
63. Hickel, W.; Duda, G.; Jurich, M.; Krohl, T,; Rockford, K.; Stegeman, G. I.; Swalen, J. D.; Wegner, G.; Knoll, W. Optical Waveguides from Novel Polymeric Langmuir-Blodgett Multilayer Assemblies. *Langmuir*. **1990,** *6*,1403–1407.
64. Sekkat, Z.; Wood, J.; Geerts, Y.; Knoll, W. Surface Plasmon Investigations of Light-I Modulation in the Optical Thickness Ofmolecularly Thin Photochromic Layers. *Langmuir* **1996,** *12*, 2976–2980.

65. Spinke, J.; Liley, M.; Schmitt, F. J.; Guder, H. J.; Angermaier, L.; Knoll, W. Molecu Recognition at Self-Assembled Monolayers: Optimization of Surface Functionalization. *J. Chem. Phys.* **1993,** *99*, 7012–7019.
66. Percot, A.; Zhu, X. X.; Lafleur, M. A Simple FTIR Spectroscopic Method for the Determination of the Lower Critical Solution Temperature of N-Isopropylacrylamide Copolymers and Related Hydrogels. *J. Polym. Sci.: Polym. Phys.* **2000,** *38*, 907–915.
67. Katsumoto, Y.; Tanaka, T.; Sato, H.; Ozaki, Y. Conformational Change of Poly(N-Isopropylacrylamide) During the Coil−Globuletransition Investigated by Attenuated Total Reflection/Infrared Spectroscopy and Density Functional Theory Calculation. *J. Phys. Chem. A.* **2002,** *106*, 3429–3435.
68. Noda, I.; Dowrey, A. E.; Marcott, C.; Story, G. M.; Ozaki, Y. Generalized Two-Dimensional Correlation Spectroscopy. *J. Appl. Spectrosc.* **2000,** *54*, 236–248.
69. Noda, I. Generalized Two-Dimensional Correlation Method Applicable to Infrared, Raman, and Other Types of Spectroscopy. *J. Appl. Spectrosc.* **1993,** *47*, 1329–1336.
70. Guo, Y.; Peng, Y.; Wu, P. A Two-Dimensional Correlation ATR-FTIR Study of Poly (Vinyl Methyl Ether) Water Solution. *J. Mol. Struct.* **2008,** *875*, 486–492.
71. Heskins, M.; Guillet, J. E. Solution Properties of Poly (N-isopropylacrylamide). *J. Macromol. Sci.: Pure Appl. Chem.* **1969,** *2*, 1441–1455.
72. Grinberg, A.; Grosberg, Y.; Tanaka, T. Studies of the Thermal Volume Transition of Poly(N-Isopropylacrylamide) Hydrogels by High-Sensitivity Differential Scanning Microcalorimetry. 2. Thermodynamic Functions. *Macromolecules.* **2000,** *33*, 8685–8692.
73. Chearúil, F. N.; Corrigan, O. I. Thermosensitivity and Release from Poly N-Isopropylacrylamide–Polylactide Copolymers. *Int. J. Pharm.* **2009,** *366*, 21–30.
74. Meraa, A. E.; Goodwin, M.; Pike, J. K.; Wynne, K. J. Synthesis, Characterization and Surface Analysis Using Dynamic Contact Angle Measurements of Graft Copolymers: Poly(methyl methacrylate)-g-poly(dimethylsiloxane) and Poly(methyl methacrylate)-poly(trifluoropropylmethylsiloxane). *Polymer* **1999,** *40*, 419–427.
75. Letellier, P.; Mayaffre, A.; Turmine, M. Drop Size Effect on Contact Angle Explained by Nonextensive Thermodynamics. Young's Equation Revisited. *J. Colloid Interface Sci.* **2007,** *314*, 604–614.
76. Arenas, E.; Bucio, E.; Burillo, G.; Lopez, G. P. Radiation Grafting of Poly(ethylene glycol) Methacrylate Onto Poly(vinyl chloride) Tubes. *Des. Monomors Polym.* **2007,** *10*, 459–467.
77. Meléndez-Ortiz, H. I.; Bucio, E.; Isoshima, T.; Hara, M. Surface Characterization of Binary Graft Copolymers (PP-g-DMAEMA)-g-NIPAAm and (PP-g-4VP)-g-NIPAAm by Using SEM and AFM. *Smart Coatings III* **2010,** *1050*, 107–120.

CHAPTER 5

Green "Bioplastics" in Solving the Global Problem of a Polyethylene Dump

TATIANA GR. VOLOVA[1,2*], EKATERINA I. SHISHATSKAYA[1,2*], and SABU THOMAS[1,3]

[1]*Siberian Federal University, 79 Svobodnyi Av., Krasnoyarsk 660041, Russia*

[2]*Institute of Biophysics SB RAS, Federal Research Center "Krasnoyarsk Science Center SB RAS," 50/50 Akademgorodok, Krasnoyarsk 660036, Russia*

[3]*International and Interuniversity Centre for Nano Science and Nano Technology, Mahatma Gandhi University, Kottayam, Kerala, India*

**Corresponding author. E-mail: volova45@mail.ru; shishatskaya@inbox.ru*

ABSTRACT

Development and use of new, environmentally friendly materials, which will be able to be degraded in the environment without producing toxic compounds, thus joining the global material cycles, is among the priorities for critical technologies of the 21st century. A major concern is that the annual production of synthetic plastics has exceeded 350 million tons; in developed countries, no more than 20% of them are recycled and they largely accumulate in landfills. Up to 10,000 ha of the land, including arable fields, is annually occupied by new landfills. Plastic waste ruins municipal sewage and drainage systems and pollutes water environments. Greenpeace reports that up to 10% of the plastic produced every year gets into the Global Ocean. Increasing requirements for environmental protection and reasonable concern in connection with the accumulation of synthetic plastics waste in the biosphere actualize research and transition to biotechnological processes for the production of degradable polymers. Among the promising "green," plastics are polyhydroxyalkanoates (PHAs), degradable polyesters of monocarboxylic acids synthesized by various microorganisms. PHA is a family of polymers of various chemical structure having different basic physicochemical properties. These polymers are promising materials for fabricating biomedical devices, degradable packaging for food and drinks, personal hygiene products, and devices and formulations for municipal engineering and agriculture.

Researchers of the Institute of Biophysics SB RAS and Siberian Federal University tested different modes of cultivation of hydrogen-oxidizing bacteria on the gaseous substrate, such as autotrophic synthesis of protein or poly (3-hydroxybutyrate) and PHA copolymers using electrolytic hydrogen; and synthesis gas; sugars, organic and fatty acids, vegetable oil. Physicochemical properties and mechanical properties of PHA having different chemical composition were studied. The laws governing the destruction of polymers in soil and aquatic natural ecosystems are investigated. Polymer products of various types for various applications were obtained—for medicine and technical purposes, as destructible packaging and containers. A new line of research in bioplastics has been formulated and deployed to create eco-friendly agro preparations of a new generation. A family of agricultural preparations forms using biodegradable PHAs as a basis was constructed and characterized. In soil ecosystems with model weed plants, as well as in crops of cultivated plants infected with pathogens of root rot or weeds, the effectiveness of pre-emergence application of the developed long-term forms of pesticides deposited in PHA was shown. It opens prospects for the creation of long-term and targeted new generation pesticides to reduce application rates and the uncontrolled spread of xenobiotics in the biosphere.

5.1 INTRODUCTION

Traditional use of products of chemical synthesis derived from nonrenewable natural resources leads to excessive and increasing accumulation of nonrecyclable wastes, coming into conflict with environmental protection activities and posing a global environmental problem. One way out of this conundrum is to expand the use of tools and methods of biotechnology, which, on the one hand, protect beneficial biota and enhance productivity in agriculture and, on the other, reduce toxic impacts on individual ecosystems and the entire biosphere.

Pollution of the planet by waste and the high rates of consumption of natural resources lead to the failure of the biosphere's self-regulating mechanism, entailing unpredictable consequences. Academician N.N. Moiseev wrote at the end of the 1970s: "The main problem of global ecology is sustainability of the biosphere; the loss of its sustainability actually means elimination of humankind on Earth." But mankind has always disturbed the balance in the biosphere and has always had an environmental problem.

Now 21st century pollution in many regions has reached a critical limit.

Intensive farming involves the use of enormous amounts of various chemicals to control pests, weeds, and pathogens causing diseases of crops. However, no more than 10% of the pesticides applied reach their target; most of these substances accumulate in biological objects, contaminate soils and water environments, kill beneficial living organisms, and upset the balance of natural ecosystems. The newest trend in research is development and agricultural use of environmentally safe new-generation pesticides with targeted and controlled release of active ingredients embedded in biodegradable matrices or covered with biodegradable coatings, which are degraded in soil and other biological media by soil micro flora to form products that are harmless to living and nonliving nature and which are gradually released into the environment. The use of such formulations can reduce the amounts of the chemicals applied to soil and enable their sustained and controlled delivery over a plant growing season, preventing sharp releases into the environment that occur in the case of using free pesticides. The main condition for constructing such formulations is the availability of appropriate materials with the following properties: degradability, ecological compatibility with the environment and global biosphere cycles; safety for living and nonliving nature; long-term presence in the natural environment (for weeks and months) and controlled degradation followed by formation of nontoxic products; chemical compatibility with pesticides and fertilizers; process ability by available methods that are compatible with pesticide and fertilizer production technologies.

Development and use of new, environmentally friendly materials, which will be able to be degraded in the environment without producing toxic compounds, thus joining the global material cycles, is among the priorities for critical technologies of the 21st century. Accumulation of synthetic plastics in the biosphere is a global ecological problem. The production of synthetic plastics has exceeded 350 million tons in year; 50% it is packaging. Cycle of synthetic plastics it is "from the cradle to the grave." More than 99% of plastics are produced from chemicals derived from oil, natural gas, and coal, and all of which are dirty, nonrenewable resources. Researchers estimate that more than 8.3 billion tonnes of chemical plastic has been produced since the early 1950s: only 10–15% of all plastic waste ever produced has been recycled; about 5% has been incinerated, while the rest >80% has accumulated in landfills, dumps, or the natural environment.

Up to 10,000 ha of the land, including arable fields, is annually occupied by new landfills. Plastic waste ruins municipal sewage and drainage systems and pollutes water environments. Greenpeace reports that up to 10% of the plastic produced every year gets into the Global Ocean. Ocean is the accumulator of all wastes. Rivers carry plastic waste to the sea, making them major contributors to ocean pollution. 10 of the biggest rivers (Niger, Nile, Indus, Meghna, Ganges, Amur, Hai He, Huang He, Mecong, et al.) alone carry more than 90% of the plastic waste that ends up in the oceans. The accumulation of plastic in the ocean is dangerous for marine biota and reduces water quality. In the oceans as a result of the accumulation of plastic waste formed "plastic islands." If people do not stop using disposable plastic products, by 2050 there will be more plastic in the oceans than fish.

"If you can refuse from plastic—then refuse"—this is the motto of the UN report on combating plastic waste 5 June 2018.

The possible ways to replace plastics:

- Obtaining plastics from natural materials
- Giving degradable synthetic plastics
- Chemical synthesis of biodegradable polymers of monocarboxylic acids (polylactide, polyglycolide)
- Biological synthesis of bioplastic (PHA).

Increasing requirements for environmental protection and reasonable concern in connection with the accumulation of synthetic plastics waste in the biosphere actualize research and transition to biotechnological processes for the production of degradable polymers. Among the promising "green" plastics are PHAs, degradable polyesters of monocarboxylic acids synthesized by various microorganisms. PHA is a family of polymers of various chemical structure having different basic physicochemical properties. These polymers are promising materials for fabricating biomedical devices, degradable packaging for food and drinks, personal hygiene products, and devices and formulations for municipal engineering and agriculture (Volova, 2004; 2009; Volova et al., 2013; 2017; 2019).

The chapter presents the results of the study of "green" plastics—PHAs and their potential as a biotechnological analog of nondestructible polyolefins.

5.2 MATERIALS AND METHODS

Chemoorganotrophic hydrogen-oxidizing bacteria taxa *Ralstonia eutropha* B 5786; *R.eutropha* B 8562 and *Cupriavidus eutrophus* B-10646 have been investigated as a producer of PHA. Bacteria were cultivated in developmental regimes in periodic culture with the use of various fermentation equipment, from glass flasks in volumes from 1 L to automated fermentation complexes with apparatus from 15 to 150 L. The variation of the C-substrate composition and the use of precursor substrates (valerate, hexanoate, butyrolactone) made it possible to synthesize polymers with different sets and ratios of monomers.

With the use of PHA in various phase states, polymer products of various structures and geometries were obtained in the form of films, fibers, 3D forms, and thinning. To study the properties of polymers used modern physico-chemical methods, regularities of PHA degradation were studied under natural conditions in various soil and aquatic ecosystems. Original research results on the properties of herbicides and fungicides fertilizers deposited in a degradable polymer base and the effectiveness of the use of these formulations in laboratory ecosystems with higher plants infected with fusariosis and weeds are presented in the monograph.

5.3 RESULTS AND DISCUSSION

In Institute of Biophysics SB RAS and Siberian Federal University, a fundamental groundwork has been created that ensured the development and implementation of polymer synthesis technologies, the design and application of high-tech products made from them for various applications, biomedicine, experimental pharmacology, agriculture and public utilities, etc.

5.3.1 PHA SYNTHESIS

The laws of the constructive metabolism of the collection of hydrogen-oxidizing bacterial strains were studied; the technologies for controlled synthesis of PHA of different composition with high yields on various substrates (sugar, carbohydrate, $CO_2 + O_2 + H_2$, synthesis gas, vegetable oils) were developed and implemented (Volova, 2004; 2009).

This knowledge was used as a basis for developing an effective process of production of polymers with various chemical structures in a first Russian pilot facility. The authors' collection of microbial strains was used to develop processes of PHA synthesis with high yields (up to 80–90%) of polymers with different chemical structure: a homopolymer of 3-hydroxybutyric acid [P(3HB)]; copolymers of 3-hydroxybutyric acid and 4-hydroxybutyric acid [P(3HB-*co*-4HB)] containing 6–51 mol.% 4HB; copolymers of 3-hydroxybutyric acid and 3-hydroxyvaleric acid [P(3HB-*co*-3HV)] containing 3-83 mol.% 3HV; and copolymers of 3-hydroxybutyric acid and 3-hydroxyhexanoic acid [P(3HB-*co*-3HHx)] containing 50 mol.% 3HHx. The study compared physicochemical properties of poly(3-hydroxybutyrate), poly (3-hydroxybutyrate-*co*-4-hydroxybutyrate), poly(3-hydroxybutyrate-*co*-3-hydroxyvalerate), and poly (3-hydroxybutyrate-*co*-3-hydroxyhexanoate) containing different monomer fractions, which were determined by IR spectroscopy, X-ray structure analysis, DSC, gel permeation chromatography, and SEM. Incorporation of 4-hydroxybutyrate, 3-hydroxyvalerate, or 3-hydroxyhexanoate into the poly(3-hydroxybutyrate) chain reduced the melting temperature and the degree of crystallinity of the copolymers as compared to the corresponding parameters of poly(3-hydroxybutyrate) (Volova et al., 2008). PHA chemical composition influenced the structure and properties of the surface of polymer films and their physico-mechanical properties. The study showed that by varying the composition of PHAs and their monomer fractions, one can produce materials and devices that would considerably differ in their basic properties (Volova et al., 2013; 2017).

During implementation of the mega projects supported by the Russian Government (Orders No. 219 and 220 of April 9, 2010), the team of researchers at the Siberian Federal University widened the range of 2- and 3-component PHAs with different chemical structures that contained major fractions of short- and medium-chain-length monomer units (Volova et al., 2013; 2017) and established a new high-productivity pilot production (PP) facility equipped with an automatic fermentation system (Bioengineering, Switzerland). The trademark "BIOPLASTOTAN" was registered for PHAs of different chemical compositions and PHA based products..

The process of PHA synthesis was scaled and studied under PP conditions. PP includes: units for media and inoculum preparation, unit for fermentation, unit for polymer extraction and purification. The PP fermentation unit includes a steam generator (Biotron, South Korea) for sterilizing fermenters and communications, a compressor (Remeza, Belarus) for air supply, a 30-L seed culture fermenter, a 150-L production fermenter, an ultrafiltration unit (Vladisart, Russia) to concentrate the culture, and a unit for cool dehumidification of the condensed bacterial suspension (LP10R ILSHIN C, South Korea) (Fig. 5.1).

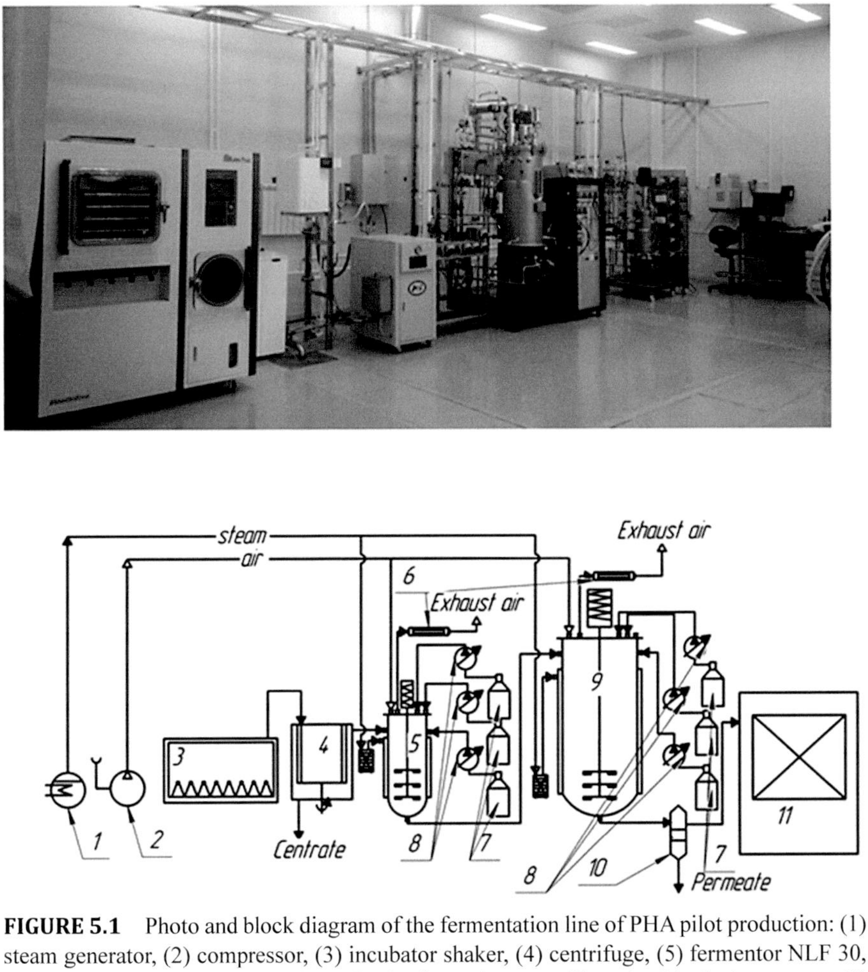

FIGURE 5.1 Photo and block diagram of the fermentation line of PHA pilot production: (1) steam generator, (2) compressor, (3) incubator shaker, (4) centrifuge, (5) fermentor NLF 30, (6) exhaust air cooler, (7) containers for feeding substrates, (8) peristaltic dosing pumps, (9) pilot scale fermentor P150, (10) ultrafiltration plant, and (11) freeze drying.

PHA of different chemical composition was synthesized, properties were studied, methods for producing polymer products were tested. The first among the isolated PHAs and the most fully characterized one is poly-3-hydroxybutyrate P(3HB). P3HB is a homopolymer of D(-)-3-b-hydroxybutyric acid, an isotactic polyester with regular ($C_4H_6O_2$) units. The melting temperature of P(3HB) varies between 160°C and 185°C, and the T_m proper is 176°C–182°C. The thermal decomposition temperature of P(3HB) lies within the range between 275°C and 280°C. The data obtained by X-ray structure analysis suggest a conclusion that the crystalline region is predominant in P(3HB). The degree of crystallinity of different P(3HB) samples is only slightly influenced by the conditions of polymer production, ranging between 70% and 80%. A disadvantage of P(3HB) is that it cannot be strain-crystallized, and, thus, it is very difficult to process (Fig. 5.2).

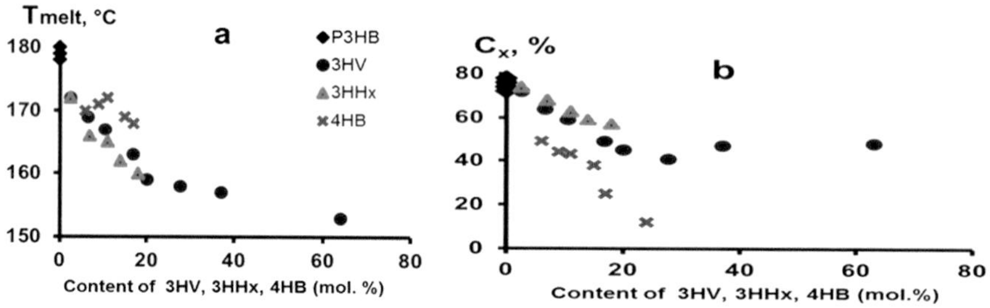

FIGURE 5.2 The effect of the composition of monomers in PHA (a) on the melting temperature and (b) degree of crystallinity.

The resulting products show low impact strength and hardness and are prone to ageing. The physicochemical properties of PHA copolymers are more diverse, and, thus, they show greater promise. However, production of PHA copolymers is a very complicated technological task, and it cannot be achieved without fundamental knowledge of the structural-functional organization of the PHA cellular cycle and the relationship between the physicochemical properties of these polymers and their chemical composition. The T_m and T_d of PHA copolymers are lower than those of P(3HB). The melting temperature of 3-hydroxybutyrate/3-hydroxyvalerate copolymers is lower than the melting peak of homogeneous P3HB, and this difference increases with an increase in the 3HV molar fraction. The melting and thermal decomposition temperatures of the copolymers consisting of 3-hydroxyhexanoate and 4-hydroxybutyrate monomer units are also lower than those of P(3HB). The low crystallization temperature of the homogeneous P(3HB) pose an obstacle to melt processing of this polymer. A possible way to increase the crystallization temperature of PHA is to synthesize 3HB copolymers tailored with 3HV and other monomers. The decrease in the melting temperatures and thermal decomposition temperatures of PHA copolymers does not reduce the difference between these parameters. Thus, different types of PHA copolymers retain their significant property—thermoplasticity. The degrees of crystallinity of P(3HB/3HV), P(3HB/3HHx), P(3HB/4HB) are lower than that of P(3HB)—20–50 (Volova, 2004; Volova et al., 2013; 2017).

5.3.2 PHA PROCESSING

Methods of solutions, emulsions, powders, and rasplavovs of PHAs in various products were developed. Subsequent studies were aimed at construction of various special devices, implants, and polymer constructs for tissue engineering and reconstructive medicine (Fig. 5.3). Interactions of fully resorbable surgical implants and devices (sutures, membranes, 3D scaffolds, microparticles, etc.) with cells and tissues were investigated to evaluate biocompatibility and functionality of these products and decide whether the novel material was suitable for medical applications. A number of items were produced from high-purity medical-grade polymers. Based on the knowledge of properties of PHA solutions, melts, emulsions, and powders, the research team determined parameters for processing of the polymers into special products such as sutures, films, 3D scaffolds, microparticles, and implants. The structure of polymer scaffolds was modified using chemical and physical methods, and functional scaffolds were prepared in the form of films and membranes. These scaffolds are good candidates for tissue engineering. PHAs with different compositions and special devices fabricated from them have been trademarked as "Bioplastotan." Research results have been reported in numerous journal papers and in several books (Volova, 2004; 2009; Volova et al., 2013; 2017).

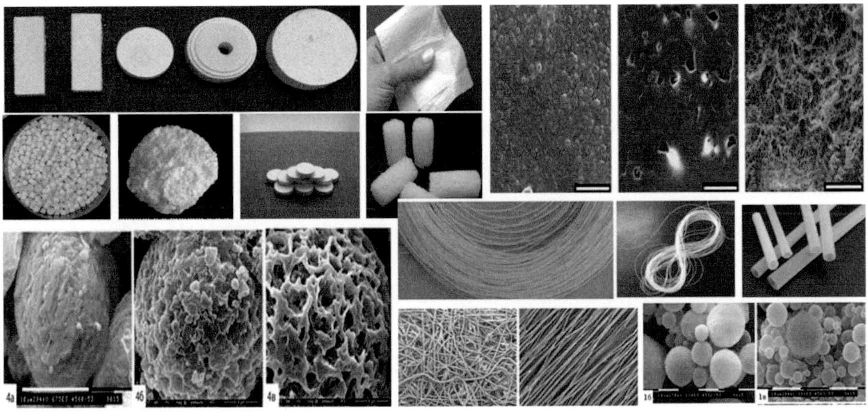

FIGURE 5.3 PHA successfully used to produce various types of polymer products.

PHAs show the greatest potential in medicine and pharmaceutics. The mild immune response to PHA implants and the sufficient duration of PHA degradation in biological media make these polymers attractive candidates for use as drug carriers in controlled-release drug delivery systems, implants and grafts for tissue and organ regeneration, materials for tissue engineering and designing of bioartificial organs. Owing to their biological and physicochemical properties, polymer devices can be regarded as promising candidates for various areas of reconstructive medicine. An experimental study of PHA monofilament suture material showed that it has all necessary physical and mechanical properties to be used in surgery as fully degradable stents for reconstruction of bile ducts. Three-dimensional PHA-hydroxyapatite implants have been constructed for bone tissue repair. Polymer microparticles are promising drug carriers that can be administered in vivo via different routes and function as sustained-release delivery systems for antiproliferative drugs.

Since then, the research team has considerably widened the scope of its PHA studies by synthesizing PHAs with different chemical compositions and by designing and studying experimental films, barrier membranes, granules, filling materials, ultrafine fibers produced by electrospinning, solid and porous 3D implants for bone tissue defect repair, tubular biliary stents, mesh implants modified by PHA coating, microparticles for drug delivery, etc. In cooperation with the V.F. Voino-Yasenetsky Krasnoyarsk Federal Medical University, the research team has conducted pioneering clinical trials. Results have been covered by several Russian patents, reported in papers published in peer-reviewed Russian and international journals, summarized in books and reviews (Volova et al., 2013; 2017).

5.3.3 BIODEGRADATION OF PHAS IN NATURAL ENVIRONMENTS

PHAs, representing the developing industry of degradable bioplastics, are good candidates to gradually replace synthetic polymers. PHA biodegradation is a complex multistage process, which is influenced by a number of factors, including PHA chemical composition and properties, the type of the polymer product and the technique employed to fabricate it, climate and weather, and the structure of the microbial community. PHA biodegradation is to a great extent determined by the composition and metabolic activity of the microbial community. As the outputs of PHAs increase, studies examining degradation of these polymers in natural environments acquire increasing significance. Results obtained in laboratory experiments cannot be used to construct prognostic models and predict PHA behavior and degradation in diverse and changeable natural ecosystems.

Analysis of the available literature shows that rather few authors reported integrated studies of various aspects of PHA degradation, which is a very complex process. Most of the studies were performed in laboratory, and they mainly addressed the mechanism of interaction between the PHA supramolecular structure and PHA-depolymerizing enzymes, the structure and molecular organization of various depolymerases and microorganisms secreting extracellular PHA depolymerases.

An important question is the pattern of polymer breakdown in the natural environment. Extensive pioneering research into PHA biodegradation behavior in natural soil ecosystems was performed at the Siberian Federal University and Institute of Biophysics SB RAS.

We studied the kinetics and laws of the degradation of PHA in natural ecosystems in various regions and received answers to key questions of the PHA biodegradation process:

What microorganisms are the most effective PHA degraders?

- How do the PHA properties change during degradation?
- How do the physico-chemical conditions of the environment (temperature, pH, oxygen availability, salinity, etc.) affect this process?
- How will the process of PHA degradation be affected by weather and climate of different regions (Volova, 2012; Volova et al., 2013; 2019).

PHA degradation influenced the total counts of microorganisms and composition of soil microflora. The microbial community that formed on the polymer surface and the soil microbial community differed in the composition and percentages of the species. By employing the clear zone technique, we, for the first time, showed that each of the PHA types studied had specific degraders. PHA degradation behavior was studied in different environments: Siberian soils under broadleaved and coniferous trees, tropical soils (in the environs of Hanoi and Nha Trang), seawater (the South China Sea), a brackish lake (Lake Shira), and freshwater recreational water bodies in Siberia. Those studies showed that degradation occurred at different rates depending on the polymer composition, shape of the specimen (film or 3D construct), climate and weather conditions, and microbial community composition. The time over which the polymer loses 50% of its mass may vary between 68.5 and 270 days in Siberian soils, between 16 and 380 days in tropical soils of Vietnam, between 73 and 324 days in the brackish lake (Shira), between 127 and 220 days in the seawater of the South China Sea, and between 17 and 65.9 days in freshwater lakes (Volova, 2012; Volova et al., 2019).

The study addresses degradation of PHAs with different chemical compositions—the polymer of 3-hydroxybutyric acid [P(3HB)] and copolymers of P(3HB) with 3-hydroxyvalerate [P(3HB/3HV)], 4-hydroxybutyrate [P(3HB/4HB)], and 3-hydroxyhexanoate [P(3HB/3HHx)] (10-12 mol.%)—in the agro-transformed field soil of the temperate zone. Based on their degradation rates at 21°C and 28°C, polymers can be ranked as follows: P(3HB/4HB) > P(3HB/3HHx) > P(3HB/3HV) > P(3HB). The microbial community on the surface of the polymers differs from the microbial community of the soil with PHA specimens in the composition and percentages of species. Thirty-five isolates of bacteria of 16 genera were identified as PHA-degraders by the clear zone technique, and each of the PHA had both specific and common degraders. P(3HB) was degraded by bacteria of the genera *Mitsuaria, Chitinophaga*, and *Acidovorax*, which were not among the degraders of the three other PHA types. *Roseateles depolymerans*, *Streptomyces gardneri*, and *Cupriavidus* sp. were specific degraders of P(3HB/4HB). *Roseomonas massiliae* and *Delftia acidovorans* degraded P(3HB/3HV), and *Pseudoxanthomonas sp., Pseudomonas fluorescens, Ensifer adhaerens*, and *Bacillus pumilus* were specific P(3HB/3HHx) degraders. All four PHA types were degraded by *Streptomyces.*

We used a comprehensive approach to studying the biodegradation of PHA in nature. We studied this process in different climates and soils, taking into account the diversity of soil microbial communities, shapes of polymer samples and methods of their preparation, and the chemical composition of the PHAs tested. One part of the studies addressed degradation of PHAs with different chemical structures in the form of film discs and pressed pellets by soil microorganisms inhabiting the rhizosphere of coniferous and broadleaved trees under varying soil temperature conditions. Experiments were performed under natural conditions, in the arboretum at the V.N. Sukachev Institute of Forest SB RAS (Krasnoyarsk) during two field seasons, which differed in temperature conditions. The first field season lasted from 2 July to 19 October 2007; the second—from 7 June to 7 September 2010. The second field season was preceded by severe winter, with the average winter temperature −22.1°C (in the second half of January and the first part of February, the temperature fell to −40°C), which was 13.5°C lower than the average temperature of the winter preceding the first field season (−8.6°C). In addition to that, the spring temperature rise occurred later in the year of the second experiment; the arboretum had been covered with snow until late April. The soddy-carbonate soil of the arboretum consists of a 10–15 to 30–40 cm thick humus layer and the underlying carbonate rock.

PHA degradation behavior was influenced by properties of the soils under the trees and characteristics of the microbial communities. In 2007, in the soil under the larch, which was moister and housed more microorganisms, PHA degradation rates were higher than those recorded under the birch. By the end of the experiment, the residual mass of P(3HB) specimens had decreased to 45% of their initial mass, and the residual mass of P(3HB-co-3HV) specimens—to 22%; the half-lives of these polymers were 83 d and 68.5 d and their average mass losses for the field season 0.325 and 0.44 mg/d, respectively. In the soil of the birch rhizosphere, degradation rates of both PHA types were lower, in spite of the great variety of the fungi present in this soil. At day 109 of the exposure, the residual masses of P(3HB) and P(3HB-co-3HV) specimens amounted to 84% and 74% of their initial masses, respectively, with the mass losses of the homopolymer and the copolymer 0.097 and 0.15 mg/d. In 2010, PHA degradation rates were lower than during the 2007 field season. At the end of the field experiment, the residual mass of the specimens amounted to 89.9% and 74% for P(3PHB) and P(3HB-co-3HV) specimens buried under the larch and to 91.4% and 89% for the specimens buried under the birch. As in 2010, the mass loss was so small, we failed to find any reliable differences in the degradation of the two PHAs used in this study (Fig. 5.4).

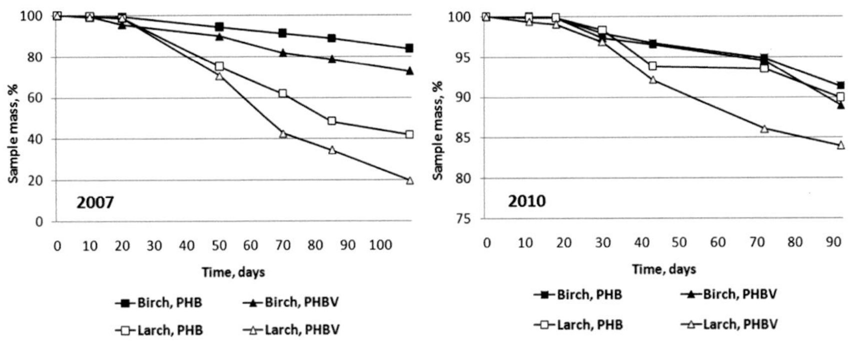

FIGURE 5.4 Dynamics of the mass of polymer specimens in Siberian soil.

The integrated study of biodegradation of two PHA types by soil microbial communities of different structures performed in Siberia conditions showed that PHA degradation is influenced by both polymer chemical composition and soil parameters: temperature, moisture content, and composition of microbial community. Twenty bacterial and eight fungal species have been identified as true PHA degraders. Under the study conditions, representatives of the bacterial genera *Variovorax, Stenotrophomonas, Acinetobacter, Pseudomonas,* and *Bacillus* have been identified as major PHA degraders. Populations of soil micromycetes that grew on polymer specimens were dominated by such PHA degraders as *Penicillium sp.* BP-1, *Penicillium sp.* BP-2, and *Paecilomyces lilacinus,* which amounted to 80% (Table 5.1).

TABLE 5.1 The Occurrence of Microorganisms in Samples of Siberian and Tropical Soils.

Soil samples		**PHA degrading microorganisms**	
		Bacteria	**Fungi**
Siberian soil	Larch	*Acientobacter sp.* *Acientobacter schindleri* *Bacillus sp.* *Pseudomonas sp.* *Stenotrophomonas maltophilia* *Variovorax paradoxus*	*Acremonium butyri* *Penicillium sp.* *Purpureocillium lilacinum* *Zygosporium masonii*
	Birch	*Acientobacter sp.* *Bacillus sp.* *Pseudomonas sp.* *Stenotrophomonas rhizophilia* *Variovorax paradoxus*	*Penicillium sp.* *Purpureocillium lilacinum* *Verticillium lateritium*
Vietnamese soil	Hanoi	*Burkholderia sp.* *Nocardiopsis sp.* *Streptomyces sp.*	*Acremonium recifei* *Gongronella butleri* *Penicillium sp.* *Purpureocillium lilacinum* *Trichoderma pseudokoningii*
	Nha Trang	*Bacillus cereus* *Burkholderia sp.* *Cupriavidus sp.* *Mycobacterium sp.* *Streptomyces sp.*	*Gongronella butleri* *Penicillium oxalicum* *Penicillium sp.*

Another study of PHA degradation in the soil was performed in the tropics. Biodegradation of PHAs of two types—poly(3-hydroxybutyrate) (P(3HB)) and poly(3-hydroxybutyrate-co-3-hydroxyvalerate) (P(3HB-co-3HV))—was analyzed in soils at field laboratories in the environs of Hanoi (Vietnam) and Nha Trang (Vietnam) (Fig. 5.5).

The air and soil temperatures and humidity in both study sites were similar throughout the study season. Precipitation at Hanoi was, however, almost an order of magnitude higher than in Nha Trang. In Nha Trang, PHAs were degraded at lower rates because of lower precipitation amounts in this area in summer. PHAs of all types were degraded at higher rates in the soil of the study site at Hanoi.

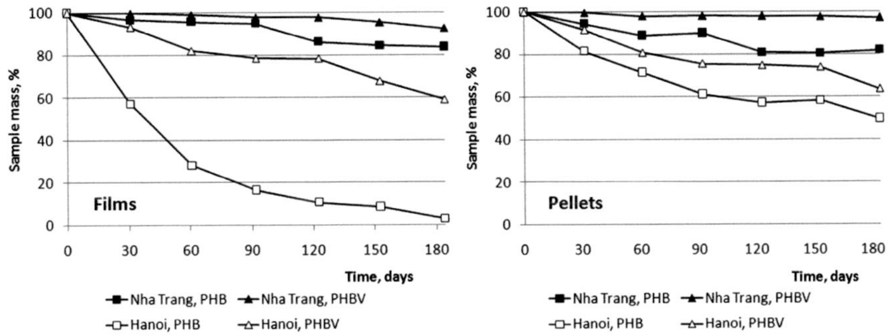

FIGURE 5.5 Dynamics of the mass of polymer specimens in Vietnamese soil.

PHA films were more prone to degradation than pressed pellets. At the end of the experiment (after 184 days of soil exposure), degradation of P(3HB) films reached more than 97%, and P(3HB-co-3HV) films were 33% degraded, while the pressed pellets were 42% and 23% degraded, respectively. In the more arid area (Nha Trang), the mass loss of P(3HB) and P(3HB-co-3HV) films was 16% and 7% and that of the pressed pellets—18% and 3% (Fig. 5.5).

Microbial communities of the soils in the two study areas in Vietnam were significantly different. Microbial populations of the Hanoi soil were dominated by *Acinetobacter calcoaceticus, Arthrobacter artocyaneus, Bacillus aerophilus, Bacillus megaterium, Bacillus sp., Brevibacillus agri, Brevibacillus invocatus, Chromobacterium violaceum, Cupriavidus gilardii, Mycobacterium fortuitum, Ochrobactrum anthropi, Staphylococcus arlettae, Staphylococcus haemoliticus, Staphylococcus pasteuri, Pseudomonas acephalitica,* and *Rodococcus equi;* while the major species in the Nha Trang soil were *Bacillus cereus, Bacillus megaterium, Bacillus mycoides, Brevibacillus agri, Gordonia terrari,* and *Microbacterium paraoxydans.* Determination of the species composition of microbial communities showed that PHA degrading bacteria were dominated by Gram-negative rods of *Burkholderia sp.;* they were isolated from the samples of both study areas. Actinobacteria of the genus *Streptomyces* were also present in the samples of both areas. Other PHA degraders isolated from the samples at Nha Trang were *Bacillus, Cupriavidus,* and *Mycobacterium spp.; Nocardiopsis* actinobacteria were isolated from the soil at Hanoi.

Gongronella butleri and *Penicillium sp.* were the fungi found in both study areas. *Acremonium recifei, Paecilomyces lilacinus,* and *Trichoderma pseudokoningii* were only isolated from the soil at Hanoi. Thus, polymer biodegradation in soils—both Siberian and tropical—is performed by bacteria and fungi. Bacteria inhabiting these soils belonged to different genera (except Bacillus), while most of the fungi in both Siberian and Vietnamese soils were represented by *Penicillium, Paecilomyces,* and *Acremonium* (Table 5.1).

Studies of PHA degradation in different soils showed that PHA biodegradation is influenced by the chemical structure of the polymer, its geometry and the technique used to process it; climate and weather, the type of the natural ecosystem and its microbial component in particular, as the factor determining the mechanism of PHA biodegradation: preferential attack of the amorphous regions of the polymer or equal degradation of both crystalline and amorphous phases. PHA degrading microorganisms that dominate microbial populations in some soil ecosystems have been isolated and identified.

5.3.4 POTENTIAL APPLICATION OF BIODEGRADABLE PHAS IN DESIGNING OF NEW GENERATION FORMULATIONS OF AGROCHEMICALS

The wide use of products of chemical synthesis based on nonrenewable resources has led to an excessive and progressive increase in the amounts of unrecycled wastes, causing environmental concerns and posing a global environmental problem. A possible solution is the wider use of biotechnological tools and methods, which, on the one hand, protect useful biota and increase agricultural production and, on the other, decrease toxic impact on different ecosystems and the entire biosphere. An important scientific task is to provide a fundamental basis for construction and agricultural use of new-generation agrochemicals in order to decrease the risk from uncontrolled spreading and accumulation of chemical products of the techno sphere in the biosphere. The following results were obtained:

- The knowledge of conditions of interactions between chemicals and PHAs in different phase states and, based on this, development of methods of loading of the chemicals intended for suppressing agents of plant diseases and killing weeds and fertilizers into variously shaped polymer matrices.

- A series of experimental slow-release formulations produced using different methods; results of studying their structure and physical/mechanical properties.
- Knowledge of the release kinetics of the active ingredients from the polymer matrix obtained in laboratory soil micro ecosystems as dependent on the geometry of formulations, concentrations of the chemicals in them, chemical composition of the soil, and the type of soil microbial community.
- Positive evaluation of the effectiveness of using the slow-release chemical formulations in laboratory conditions: in soil ecosystems with known properties containing higher plants infected by plant pathogens and weeds (Volova et al., 2019).

During initial stage, based on the review of the modern literature and preliminary studies, we selected the agrochemicals that were sufficiently effective, could be used on a large scale, were suitable for soil applications, were compatible with the polymeric matrix in different phase states, were stable in nonpolar solvent solutions, and could be analyzed by spectrophotometric methods (chromatography-mass spectrometry, HPLC, and IR spectroscopy). These were herbicides Magnum Super and Sencor Ultra, fungicide Vial Trust (tebuconazole), and nitrogen fertilizers—granular urea and ammonium nitrate. Then, we prepared polymer/active ingredient two-phase systems in the form of solutions, emulsions, and powders, which were investigated by using HPLC, DSC, X-Ray, and IR spectroscopy.

These polymer/active ingredient two-phase systems in the form of solutions, emulsions, and powders were used to construct slow-release formulations of the selected chemicals, which were shaped as microparticles, microgranules, films, and pellets and were loaded with different amounts of agrochemicals. These were:

- Herbicide (metribuzin, 2,4-chloroacetic acid, tribenuron) formulations loaded with the active ingredient to different extents in the form of films prepared from the solution of the polymer of 3-hydroxybutyric acid [P(3HB)], granules prepared from the polymer solution, 3D constructs prepared from a mixture of powdered polymer and MET, and MET-loaded microparticles prepared by microemulsion technique.
- Formulations of fungicide tebuconazole in the form of films, microgranules, and 3D pressed constructs, loaded with the active ingredient to different extents.

- A series of slow-release formulations of nitrogen fertilizer based on urea and ammonium nitrate in the form of films, 3D constructs, core/shell polymer-coated urea granules, and pressed 3D constructs coated with several layers of the polymer.

Investigation of the initial substances (polymer and chemicals) and the experimental formulations by methods of IR spectroscopy, DSC, X-Ray, and HPLC showed that loading of the polymer matrix with the chemicals did not produce any noticeable effect on the physicochemical properties of the polymer and, hence, its performance and that no chemical binding of the components occurred and the constructed systems were physical blends.

To evaluate the experimental slow-release formulations, we constructed and characterized laboratory soil microecosystems with two types of agro-transformed soils (field and garden soils). We examined the chemical composition of the soils and the structure of soil microbial communities and revealed the dominant microbial species, including primary degraders of PHAs of different chemical compositions. We studied the relationships between the release profiles of the fungicide, herbicide, and nitrogen fertilizer and the geometry of the forms, the type of the embedded chemical, and the degree of loading of the matrix, taking into account degradation rate of the polymer matrix. The embedding of the chemicals into the degradable PHAs enabled slow and burst-free release of the active ingredients into soil and, hence, long-lasting effect of the agrochemicals.

To influence PHA degradation and to increase the availability of these polymers, we prepared PHA composites with polyethylene glycol, poly-caprolactone, and birch wood chips. The composition of the matrix had a substantial effect on release of the agrochemicals. By varying the composition of the matrix and by coating nitrogen fertilizer with polymer, one can regulate release rates of the active ingredient within a wide range. The outcomes of the studies conducted during implementation of the project were development of the methods and techniques for construction of the novel formulations of agrochemicals based on degradable PHAs. The studies addressed the effects of various factors (geometry of the form, the degree of loading, the chemical composition of the polymer matrix, the type of soil and its microbial community) on degradation of the formulations in the soil and release profiles of the active ingredients and determined the major factors that could be used to regulate those processes. The results proved that these formulations were suitable for slow release of herbicides, fungicides, and nitrogen fertilizers.

In the final stage of the project implementation, we studied the effectiveness of the formulations developed during the research in laboratory ecosystems with plants, which contained the following model weeds and crops:

- Weeds (a perennial grass species *Agrostis stolonifera*—creeping bentgrass, *Setaria macrocheata*—foxtail millet, *Chenopodium album*—lamb's quarters, *Melilotus albus*—white sweet clover, *Amaránthus retrofléxus*—red-root amaranth);
- Crops (lettuce *Latuca sativa*, wheat *Triticum aestivum*);
- Soil microecosystems infected by plant pathogens *F. moniliforme* and *F. solani*—root rot agents;
- Crop *Triticum aestivum* infected by *Fusarium moniliforme* and weeds.

For studying herbicidal activity, five types of weed plants (*Agrostis stolonifera, Setaria macrocheata, Chenopodium album, Melilotus albus,* and *Amaranthus retroflexus*) were exposed to experimental formulations P(3HB)/MET and compared with free form metribuzin in the commercial pesticide Sencor Ultra. In the herbicide-treated ecosystems, the density of weed plants and the aboveground biomass were significantly reduced. As a result, complete suppression of the growth of weeds was achieved (Fig. 5.6).

FIGURE 5.6 The growth suppression of weeds *Agrostis stolonifera* (1) and *Setaria macrocheata* (2) by different metribuzin formulations: control without herbicide (a), Sencor Ultra (b), P(3HB)/MET-films with 10% (c), and 50% (d) of metribuzin.

During the first 10 days of the experiment, the efficiency of P(3HB)/MET formulations was lower or comparable to that of the commercial pesticide. However, after 30 days, a prolonged effect and higher activity of the experimental formulations as compared to Sencor Ultra were observed.

A similar effect was observed in co-cultivation of wheat (*Triticum aestivum*) and weeds (*Melilotus albus*). After 50 days, there was no growth of weeds in all versions of metribuzin formulations, both embedded and free forms (Fig. 5.7).

An analysis of fungicidal action of P(3HB)/TEB formulations on phytopathogenic fungi *Fusarium verticillioides* (former *F. moniliforme*) showed an effective suppression of them in soil during the 30 days of the experiment, at the rate comparable with the effect of the commercial fungicide Raxil Ultra at similar concentrations of TEB (Fig. 5.8). A slight decrease in the total number of fungi in soil samples treated by P(3HB)/TEB formulations was also observed, while the number of bacteria did not change significantly.

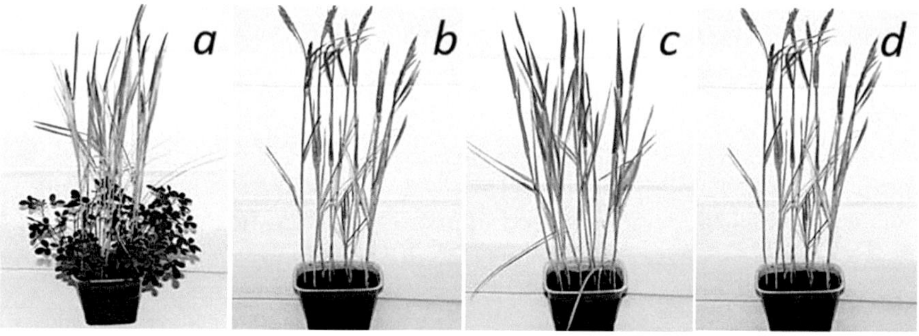

FIGURE 5.7 The growth suppression of *Melilotus albus* by different metribuzin formulations: control without (a) herbicide, (b) Sencor Ultra, (c) P(3HB)/MET-films, and (d) P(3HB)/MET-microparticles.

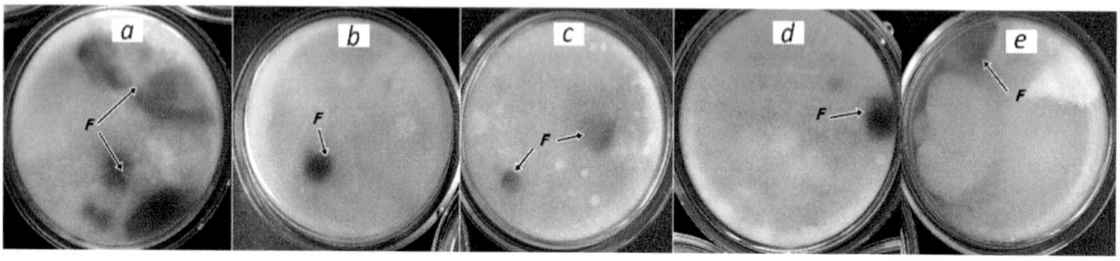

FIGURE 5.8 The growth of *Fusarium verticillioides* colonies; (a) control, (b) Raksil Ultra, (c) P(3HB)/TEB-films, (d) P(3HB)/TEB-pellets, (e) P(3HB)/TEB-granules. Arrows indicate colonies of Fusarium.

The efficiency of the experimental forms of tebuconasole P(3HB)/TEB was assessed by infection rate of fungal root rot pathogens in wheat roots. It was shown that the proportion of roots infected with *Fusarium*, *Alternaria*, and *Bipolaris* fungi decreased by 10% when embedded tebuconazole was used. Furthermore, the fungicidal effect was prolonged and, after 30 days, was comparable to the effect of the commercial fungicide Raxil Ultra (Fig. 5.9). Experimental forms of the fungicide showed a strongly pronounced healing effect on the wheat root system in the case of additional contamination of soil with *F. verticillioides* spores in concentrations of up to 1×10^6 g^{-1}. After 30 days, the degree of infected roots was reduced by 34%, and the fungicidal effect exceeded the effect of the commercial fungicide.

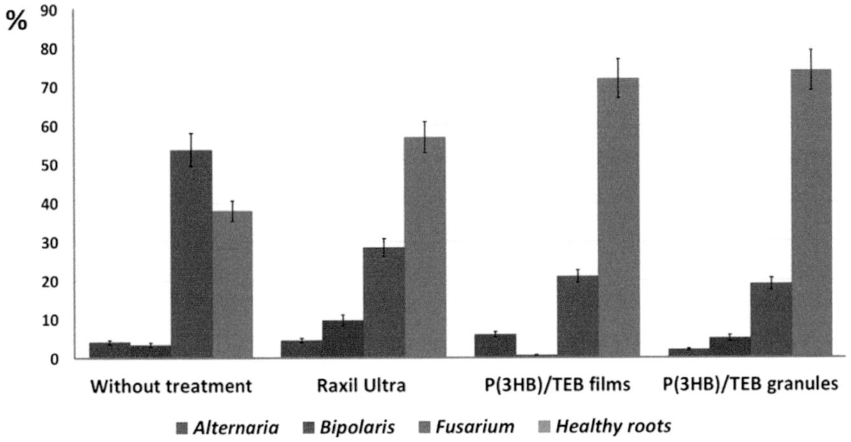

FIGURE 5.9 The degree of infected roots (%) of *Triticum aestivum* at 30 days after tebuconazole treatment.

The studies showed that the slow-release herbicide and fungicide formulations developed during implementation of this project were effective against weeds and plant pathogens and that their effects were comparable with or stronger than those of free forms of agrochemicals. Gradual release of nitrogen from the experimental nitrogen fertilizer formulations significantly reduces nitrogen losses (Volova et al., 2019). Positive results of the research project suggest that the use of natural polyesters, PHAs, as degradable polymer matrix for constructing slow-release formulations of agrochemicals is a useful approach and that the next step should be field tests of the experimental formulations.

5.4 CONCLUSION

Available results allow us to summarize:

- Biotechnological processes make it possible to obtain new generation materials that do not accumulate in the biosphere.
- PHA and polymeric products from them can be obtained from various raw materials, including industrial wastes.
- PHA is a good candidate for replacement of polyethylene and polypropylene as a degradable package and for products for agriculture.

The development of this line of research will help mitigate the risk of accumulation and uncontrolled spread of xenobiotics in the environment and replace dead-end synthetic plastics by degradable materials capable of joining the biospheric cycles.

ACKNOWLEDGMENT

This study was financially supported by Mega-Grant "Agro preparations of the new generation: a strategy of construction and realization" (Agreement No 074-02-2018-328) in accordance with Resolution No 220 of the Government of the Russian Federation.

KEYWORDS

- **plastic waste**
- **ecology**
- **"green" bioplastics**
- **polyhydroxyalkanoates**
- **properties**
- **biodegradability**
- **application**

REFERENCES

1. Prudnikova, S. V.; Volova, T. G. *The Environmental Role of Polyhydroxyalkanoates: Patterns of Biodegradation in the Natural Environment and Interaction with Microorganisms*/Krasnoyarsk: Krasnoyarsk writer, 2012; p 197. ISBN 978-5-98997-062-9.

2. Volova, T. G. *Polyhydroxyalkanoates—Plastic Materials of the 21st Century: Production, Properties, Application.* Nova Science Pub. Inc.: New York, 2004; p 282. ISBN 1-59033-992-4.
3. Volova, T. G. *Hydrogen-Based Biosynthesis.* Nova Science Pub. Inc. NY, USA, 2009; p 287. ISBN 978-1-60876-053-4.
4. Volova, T. G.; Shishatskaya, E. I.; Sinskey, A. J. *Degradable Polymers: Production, Properties and Applications.* Nova Science Pub. Inc.: New York, 2013c; p 380. ISBN 978-1-62257-832-0.
5. Volova, T. G.; Zhila, N. O.; Prudnikova, S. V.; Boyandin, A. N.; Shishatskaya, E. I. *Fundamental Principles of Design and Application of New Generation Agricultural Products;* Krasnoyarsk, 2016d; p 220. ISBN: 978-5-906740-01-4.
6. Volova, T. G.; Shishatskaya, E. I.; Prudnikova S.V.; Zhila, N. O; Boyandin, A. N. *New Generation Formulations of Agrochemicals: Current Trends and Future Priorities.* Toronto-Canada: CRC/Taylor&Francis/Apple Academic Press. p 250 ISBN: 9781771887496//DOI: 10.1201/9780429433610.

CHAPTER 6

Potency of Endophytic Fungi Isolated from Medicinal Plants in Egypt as a Resource for Anticancer Compounds

AHMED M. ABDEL-AZEEM[1*], SHERIF M. ZAKI[2], FATMA M. SALEM[3], WALEED F. KHALIL[4], and SHERIF Y. SALEH[5]

[1]*Department of Botany, Faculty of Science, Suez Canal University, Ismailia 41522, Egypt*

[2]*Department of Microbiology, Faculty of Science, Ain Shams University, Abbassia-Cairo 11566, Egypt*

[3]*Department of Plant Protection, Desert Research Center, El-Matarya, Cairo 11753, Egypt*

[4]*Department of Pharmacology, Faculty of Veterinary Medicine, Suez Canal University, Ismailia 41522, Egypt*

[5]*Department of Biochemistry, Faculty of Veterinary Medicine, Suez Canal University, Ismailia,41522, Egypt*

[*]*Corresponding author. E-mail: ahmed_abdelazeem@science.suez.edu.eg; zemo3000@yahoo.com*

ABSTRACT

This study aimed to bioprospect endophytic fungi producing anticancer compounds hosted ethnobotanical plant species inhabiting a protected area in Egypt. A total of 58 fungal endophytic species belonging to 32 genera were isolated from 2 medicinal plant species inhabiting Saint Katherine Protectorate. *Trichothecium roseum* and *Stachybotrys chartarum* were the most frequently isolated species and were molecularly identified via comparison of their 18S rRNA sequences with those deposited in the GenBank database. Fungi were cultivated on half-strength potato dextrose broth for 15 days at 28°C on a rotary shaker at 180 rpm and were then extracted in ethyl acetate (EtOAc). Acute toxicity and brine shrimp lethality bioassays were performed to evaluate the cytotoxicity of the aqueous and EtOAc extracts of the *T. roseum* (GenBank accession number MF399479) and *S. chartarum* (MF399480) strains. Both taxa exhibited considerable cytotoxic activities against *Artemia salina*, and their effects on MCF-7 breast cancer cell proliferation and apoptosis were investigated in mice. The elevated tumor volumes and cell counts of the tumor-bearing mice decreased significantly after treatment with the *T. roseum* and *S. chartarum* extracts. A significant decrease in the cell viability of MCF-7 cells was recorded for 550 to 714 mg/kg of the EtOAc extract and for 169 to 178 mg/kg of the aqueous extract of *T. roseum*. Furthermore, the lethal concentration (LC_{50}) value was 2.94 μg/mL for the EtOAc extract and 1.63 μg/mL for the aqueous extract, as only 50% of the brine shrimp were viable after 24 h of treatment at these concentrations. The present study revealed that the secondary metabolites of a native isolate of *T. roseum* (MF399479) hosted by *Achillea fragrantissima* have a direct inhibitory effect on *Ehrlich* ascites carcinoma in a mouse model. This type of carcinoma is undifferentiated and originally hyperdiploid and does not exhibit tumor-specific transplantation antigen; moreover, it demonstrates high transplantable capability with no regression and rapid proliferation, exhibiting 100% malignancy, which leads to a shorter life span.

6.1 INTRODUCTION

The prospect of fungal endophytes producing active metabolites that may be effective candidates for the treatment of human conditions has attracted the attention of the scientific community in many countries (Abdel-Azeem et al., 2016). Endophytic fungi were recently indicated to produce 51% of previously unknown bioactive substances. The search for new sources of effective natural compounds as novel cancer therapeutic agents is of great importance due to the number of annual cancer deaths worldwide and the high cost and serious side effects of available cancer therapies (Kharwar et al., 2011). The unique environments, substantial diversity, ethnobotanical history, and endemicity of endophyte host plants should be considered during study selection processes (Salem and Abdel-Azeem, 2014). The mountainous region of southern Sinai exhibits greater biodiversity than the rest of Egypt, and 4350 km^2 of this area was declared a Protectorate in 1996 (Abdel-Azeem and Salem, 2015). Approximately, 170 plant species that inhabit south Sinai are used traditionally in folk medicine (Fayed and Shaltout, 2004).

This study investigated the capability of endophytic mycobionts hosted by two species of medicinal plants in to produce anticancer metabolites. Furthermore, the potential inhibitory effects of these compounds on Michigan Cancer Foundation-7 (MCF-7) breast cancer cell proliferation and apoptosis in mice were explored.

6.2 MATERIALS AND METHODS

6.2.1 SAMPLING OF MEDICINAL PLANTS

Two plant species, *Achillea fragrantissima* (Forssk) Sch. Bip and *Origanum syriacum* L., were collected from 12 sites representing different elevation wadis (1290 m above sea level (m.a.s.l.) up to 2300 m.a.s.l.) in Saint Katherine Protectorate, Sinai, Egypt, following the ethical rules of the protectorates. Samples of aerial parts from each plant species were collected in sterile polyethylene bags, closed using rubber bands, and transferred to the laboratory until plating.

6.2.2 ISOLATION AND IDENTIFICATION OF ENDOPHYTES

The aerial parts of collected plant samples were washed in running water, cut into small pieces, immersed in 75% ethanol for 1 min, dipped in sterile distilled water twice then 0.05 g/mL of a sodium hypochlorite solution for 3–5 min, and rinsed three times in sterile distilled water (Abdel-Azeem and Salem, 2012). The surface-sterilized segments were cultured on potato dextrose agar medium (PDA, Difco™) amended with 0.05 g/L rose bengal and 150 mg/L chloramphenicol. Petri dishes were sealed using Parafilm™ and incubated at 28±1°C for 7–21 days. Growing fungi were purified for identification.

6.2.2.1 PHENOTYPIC IDENTIFICATION

Phenotypic identification of endophytic fungal isolates was primarily based on the relevant identification keys for *Penicillium* (Pitt, 1979), *Aspergillus* (Klich, 2002), dematiaceous hyphomycetes (Ellis, 1976, 1971), *Fusarium* (Leslie and Summerell, 2006), miscellaneous fungi (Domsch et al., 2007), soil ascomycetes (Guarro Safont, 2012), *Chaetomium* (Doveri, 2013), and *Alternaria* (Simmons, 2007). The names of fungal taxa have been shortened in accordance to (Kirk and Ansell, 1992), and their systematic arrangement followed Ainsworth and Bisby's Dictionary of the Fungi (Kirk et al., 2008). All reported taxa were checked against Index Fungorum website database (Kirk, 2017) for name corrections, authorities, and taxonomic assignments.

6.2.2.2 MOLECULAR IDENTIFICATION

As the most frequent taxa, both *Stachybotrys chartarum* (Ehrenb.) S. Hughes and *Trichothecium roseum* (Pers.) Link were identified via comparison of their partial 18S rDNA sequences with reference strains data deposited in GenBank using BLAST homology searches on the NCBI website.

Fungal isolates were grown on PDA. DNA extraction, amplification and sequencing of the 18S rDNA were described previously (White et al., 1990; Zaki et al., 2013).

6.2.3 EXTRACTION OF ACTIVE METABOLITES

Selected isolates of *T. roseum* and *S. chartarum* were grown in 2-liter standard flasks containing 500 mL of potato dextrose broth (PDB) on a rotary shaker incubator at 180 rpm/min for 15 days at 28°C. The culture fluid was passed through two layers of filter paper to remove solids, and the metabolites were extracted using ethyl acetate (EtOAc) as the organic solvent. Crude fermentation broths were blended thoroughly and centrifuged at 3000 rpm for 10 min. Supernatants were concentrated to 10% of their original volume via rotary evaporation at 49°C. The concentrated broths were passed through a filtration membrane (d = 0.22–l m) and reconstituted in 5% dimethylsulfoxide (DMSO, Merck) in ethanol (v/v) to 5 mg/mL prior to toxicity and bioactivity evaluations.

6.2.4 TOXICOLOGICAL BIOASSAYS

6.2.4.1 BRINE SHRIMP LETHALITY BIOASSAY

Brine shrimp lethality bioassays were used to evaluate the cytotoxicity of the aqueous and ethyl acetate extracts of *T. roseum* and *S. chartarum* strains according to (Wakawa and Fasihuddin, 2016). A sample of each extract (8 mg) was dissolved in DMSO, and solutions of varying concentrations (800, 400, 200, 100, 50, 25, 12.5, 6.25, 3.13, 1.56, and 0.78 μg/mL) were obtained via serial dilution using simulated seawater. Each concentration was investigated in triplicate. The solutions were included with premarked vials loaded with 10 live nauplii in 5 mL of simulated sea water. The vials were examined just after one day with a magnification glass, and the amount of surviving nauplii in every vial was measured. The mortality endpoint of this bioassay was described as the lack of controlled onward movement throughout half a minute of observation.[3] Vials containing DMSO and the extraction (500 μL) were set as controls. Vincristine sulfate (VS) was adopted as a positive control. Sea salt (Sigma 9883) was applied in activity tests. The complete quantity of shrimp in every container was measured and recorded. The death percentage and lethal concentration (LC) were determined. The LC_{50} after 24 h was obtained using a plot of the percentage of brine shrimp killed against the logarithm of the extract concentration (toxicant concentration). The best concentration was obtained from the curve data using regression analysis compared to the positive control (VS LC_{50} = 0.52 μg/mL). The following regression equation was used: $Y = f(X,ß) = a + bx$ according to (Salem and Abdel-Azeem, 2014).

6.2.4.2 ACUTE TOXICITY BIOASSAY

The acute intraperitoneal toxicity of the fungal extracts was determined by calculation of the lethal dose that kills 50% of animals (LD_{50}) using up and down techniques according to a previously described method (Bruce, 1985) in 12 albino mice per extract type. Fungal extracts were administered intraperitoneally (i.p.) in graded doses of 422.5, 550, 714, and 928 mg of extract/kg body weight for the *S. chartarum* EtOAc extract (S1); 100, 130, 169, and 219.7 mg of extract/kg body weight for the *S. chartarum* aqueous extract (AQ1); 500, 650, 845, and 1099 mg of extract/kg body weight for the *T. roseum* EtOAc extract (S2); and 500, 650, 845, 1099, and 1428 mg of extract/kg body weight for the *T. roseum* aqueous extract (AQ2). Mortality rates were recorded within the first 24 h after administration. Doses were selected and adjusted using a constant multiplicative factor of 1.3 for this test. For each successive animal, the dose was adjusted down or up depending on the previous result.

6.2.5 ANIMALS AND ANTICANCER ACTIVITY BIOASSAY

The Michigan Cancer Foundation-7 human breast tumor cell line (MCF-7) was supplied by the Egyptian Institute of Cancer in Cairo, Egypt, and was kept in a female of Swiss albino mice. The in vivo antitumor activity of fermentation broths was tested using a protocol approved by the Scientific Research Ethics Committee of the Veterinary Medicine Faculty at Suez Canal University. Animals were housed as described by standard animal care requirements in groups up to 10 individuals per cage. They maintained under pathogen-free environment on a 12/12 h light/dark system. Seventy mice of Swiss albino were divided into 7 groups (10 animals/group). Each mouse, except those in the negative control group, received an i.p. injection of 0.2 mL of a cell suspension containing 2×10^6 Ehrlich ascites carcinoma (EAC) on day 0. Group I mice served as a negative control. Group II mice were left untreated as a positive control. Group III mice were treated with 5-fluorouracil (5-FU) as a standard reference drug. Animal groups IV, V, VI, and VII were subcutaneously injected with 0.1 mL of OS (28 mg/kg b.wt.), AQS (9 mg/kg b.wt.), OT (45 mg/kg b.wt.) and AQT (58 mg/kg b.wt.), respectively, 24 h after tumor cell inoculation. The abdominal circumferences of the mice were measured after fungal extract injection using a measuring tape. Animals were weighed immediately before tumor inoculation and twice weekly after tumor inoculation. Five mice from each group were sacrificed via cervical dislocation 13 days after injection with the fungal extracts, and changes in the body weight, ascites tumor volume, and viable tumor cell count of the mice were measured and statistically analyzed. Median survival time (MST) and the percent increase in life span (ILS%) for the remaining mice in each group were observed and calculated (Sur and Ganguly, 1994).

6.2.6 BIOCHEMICAL BIOASSAY

Ascetic fluid and serum samples were collected from the animals treated with fungal extracts and sent directly to the laboratory for analysis. The serum-ascites albumin gradient (SAAG) was used in the differential diagnosis of ascites, particularly with reference to the prediction of portal hypertension, and both fluid and serum samples were required for this analysis. The following examinations were performed: physical parameter history and clinical examination and abdominal (Nyland and Mattoon, 2002) and biochemical examinations (Rudloff, 2005). Blood was immediately transferred to test tubes and maintained at room temperature for 30 min according to the manufacturer's protocol. Blood was centrifuged at $1200 \times g$ for 20 min. Serum assays for albumin, alanine amino transferase (ALT), creatinine, C-reactive protein (CRP), alpha-fetoprotein (AFP) and carcinoembryonic antigen (CEA) were measured using photometric or enzyme-linked immunosorbent assay (ELISA) methods according to the manufacturers' instructions.

6.2.7 STATISTICAL ANALYSIS

Data obtained from the endophytic fungal extract experiments were subjected to statistical analyses, including descriptive statistics and graphical presentations using Excel software (Microsoft Office-XP Package 2002). The results of the bioassays are mentioned as the mean value ± standard error of mean. Comparisons were carried out inside treatments with the analysis of variance (ANOVA).). The mean values obtained for the different groups were compared using one-way ANOVA followed by Duncan's multiple ranges (Duncan, 1955), and the least significant difference test was used to determine the significant difference between the means of each parameter. Student's t-test was used for some parameters. The obtained results are expressed as means ± S.E.M.

6.3 RESULTS

6.3.1 SPECIES RICHNESS OF THE ISOLATED ENDOPHYTES

Twenty-four endophytic species belonging to 14 genera were recovered from *Achillea fragrantissima*, and *Trichothecium roseum* was the most prevalent endophyte isolate. Thirty-four species belonging to 16 genera were isolated from *Origanum syriacum*, and *Stachybotrys chartarum* was the most prevalent endophyte isolate (Table 6.1). The isolated endophytic species were deposited in the Suez Canal University Fungarium (SCUF). *T. roseum* (SCUF23A) and *S. chartarum* (SCUF4E) were screened for anticancer metabolite production.

6.3.2 PHENOTYPIC IDENTIFICATION OF THE TESTED ENDOPHYTES

T. roseum is a fast growing fungus reached about 9 cm in diameter within 10 days at room temperature on PDA. Colonies were flat, powdery texture, and initially white, but they became rosy, pink, or orange with age. Conidiophores were long, erect, septated, 150–260 μm long and 4–5 μm wide. Conidia were hyaline to brightly colored, smooth, 2-celled, ovoid to ellipsoid, with a protuberant hilum, 8.75–13.75 (−16) × 5–6.25 (−7.5) μm in size (Fig. 6.1 A).

TABLE 6.1 Frequency of Recovered Taxa Hosted *Achillea fragrantissima* and *Origanum syriacum*.

Fungal species	Isolation frequency	
	A. fragrantissima	O. syriacum
Acremonium alternatum Link	0	2
A. murorum (Corda) W. Gams	1	1
A. rutilum W. Gams	1	1
Alternaria alternata (Fr.) Keissl.	2	2
A. atra (Preuss) Woudenb. & Crous	2	2
A. tenuissima (Kunze) Wiltshire	1	1
Aspergillus bisporus Kwon-Chung & Fennell	1	0
A. candidus Link	0	2
A. flavus Link	1	2
A. fumigatus Fresen.	1	0
A. japonicus Saito	0	1
A. niger Tiegh.	2	2
A. terreus Thom	3	0
Chaetomium bostrychodes Zopf	1	1
Ch. globosum Knuze	3	1
Ch. hamadae (Udagawa) Arx	0	1
Ch. nigricolor L.M. Ames	0	1
Ch. perlucidum Sergeeva	0	1
Ch. piluliferum J. Daniels	0	1
Ch. senegalense L.M. Ames	1	0
Ch. subspirilliferum Sergeeva	1	0
Cladosporium herbarum (Pers.) Link	1	0
Curvularia lunata (Wakker) Boedijn	2	2
Drechslera australiensis Bugnic. Ex Subram. & B.L. Jain	0	1
D. bicolor (Mitra) Subram. & B.L. Jain	0	2
D. hawaiiensis Bugnic. ex Subram. & B.L. Jain	0	2
Embellisia phragmospora (Emden) E.G. Simmons	1	1
Eurotium amstelodami L. Mangin	1	0
E. chevalieri L. Mangin	0	1
Fusarium oxysporum E.F. Sm. & Swingle	1	1
F. solani (Mart.) Sacc.	0	1
Melanospora caprina (Fr.) Sacc.	1	0
Microascus trigonosporus C.W. Emmons & B.O. Dodge	0	2

TABLE 6.1 *(Continued)*

Fungal species	Isolation frequency	
	A. fragrantissima	**O. syriacum**
Nigrospora oryzae (Berk. & Broome) Petch	2	2
Penicillium chrysogenum Thom	1	3
P. rubrum Stoll	0	1
Sarocladium strictum (W. Gams) Summerb.	1	1
Sordaria fimicola (Roberge ex Desm.) Ces. & De Not.	0	1
S. humana (Fuckel) G. Winter	0	1
Stachybotrys chartarum (Ehrenb.) S. Hughes	0	6
Trichoderma harzianum Rifai	0	1
T. viride Pers.	0	1
Trichothecium roseum (Pers.) Link	7	0

Colonies of *S. chartarum* reached 1.5 cm in diameter within 5 days on PDA at 25°C and were covered with a dark powdery bloom of conidial masses with the reverse uncolored. Conidiophores were simple or irregularly branched and were generally approximately 100 μm, but they could reach 1000 μm long and 3–6 μm wide; furthermore, the conidiophores were hyaline at the base and could become dark olivaceous and were sometimes roughened toward the apex, bearing clusters of 4–10 phialides. Phialides were obovate or ellipsoidal and hyaline, though they could later become olivaceous (9–14 × 4–6 μm). Conidia were ellipsoidal and aggregated in slimy masses; at first hyaline, the conidia became dark olivaceous-gray ± opaque at maturity and exhibited an almost smooth-walled to coarsely roughened appearance with warts and ridges, 6.5–8.75 (−10) × 4–6 (−7.5) μm (Fig. 6.1 B).

6.3.3 MOLECULAR IDENTIFICATION OF TESTED ENDOPHYTES

The partial 18S rDNA sequences of the *T. roseum* (SCUF23A) and *S. chartarum* (SCUF4E) isolates were 1012 and 1344 bp, respectively. The NCBI repository was utilized to distinguish the isolates by means of BLAST searches of the acquired 18S rDNA sequences. The 18S rDNA partial sequences of the isolated cultures were more than 99% identical to the 18S partial sequences of their respective reference strains deposited in GenBank.

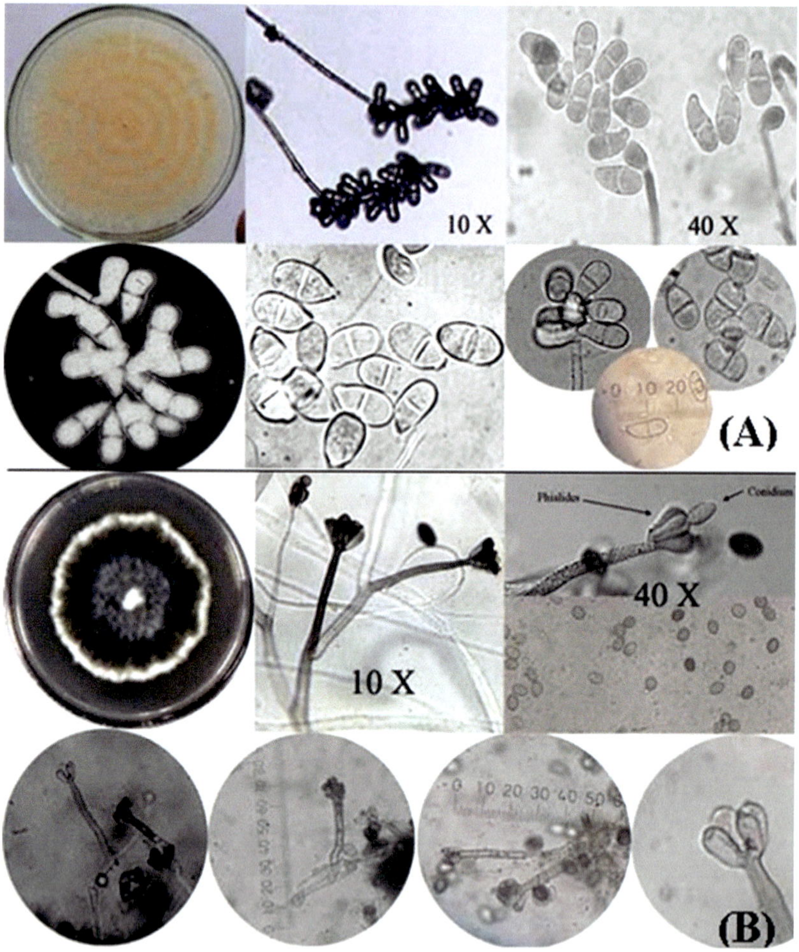

FIGURE 6.1 (A) *Trichothecium roseum* (SCUF23A-MF399479) and (B) *Stachybotrys chartarum* (SCUF4E- MF399480).

The nucleotide sequence data for the SCUF23A and SCUF4E isolates from the current study were registered in the NCBI GenBank repository as MF399479 and MF399480, respectively.

6.3.4 CYTOTOXIC ACTIVITY OF THE FUNGAL EXTRACTS

The cytotoxic activity showed by fungal extract fractions was appealing; these results obviously reveal the existence of powerful bioactive molecules.

The LC_{50} values of the brine shrimp lethality bioassays were 5.64, 4.13, 2.94, and 2.13 µg/mL for the *S. chartarum* EtOAc extract (OS) and aqueous extract (AQS) and the *T. roseum* EtOAc *extract* (OT) and aqueous extract (AQT), respectively (Table 6.2, Fig. 6.2). The LD_{50} values of the acute toxicity bioassays were 891.25, 1158, 565, and 178 mg/kg for the *S. chartarum* EtOAc extract (OS) and aqueous extract (AQS) and the *T. roseum* EtOAc *extract* (OT) and aqueous extract (AQT), respectively (Fig. 6.3).

TABLE 6.2 Brine Shrimp Lethality Bioassay Results.

Sample	**LC_{50} (µg/mL)**	**Regression equation**	**R^2**
S. chartarum EtOAc extract (OS)	5.64	y = 0.0394 x + 0.8133	0.5987
S. chartarum aqueous extract (AQS)	4.13	y = 0.0449 x + 0.5736	0.8004
T. roseum EtOAc extract (OT)	2.94	y = 0.0237 x + 0.1325	0.9761
T. roseum aqueous extract (AQT)	1.63	y = 0.0234 x + (−0.0915)	0.9549
Vincristine sulfate (Std.)	0.52	y = 33.256 x + 58.740	0.9580
DMSO	2.13	y = 0.0234 x + (−0.0915)	0.9549

FIGURE 6.2 Graph showing the relationship between % mortality of *Artemia salina* and the log concentration of fungal extracts.

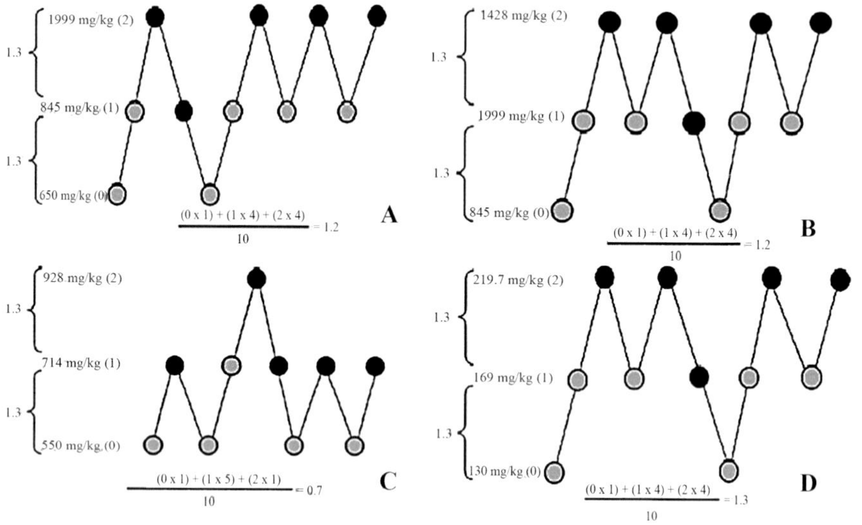

FIGURE 6.3 Schematic drawing showing the up-and-down method for LD_{50} determination of the (A) *S. chartarum* EtOAc extract (OS), (B) *S. chartarum* aqueous extract (AQS), (C) *T. roseum* EtOAc *extract* (OT), and (D) *T. roseum* aqueous extract (AQT) in mice.

6.3.5 ANTICANCER ACTIVITY OF FUNGAL METABOLITES

The EtOAc (OT) and aqueous extracts (AQT) of *T. roseum* exhibited high anticancer activity and decreased tumor volumes significantly compared to those of the 5-FU-treated group of mice. The EtOAc (OS) and aqueous extracts (AQS) of *S. chartarum* exhibited less activity than either of these groups (Fig. 6.4 and Table 6.3). Significant differences were observed in the body weights of mice treated with 5-FU or fungal extracts in comparison with those of the untreated positive control group. These differences were highest in the 5-FU group, followed by the EtOAc (OT) and aqueous extracts (AQT) of *T. roseum*; the EtOAc (OS) and aqueous extracts (AQS) of *S. chartarum* exhibited the least activity (Fig. 6.5). Treatments with *T. roseum* extracts significantly ($p < 0.001$) decreased the abdominal circumferences of the mice after 14 days of tumor growth evaluation, and they also increased the MST and ILS%. Treatments with *S. chartarum extracts exhibited non-significant results* (Fig. 6.6 and Table 6.4).

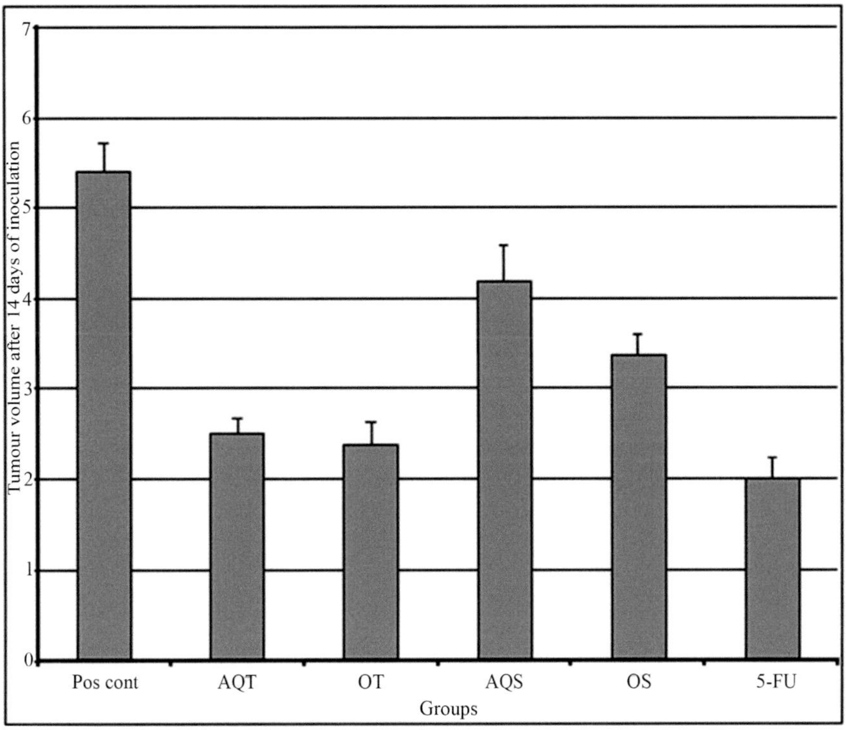

FIGURE 6.4 Tumor volume at 14 days after inoculation.

TABLE 6.3 Tumor Volume at 14 Days after Tumor Transplantation (mL).

			Ascites volume (mL)			
Mouse No.	**Positive control**	**OT**	**AQT**	**OS**	**AQS**	**5-FU**
1	4.7	2.0	2.3	3.6	5.5	2.1
2	6.2	1.6	2.6	4.2	3.3	1.8
3	4.8	2.9	3.1	3.1	4.1	1.3
4	5.2	2.6	2.1	2.9	3.4	2.4
5	6.1	2.8	2.4	3.0	4.6	2.5
Mean ± SE	5.40 ± 0.32[c]	2.38±0.25[a]	2.50±0.17[ab]	3.36 ± 0.24	4.18 ± 0.41	2.02 ± 0.22[a]
SD	0.71	0.56	0.38	0.54	0.91	0.49
ANOVA single factor		F-ratio = 21.51			*p*-value= <0.0001***	

*Significant at $p < 0.05$; n.s. non-significant difference according to one-way analysis of variance (ANOVA). Means followed by different letters are significantly different according Duncan's multiple comparisons.

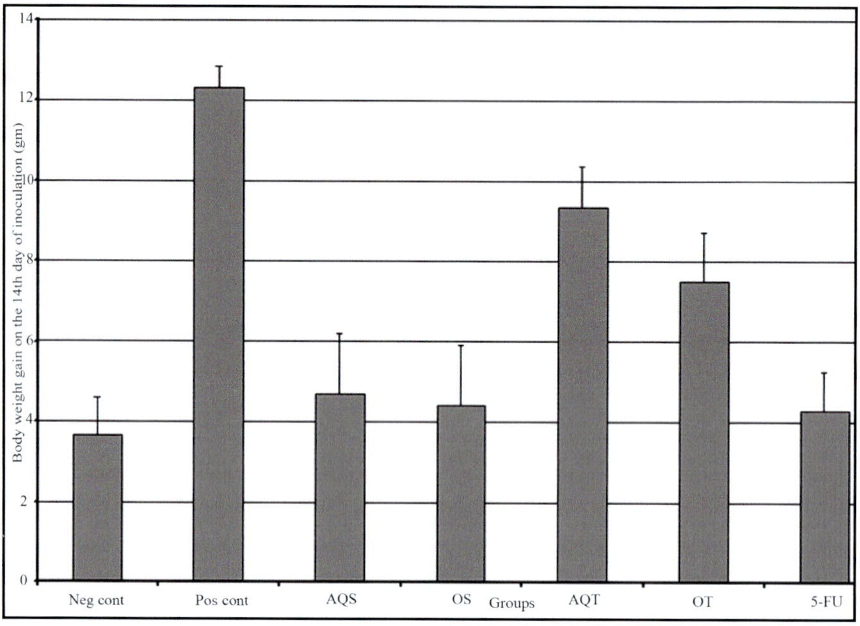

FIGURE 6.5 Body weight at 14 days after inoculation.

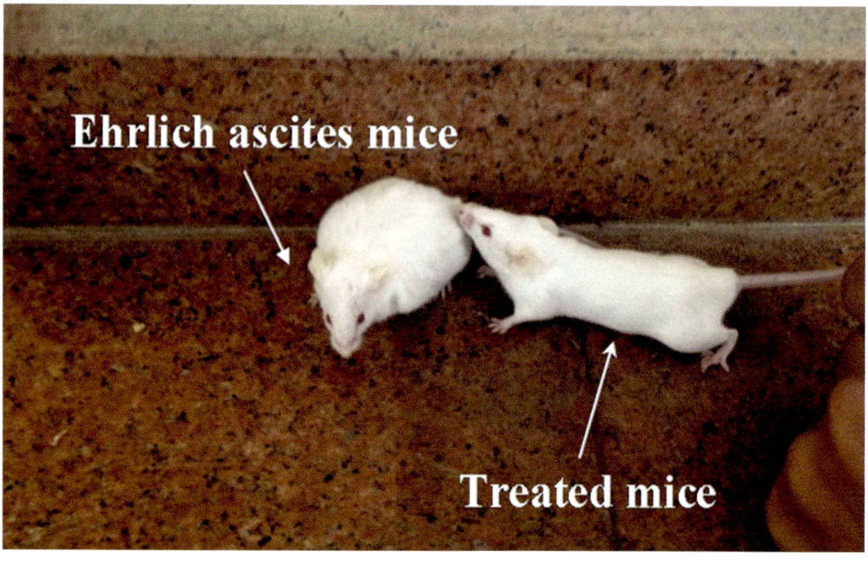

FIGURE 6.6 Mice treated with *Trichothecium roseum* (SCUF23A-MF399479) extracts and untreated mice.

TABLE 6.4 Effect of Fungal Extracts on Median Survival Time and Percent Increase in Life Span.

Parameter	Positive control	OS	AQS	OT	AQT	5-FU	ANOVA	
							F-ratio	p-value
MST	$15.40^{a} \pm 0.51$	$14.80^{a} \pm 0.83$	$25.40^{b} \pm 1.96$	$15.80^{a} \pm 0.58$	$22.40^{b} \pm 1.53$	$42.33^{c} \pm 1.78$	57.21	$<0.001^{***}$
ILS %	0.00	−3.23	0.00	70.97	48.39	174.19		

*Significant at $p<0.05$; n.s. non-significant difference according to one-way analysis of variance (ANOVA). Means followed by different letters are significantly different according Duncan's multiple comparisons.

6.3.6 BIOCHEMICAL FINDINGS

The four fungal extracts exhibited anticancer activity greater than 30% (Table 6.5). Maximum inhibition (63.89%) was recorded for the aqueous extract of *T. roseum*, followed by its EtOAc extract (53.48%).

6.4 DISCUSSION

Endophytic fungi are scientifically important due to their production of novel anticancer compounds or their analogues in greater yields than those produced by higher plants as well as their greater ability to be manipulated physiochemically or genetically (Wang et al., 2015). Plants with an ethnobotanical history are unique biotopes that served as the chief sources of medicinal compounds for thousands of years, and their use can be attributed to their endophyte populations more than to plant biochemistry itself (Gupta et al., 2015). Fifty-eight fungal endophytic species belonging to 32 genera were isolated from 2 medicinal plant species that are used in folk medicine for cancer treatment and inhabit a protectorate in Egypt. *T. roseum* and *S. chartarum* were the most frequently isolated endophytic fungal species, and they were screened for anticancer metabolite production.

Both taxa exhibited considerable cytotoxic activity against *Artemia salina*, and their effects on MCF-7 breast cancer cell proliferation and apoptosis were investigated in mice. Elevated tumor volumes and cell counts in tumor-bearing mice decreased significantly after treatment with extracts of *T. roseum* and *S. chartarum*, which further provides scientific evidence of their antitumor potential. The most promising results were obtained with metabolites from *T. roseum*, and a significant decrease in the cell viability of MCF-7 was recorded for 550–714 mg/kg of its EtOAc extract and for

TABLE 6.5 Biochemical Parameters Estimated after Application of Fungal Metabolites ± SE.

Groups	Parameters								
	ALT	**Albumin**	**Creatinine**	**CRP**	**AFP**	**CEA**	**CA19.9**	**CA12.5**	**CA15.3**
Negative control	22.20^{bc} ±1.16	3.06^{b} ±0.05	0.21^{a} ±0.01	1.06^{a} ±0.05	4.08^{a} ±0.17	2.10^{a} ±0.13	19.24^{a} ±0.65	21.18^{ab} ±0.73	11.10^{a} ±0.60
Positive control	66.40^{e} ±3.08	2.28^{a} ±0.12	0.31^{a} ±0.02	19.76^{d} ±0.67	11.30^{e} ±0.69	12.22^{e} ±0.85	45.34^{d} ±3.05	39.40^{d} ±2.34	77.80^{e} ±4.36
OT	31.40^{cd} ±3.36	3.14^{b} ±0.09	0.29^{a} ±0.02	3.24^{b} ±0.31	4.14^{a} ±0.29	3.94^{b} ±0.16	26.42^{b} ±1.43	17.22^{a} ±0.69	36.20^{c} ±2.30
AQT	15.20^{a} ±0.86	3.16^{b} ±0.07	0.21^{a} ±0.01	1.12^{a} ±0.10	6.20^{bc} ±0.44	6.16^{cd} ±0.27	22.98^{ab} ±1.73	19.10^{a} ±1.24	28.10^{bc} ±2.64
OS	84.80^{f} ±4.33	3.18^{b} ±0.09	0.42^{b} ±0.03	3.40^{b} ±0.26	9.28^{d} ±0.58	7.16^{d} ±0.44	18.94^{a} ±0.70	34.62^{d} ±2.39	49.94^{d} ±4.63
AQS	24.20^{bc} ±1.36	3.04^{b} ±0.10	0.30^{a} ±0.02	5.16^{c} ±0.36	5.28^{ab} ±0.35	5.54^{c} ±0.37	38.92^{c} ±3.63	28.64^{c} ±3.23	52.40^{d} ±2.66
5-FU	38.8^{d} ±2.82	2.56^{a} ±0.16	0.84^{c} ±0.07	6.02^{c} ±0.46	6.72^{c} ±0.27	3.10^{ab} ±0.17	22.46^{ab} ±1.43	25.18^{bc} ±1.29	21.32^{b} ±1.54
One-Way ANOVA									
F-ratio	22.36*	8.81*	15.72*	211.18*	42.45*	47.70*	23.80*	17.97*	45.82*
***p*-value**	<0.001	<0.001	<0.001	<0.001	<0.001	<0.001	<0.001	<0.001	<0.001

*Significant at $p<0.05$; n.s. non-significant difference according to one-way analysis of variance (ANOVA). Means followed by different letters are significantly different according Duncan's multiple comparisons.

169–178 mg/kg of its aqueous extract. Furthermore, the LC_{50} value was 2.94 μg/mL for the EtOAc extract and 1.63 μg/mL for the aqueous extract, as only 50% of the brine shrimp were viable after 24 h of treatment at these concentrations.

Endobionts develop significant and novel characteristics during the co-existence process with their hosts. *T. roseum* and *S. chartarum* produce numerous secondary metabolites, including trichothecenes, triprenylated phenolics, and diterpenoids (Hope, 2013). Metabolites of an endophytic native isolate of *T. roseum* (MF399479) recovered from *Achillea fragrantissima* exhibited a direct inhibitory effect on EAC in a mouse model. This type of carcinoma is undifferentiated, originally hyperdiploid and does not exhibit tumor-specific transplantation antigen (TSTA); moreover, it demonstrates high transplantable capability with no regression and rapid proliferation, exhibiting 100% malignancy, which leads to shorter life span. The current research illustrates the outstanding usage of fungal technology to generate possibly precious products (anticancer drugs), delivering powerful medical evidence for the folkloric usages of certain plant in the therapy for tumors that may be remarkable from a conservationist perspective considering that isolated native endophytic taxa are preserved in the Fungarium of the Arab Society for Fungal Conservation (ASFC). We propose additional pharmaceutical studies to purify the active metabolites of the T. roseum (MF399479) extracts evaluated in the present study. Bioprospecting of the endophytic fungal communities associated with the wealth of ethnobotanical plants inhabiting the Saint Katherine Protectorate is of excellent value and will contribute to the international scientific community by maximizing the benefits of drug discovery from natural products.

ACKNOWLEDGMENTS

We thank Prof. Abdelghafar M. Abu-Elsaoud (Botany Department, Faculty of Science, Suez Canal University) for his kind help in statistical analyses and Dr. Diaa El-Din A. Mohamed (Marine Biology Department, Faculty of Science, Suez Canal University) for his unlimited support throughout this work.

Conflict of interest

All the authors of this study declare the absence of any potential conflicts of interest.

KEYWORDS

- **anticancer metabolites**
- **bioprospecting**
- ***Stachybotrys chartarum***
- ***Trichothecium roseum***
- **Saint Katherine Protectorate**

REFERENCES

1. Abdel-Azeem, A. M.; Salem, F. M. Biodiversity of Laccase Producing Fungi in Egypt. *Mycosphere* **2012,** *3*, 900–920.
2. Abdel-Azeem, A. M.; Salem, F. M. Fungi Fimicola Aegyptiaci: I. Recent Investigations and Conservation in Arid South Sinai. *Mycosphere* **2015,** *6*, 174–194. https://doi.org/10.5943/mycosphere/6/2/8
3. Abdel-Azeem, A. M.; Zaki, S. M.; Khalil, W. F.; Makhlouf, N. A.; Farghaly, L. M. Anti-Rheumatoid Activity of Secondary Metabolites Produced by Endophytic Chaetomium globosum. *Front. Microbiol.* **2016,** *7*. https://doi.org/10.3389/fmicb.2016.01477
4. Bruce, R. An Up-and-Down Procedure for Acute Toxicity Testing. *Fundam. Appl. Toxicol.* **1985,** *5*, 151–157. https://doi.org/10.1016/0272-0590(85)90059-4
5. Domsch, K. H.; Gams, W.; Anderson, T. H. *Compendium of Soil Fungi*, 2nd ed.; Taxonomically Rev. ed.; IHW-Verl: Eching, 2007.
6. Doveri, F. An Additional Update on the Genus Chaetomium with Descriptions of Two Coprophilous Species, New to Italy. *Mycosphere* **2013,** *4*, 820–846. https://doi.org/10.5943/mycosphere/4/4/17
7. Duncan, D. B. Multiple Range and Multiple F Tests. *Biometrics* **1955,** *11*, 1. https://doi.org/10.2307/3001478
8. Ellis, M. B. Dematiaceous Hyphomycetes. Commonwealth Mycological Institute: Kew, 1971.
9. Ellis, M. B. More Dematiaceous Hyphomycetes. Commonwealth Mycological Institute: Kew, Eng., 1976.
10. Fayed, A.; Shaltout, K. Flora of Saint Catherine Protectorate, Final Report and Floristic Survey of the Mountainous Southern Sinai: Saint Katherine Protectorate (Conservation and Sustainable use of Medicinal Plants in Arid and Semi-Arid Eco-Systems Project, Egypt (GEF, UNDP) project no: 12347/12348). UNDP, 2004.
11. Guarro Safont, J., Ed. *Atlas of Soil Ascomycetes, CBS Biodiversity Series*. CBS-KNAW Fungal Biodiversity Centre: Utrecht, The Netherlands, 2012.
12. Gupta, V. K.; Mach, R. L.; Sreenivasaprasad, S., Eds. Fungal Biomolecules: Sources, Applications and Recent Developments. John Wiley & Sons, Ltd: Chichester, UK, 2015. https://doi.org/10.1002/9781118958308.

13. Hope, J. A Review of the Mechanism of Injury and Treatment Approaches for Illness Resulting from Exposure to Water-Damaged Buildings, Mold, and Mycotoxins. *Sci. World J.* **2013,** *2013*, 1–20. https://doi.org/10.1155/2013/767482.
14. Kharwar, R. N.; Mishra, A.; Gond, S. K.; Stierle, A.; Stierle, D. Anticancer Compounds Derived from Fungal Endophytes: Their Importance and Future Challenges. *Nat. Prod. Rep.* **2011,** *28*, 1208. https://doi.org/10.1039/c1np00008j
15. Kirk, P. M. Index Fungorum—Search Page [WWW Document]. Index Fungorum, 2017. http://www.indexfungorum.org/Names/Names.asp (accessed July 21, 2017).
16. Kirk, P. M.; Ansell, A. E. Authors of Fungal Names: a List of Authors of Scientific Names of Fungi, with Recommended Standard Forms of their Names, Including Abbreviations ; Index of Fungi Supplement. Internat. Mycological Inst.: Wallingford, 1992.
17. Kirk, P. M.; Cannon, P. F.; Minter, D. W.; Stalpers, J. A. *Ainsworth & Bisby's Dictionary of the Fungi*, 10th ed.; Wallingford, CT: CAB International, 2008.
18. Klich, M. A., Ed. *Identification of Common Aspergillus Species*. Centraalbureau Voor Schimmelcultures: Utrecht, 2002.
19. Leslie, J. F.; Summerell, B. A. *The Fusarium Laboratory Manual*, 1st ed.; Blackwell Pub: Ames, Iowa, 2006.
20. Nyland, T. G.; Mattoon, J. S., Eds. *Small Animal Diagnostic Ultrasound*, 2nd ed.; W. B. Saunders Co: Philadelphia, Pa, 2002.
21. Pitt, J. I. The Genus Penicillium and its Teleomorphic States Eupenicillium and Talaromyces. Academic Press: London; New York, 1979.
22. Rudloff, E. Abdominocentesis and Diagnostic Peritoneal Lavage. In *Textbook of Veterinary Internal Medicine: Diseases of the Dog and Cat*; Elsevier Saunders, 2005; pp 269–270.
23. Salem, F., Abdel-Azeem, A. Screening of Anticancer Metabolites Produced by Endophytic Fungi Of Some Medicinal Plants in Saint Katherine Protectorate. LAP Lambert Academic Publishing: Saarbrücken, 2014.
24. Simmons, E. G. Alternaria: an Identification Manual; Fully Illustrated and with Catalogue Raisonné 1796–2007, CBS Biodiversity series. Centraalbureau voor Schimmelcultures: Utrecht, 2007.
25. Sur, P.; Ganguly, D. Tea Plant Root Extract (TRE) as an Antineoplastic Agent. *Planta Med.* **1994,** *60*, 106–109. https://doi.org/10.1055/s-2006-959427
26. Wakawa, H. Y., Fasihuddin, B. A. Brine Shrimp Lethality Bioassay of Abrus Precatorius (Linn) Leaves and Root Extract. *Int. J. Pharm. Pharm. Sci.* **2016,** *9*, 179. https://doi.org/10.22159/ijpps.2017v9i1.15057
27. Wang, X.; Wang, C.; Sun, Y. T.; Sun, C. Z.; Zhang, Y.; Wang, X. H.; Zhao, K. Taxol Produced from Endophytic Fungi Induces Apoptosis in Human Breast, Cervical and Ovarian Cancer Cells. *Asian Pac. J. Cancer Prev.* **2015,** *16*, 125–131. https://doi.org/10.7314/APJCP.2015.16.1.125
28. White, T. J.; Bruns, T.; Lee, S.; Taylor, J. W. Amplification and Direct Sequencing of Fungal Ribosomal RNA Genes for Phylogenetics. *PCR Protoc. Guide Methods Appl.* **1990,** *18*, 315–322.
29. Zaki, S. M.; Elkholy, I. M.; Elkady, N. A.; Abdel-Ghany, K. Mucormycosis in Cairo, Egypt: Review of 10 Reported Cases. *Med. Mycol.* **2013,** 1–8. https://doi.org/10.3109/13693786.2013.809629

CHAPTER 7

Foodborne Pathogens and Nanoparticles as a Tool for Quality Assurance and Intervention of Foodborne Pathogens

PORTEEN KANNAN[1*], S.WILFRED RUBAN[2], and M. NITHYA QUINTOIL[3]

[1]*Department of Veterinary Public Health and Epidemiology, Madras Veterinary College, Chennai 600007, Tamil Nadu*

[2]*Department of Livestock Products Technology, Veterinary College, Bengaluru 560024, Karnataka*

[3]*Department of Veterinary Public Health, Rajiv Gandhi Institute of Veterinary Education and Research, Puducherry*

[*]*Corresponding author. E-mail: rajavet2002@gmail.com*

ABSTRACT

Food safety has emerged as an important global issue due to international trade and public health implications. In part, this reflects consumer concerns about the safety of the food they consume, particularly in developed countries. In response to the increasing number of foodborne illnesses, governments all over the world are intensifying their efforts to improve food safety. In the recent years, there have been heightened concerns about the food safety not only among scientists with an interest in food microbiology or toxicology, but also among economists and other social scientists that focus on the wider socioeconomic issues associated with a country's food supply. An added dimension is the impact of food safety regulations on global trade in agricultural and food products.

7.1 FOODBORNE DISEASES: GLOBAL SCENARIO

Foodborne diseases result from the ingestion of contaminated foods and food products and include a broad group of illnesses caused by either bacteria and bacterial toxins, viruses, zoonotic parasites, fungi and fungal toxins, pesticide residues, heavy metals, drug residues, food adulterants, or food additives, etc., which contaminate food at different points in the food production and preparation process. Globally, increase in the incidence of foodborne illnesses continue to be reported, often associated with outbreaks and food contamination that raise international concern. These reports, however, are largely adhoc and data from developing countries where populations are particularly exposed to contaminated environments, are scarce. In the absence of a systematic and comprehensive global assessment, the true burden of foodborne diseases remains unclear.

The burden of foodborne diseases continues to be a major constraint in sustained agricultural development and food security. The term foodborne diseases (also frequently referred to as food poisoning) covers illnesses acquired through consumption of contaminated food and includes foodborne infections and intoxications.[1] Worldwide, foodborne diseases are major health burden leading to high morbidity and mortality. The global burden of infectious diarrhoea involves 3–5 billion cases and nearly 1.8 million deaths annually, mainly in young children, caused by contaminated food and water. World Health Organization (WHO) reports that 20% of deaths among children under five are caused by diarrheal diseases. According to the CDC, an estimated 76 million cases of foodborne disease are reported annually in the USA with approximately 5000 deaths. In USA, diseases caused by the major pathogens alone are estimated to cost up to US $35 billion annually (1997) while in UK nearly £1.5 billion (2010), in medical costs and lost productivity. There is approximately $9.3 billion to $12.9 billion cost paid on human disease treatment caused by only six major bacterial pathogens including *Salmonella* spp., *Campylobacter jejuni*, *Escherichia coli* including O157:H7, *Listeria monocytogenes*, *Staphylococcus aureus*, and *Clostridium perfringens* (USDA, 2011) and all these are foodborne.

To date, 250 different foodborne diseases have been described and bacteria are the causative agents of two thirds of them. Among the predominant bacteria reported include *E. coli, Salmonella* spp., *Shigella* spp., *Bacillus cereus, Clostridium* spp., *Staphylococcus aureus, Vibrio* spp., *Listeria monocytogenes, Campylobacter* spp., *Yersenia* spp. *Brucella* spp., *Mycobacterium* spp. etc. Among viruses, rotavirus, norovirus, and hepatitis virus, etc. have reported to be predominant.

Food contamination creates an enormous social and economic burden on communities and their health systems. Developing countries bear the brunt of the problem due to the existence of a wide range of foodborne pathogens and more importantly, non-compliance to hygienic practices. In India, an estimated 500,000 children below 5 years age die each year due to diarrhea (UNICEF estimate 1000/day). These figures jeopardize international and India's own development efforts including the achievement of the Millennium Development Goals (MDGs). In the context of widespread poverty and malnutrition in the developing countries, programs directed toward the promotion of adequate access to food that satisfy calorie needs and minimize hunger and malnutrition have precedence over programs designed to ensure wholesomeness and quality of food. In short, the emphasis so far has been more on food adequacy rather than on food quality. Therefore, the World Health Assembly adopted a resolution (WHA 53.15) in which the WHO was asked "*to give greater emphasis on food safety…with the goal of developing suitable, integrated food safety systems for the reduction in health risk along the entire food chain, from primary producer to the consumers.*"

Foodborne infections can not only spread faster, they appear to be emerging more rapidly than ever before and are able to circumvent conventional control measures. The growing industrialization of food production catalyzes the appearance and spread of new or antibiotic-resistant pathogens. Growing international trade, migration and travel increase the spread of dangerous pathogens and contaminants in food. In today's interconnected and interdependent world, local foodborne disease outbreaks may become a potential threat to the entire globe. In 1991, cholera which was thought to have originated from contaminated seafood harvested off the coast of Peru, rapidly spread across Latin America resulting in approximately 400,000 reported cases and more than 4000 deaths in several countries. Through the globalization of food marketing and distribution, both accidentally and deliberately contaminated food products can affect the health of people in numerous countries at the same time. The identification of one single contaminated food ingredient can lead to the recall of tonnes of food products, to considerable economic losses in production and from trade embargoes, as well as damage to the tourist industry. In early 2008, an outbreak of avian influenza in Bangalore, India, led to an import ban of Indian poultry products in the Middle East, resulting in losses totaling hundreds of thousands of US Dollars to the Indian economy.

At present, the reporting and surveillance of foodborne diseases in developing countries is grossly neglected. The exact extent of the problem of foodborne diseases in developing countries including India has not been fully understood.[2] Although most of the studies showed the incidence of foodborne diseases, they either lacked data on the organisms involved or the food implicated. This may be due to the time lapse in reporting disease outbreaks as the foodborne diseases appear to be non-epidemic in nature and are most often not recognized either by the public or by the health authorities. Studies revealed that foodborne diseases are a serious health hazard and important cause of morbidity and mortality in developing countries. Most cases go unreported and scientific investigations are rarely feasible. The actual scenario of foodborne diseases can emerge only with proper emphasis on surveillance and with the establishment of a national foodborne disease surveillance system.

7.2 VARIOUS CAUSES OF FOODBORNE DISEASES

Below mentioned is the list of probable causes of foodborne disease:

- Viable pathogenic microorganisms (bacteria, viruses, fungi) or their preformed toxins.
- Pathogenic algae, parasites, protozoa, and their preformed toxins.
- Toxins naturally present or formed in some foods, for example, toxic mushrooms, some sea foods, red kidney bean poisoning, biological amines in cheese, and fermented meats, etc.
- Toxic chemicals in contaminated food and water, such as heavy metal and some pesticides.
- Allergy to or inability to utilize some normal components of food.
- Indigestion from over eating or other reasons.

7.3 TYPES OF MICROBIAL FOOD BORNE DISEASES

On the basis of mode of illnesses, food borne disease can be arbitrarily divided into three groups.

Infection

Occurs as a result of the consumption of food and water contaminated with enteropathogenic bacteria. It is necessary for the cells of enteropathogenic bacteria to remain alive in the food or water during consumption. The viable cells even if present in small numbers have the potential too establish and multiply in the digestive tract to cause the illness, for example., Salmonellosis.

Intoxication

Occurs as a consequence of ingestion of a preformed bacterial or a mold toxin due to its growth in a food. A toxin has to be present in the contaminated food. Once the microorganism have grown and produced toxin in a food, there is no need of viable cells during the consumption of the food for illness to occur, for example, Staphylococcal food poisoning.

Toxicoinfection

Illness occurs from the ingestion of a large number of viable cells of some pathogenic bacteria through contaminated food and water. Generally the bacterial cells either sporulate or die and release toxin(s) to produce the symptoms, for example, *Bacillus cereus* gastroenteritis.

7.4 VARIOUS SOURCES OF MICROBES IN FOOD

a) Foodborne bacteria of animal origin including *Listeria monocytogenes*, *Salmonella* spp, *Campylobacter spp; Yersinia enterocolitica* could be derived form poultry and other meat, eggs and dairy products. *E. coli* 0157:H7 and *Brucella spp*, from raw or ground beef and unpasteurized milk. Foodborne bacteria of soil origin include *Listeria monocytogenes, Clostridium perfringenes* from the soil could contaminate cooked meat and poultry. *Clostridium botulinum* could be found in improperly canned low acid foods (vegetables, meat, fish, poultry), smoked fish, and cooked refrigerated low acid foods (toxin is broken down by heat). Foodborne bacteria of food handlers include *Staphylococcus aureus* and *Salmonella* spp., may be found in proteins containing foods such as meat, poultry, fish, milk, and salad made with meat. Foodborne bacteria from water include *Listeria monocytogenes, Salmonella* spp., *Vibrio cholerae, Clostridium* spp. and *Aeromonas hydrophylia.* Foodborne bacteria from fruits and vegetables include *Listeria monocytogenes, Salmonella* spp.*, and Staphylococcus aureus.*

b) Foodborne viruses originate from human intestine and are shed in the faeces. These viral diseases result from fecal contamination of food directly through mishandling of food by infected persons or indirectly through sewage water contamination, for example, Hepatitis A and E, Roraviruses and Norwalk virus (Noroviruses).
c) Foodborne parasites include intestinal worms and protozoan parasites. Intestinal worms include *Trichinella* spp. round worms in pork and bear meat, *Cysticercus* spp. beef tape worm, *Anisaka* fish food worm. Protozoan parasites include *Toxoplasma gondii, Giardia (lamblia) intestinalis* and *Cryptosporidium* spp. in water or from animal feces or food or person–person.
d) Foodborne moulds include aflatoxin from peanuts, corn, cotton seed, and groundnuts.

7.5 FOODBORNE PATHOGENS OF PUBLIC HEALTH CONCERN

Many foodborne pathogens have been recognized in the past. Cases of these traditional pathogens are increasing. This increase has been attributed to better reporting systems, changes in agricultural practices associated with recycling of waste products of human and animal origin. Microbial causes of foodborne diseases which are important in the current scenario include *Listeria monocytogenes, Campylobacter jejuni, Salmonella* Enteritidis, *Salmonella* Typhimurium, *Clostridium perfringens, Clostridium botulinus Clostridium dificile, Yersinia enterocolitica, E. coli* 0157:H7 (and related *E. coli*; 0111: NM, 0104: H21 etc.), *Cryptosporidium parvum, Cyclospora cayetanensis*, Norwalk-like virus (Noroviruses), Hepatitis A and E virus, Rotavirus, etc.

7.5.1 EMERGING FOODBORNE DISEASES

7.5.1.1 BACTERIAL CAUSES

Enterohaemorrhagic E. coli (EHEC)

This pathogen produces toxins known as verotoxins. Cattle appear to be the main reservoir. Transmission to humans is principally through the consumption of contaminated foods such as raw or undercooked meat products and raw milk. Fresh-pressed apple juice or cider, yoghurt, cheese, salad vegetables, and cooked maize have also been implicated. Fecal contamination of water and foods, as well as cross-contamination during food preparation, can lead to infection, as can person-to-person contact. It is a major cause of bloody and non-bloody diarrhoea and often leads to long-term complications such as haemolytic uremic syndrome. One of the most important serotype is *Escherichia coli* O157:H7; however, there are other *E. coli* serotypes which have been reported to produce verotoxins and causes similar type of diseases. There epidemiology differs in various countries.

Enteroaggregative E. coli (EAEC)

Enteroaggregative *E. coli* (EAEC) has increasingly been recognized as an agent of a watery mucoid diarrhoea especially in children in developing as well as developed countries. It is particularly associated with persistent diarrhoea (lasting for more than 14 days), a major cause of illness and death. It is thought that EAEC adheres to the intestinal mucosa and elaborates enterotoxins and cytotoxins, which result in secretory diarrhoea and mucosal damage. Recent studies support the association of EAEC with malnutrition and growth retardation in the absence of diarrhoea.

Listeria monocytogenes

This ubiquitous microorganism has been isolated from various environments, including decaying vegetation, soil, animal feed, sewage, and water. It is resistant to diverse environmental conditions and can grow at temperatures as low as 3°C. It is found in a wide variety of raw and processed foods-such as milk and cheeses, meat (including poultry) and meat products, and seafood and fish products where it can survive and multiply rapidly during storage. *L. monocytogenes* is responsible for opportunistic infections, preferentially affecting individuals whose immune system is perturbed, including pregnant women, newborn babies, and the elderly. It primarily causes meningitis, encephalitis, or septicaemia and when pregnant women are infected, it can lead to abortion, stillbirth, or premature birth.

Salmonella spp.

This microorganism has been isolated from cattle, poultry, sheep, pigs, and horses. Antimicrobial therapy is used extensively to combat *S.* Typhimurium infection in animals and the evolution of a strain resistant to the commonly used antibiotics has made infections with *S.* Typhimurium in food animals difficult to control. The primary route by which humans acquire infection is through the consumption of a large range of contaminated foods of animal origin. *Salmonella* Enteritidis is the dominant cause of human salmonellosis in many parts of the world. Poultry, eggs and egg products, in particular, are contaminated, but the microorganism has also been found in other foodstuffs such as ice cream. Cross-contamination, undercooking, and inadequate cooling procedures promote the spread and growth of *Salmonella* during processing and handling. One important characteristic of *S.* Enteritidis is its ability to contaminate the intact egg shells. Manifestation of illness includes invasive disease and reactive arthritis.

Shigella spp.

Infection with *Shigella* spp. is a major cause of foodborne diseases, which have increased considerably during the past decades, but only a small fraction of cases are reported. *S. dysenteriae* and *S. flexneri* are the predominant species in the tropics. *S. sonnei* occurs more frequently in industrialized than in developing countries and causes milder illness than *S. dysenteriae* and *S. flexneri*. However, occasional foodborne outbreaks by antimicrobial drug-resistant *S. sonnei* have been reported from the United States, Japan, and European countries, mostly among children. During recent years, in Thailand, Vietnam, and Sri Lanka, the predominant species has shifted from *S. flexneri* to *S. sonnei*, a phenomenon possibly linked with country's level of development. As a result, *S. sonnei* outbreaks are also being reported from developing countries. In India, the scenario differed somewhat. Devastating outbreaks of dysentery by multidrug-resistant *S. dysenteriae* type 1, with high case-fatality rates, affected major parts of the country during 1984–1985 and 2002–2003. Two foodborne outbreaks of *S. sonnei* in India, 1 each from Kerala (southern part) in February 2009 and Maharashtra (western part) in February 2010, which support extension of *S. sonnei* into India.

Campylobacter jejuni

Most sporadic infections with this pathogen are associated with improper preparation or consumption of mishandled poultry products. Most *C. jejuni* outbreaks are associated with the consumption of raw milk or un-chlorinated water. Campylobacteriosis may lead to Guillain–Barré syndrome, a cause of flaccid paralysis. The reservoirs of this organism include poultry, cattle, swine, sheep, rodents, and birds.

Yersinia enterocolitica

Yersiniosis is an infectious disease caused by a bacterium of the genus *Yersinia*. In the United States, most human illness is caused by one species, *Y. enterocolitica*. Infection with *Y. enterocolitica*can cause a variety of symptoms depending on the age of the person infected. Infection with *Y. enterocolitica* occurs most often in young children. Common symptoms in children are fever, abdominal pain, and diarrhea, which is often bloody. Symptoms typically develop 4–7 days after exposure and may last 1–3 weeks or longer. In older children and adults, right-sided abdominal pain and fever may be the predominant symptoms, and may be confused with appendicitis. In a small proportion of cases, complications such as skin rash, joint pains, or spread of bacteria to the bloodstream can occur.

Vibrio spp. (V. parahemolyticus and V. vulnificus)

The consumption of raw molluscan shellfish that are contaminated with this microorganism, which is a normal inhabitant of some marine environments, often leads to primary septicemia and death. Individuals most susceptible to infection with this agent include those with chronic liver disease or chronic alcoholism, or those who are immuno-suppressed in some way.[5]

Streptococcus parasanguinis

Pure isolates of this bacterium were recovered from two sheep in Spain during a recent bacteriological survey for determining the prevalence of subclinical mastitis. As this bacterium has been associated with the development of experimental endocarditis, its presence at relatively high concentrations in apparently healthy sheep's milk may pose a health risk in persons with predisposing heart lesions.

Arcobacter butzleri: A Potential New Foodborne Pathogen

The family *Campylobacteriaceae* includes the genera *Arcobacter* and *Campylobacter*, characterized as fastidious gram-negative, non-spore-forming, motile, spiral-shaped organisms. To date, four species have been differentiated within the genus *Arcobacter: A. butzleri, A. cryaerophilus, A. skirrowii*, and *A. nitrofigilis. Arcobacter* can grow microaerobically and aerobically and has the ability to grow at 15°C, which is a distinctive feature that differentiates *Arcobacter* from *Campylobacter*. Recent evidence suggests that arcobacters, especially *Arcobacter butzleri,* may be involved in human enteric diseases and occasionally in extra-intestinal diseases. However, little is known about the mechanisms of pathogenicity and potential virulence factors of *Arcobacter* spp. There is evidence that livestock animals might be a significant reservoir of *Arcobacter* spp. Furthermore, over the last few years the organisms presence in raw meat products as well as in surface and ground water has received increasing attention. Prevalence of *Arcobacter* isolation from retail raw red meat range between 2% and 51%. Investigation revealed that *A. butzleri* and *A. cryaerophilus* are commonly present on slaughter equipment. With the available data it is not possible to assess the significance of *Arcobacter* spp. as a human pathogen or as food- and waterborne pathogen. Further studies are needed to estimate the prevalence of *Arcobacter* spp. in patients with diarrhoea and to study the potential virulence factors of *Arcobacter* spp. Furthermore, additional information on the epidemiology of these microorganisms is also necessary. In India about 10% prevalence of this potential foodborne pathogen has been reported.[3,4]

7.5.1.2 VIRAL CAUSES

Hepatitis E

The hepatitis E virus (HEV) usually enters the body through water or food, especially raw shellfish that has been contaminated by sewage. Anti-HEV activity has been determined in the serum of a number of domestic animals in areas with a high endemicity of human infection, indicating that this may be an emerging zoonosis.

Hepatitis E, whose symptoms and methods of transmission resemble hepatitis A, is caused by a virus commonly found in the Indian Ocean region, Africa, and in underdeveloped countries. Hepatitis E is rarely, if ever, responsible for causes of chronic hepatitis. Adequate sanitation and good personal hygiene reduces the risk of hepatitis A and E. Water should be boiled prior to its use if any question of safety exists. Similarly, in areas where sanitation is questionable, food should be cooked well and fruits peeled. Those planning to travel to areas where hepatitis A or E is widespread are advised to take immune globulin before leaving.

Norovirus (Norwalk virus/Norwalk-like viruses)

The original Norovirus, previously named Norwalk virus, was first identified in 1972 after an outbreak of gastrointestinal illness in Norwalk, Ohio. Later, other viruses with similar features to Norwalk viruses, and therefore called Norwalk-like viruses or "SRSVs" ("small round structured viruses") were identified, causing similar disease, and were classified together as members of the calicivirus family.

Norwalk-like viruses are a common cause of non-bacterial outbreaks of stomach upset. Viral gastroenteritis, often referred to as "winter vomiting disease" or "stomach flu," is a common illness. It should not be confused with influenza, which is commonly referred to as the "flu." The main source of the virus is stool (feces) and vomit from infected persons. The virus is most often spread from person-to-person on unwashed hands. The virus can also be spread by food, water, or ice that has been handled by a sick person. Vomiting may spread the virus through the air. The virus can survive on surfaces such as countertops or sink taps for a long time. Outbreaks of Norovirus have been caused by food especially raw or undercooked shellfish and water contaminated with the virus. These agents cause mild to moderate disease with gastrointestinal symptoms. The main symptoms are sudden onset of nausea and vomiting (common in the young), diarrhea (more common in adults), stomach pain, muscle aches, tiredness, headache and low grade fever. Symptoms generally last between 24 and 48 h. Fluid loss resulting from vomiting and diarrhea can cause serious problems, especially for the elderly and the very young.

7.5.1.3 PROTOZOAL CAUSES

Cyclospora cayetanensis

This coccidian parasite occurs in tropical waters worldwide and causes watery and sometimes explosive diarrhoea in humans. It was initially associated with waterborne transmission but has also been linked to the consumption of raspberries, lettuce, and fresh basil. The incubation period is 1 week after the ingestion of the contaminated food and the agent is shed in the feces for more than 3 weeks.

Toxoplasma gondii

The primary hosts of this protozoan are cats, and human infection takes place when contact is made with their feces. It can also occur through the ingestion of raw or undercooked meat from intermediate hosts, such as rodents, swine, cattle, goats, chicken, and birds. Toxoplasmosis in humans often produces mononucleosis-type symptoms, but transplacental infection can result in fetal death if it occurs early in pregnancy. In immunocompromised individuals infection can cause pneumonitis, myocarditis, meningoencephalitis, hepatitis, chorioretinitis or combinations of these. Cerebral toxoplasmosis is often seen in AIDS patients.

Cryptosporidium parvum

The mode of transmission of this coccidian protozoan is fecal to oral, including waterborne and foodborne means. The reservoirs include humans and domestic animals, including cattle. Oocysts can survive in the environment for long periods; they remain infective and are capable of resisting chemicals used to purify drinking water. They can, however, be removed from water supplies by filtration. Symptoms of cryptosporosis in humans include fever, diarrhoea, abdominal pain, and anorexia. The disease usually subsides within 30 days, but may be prolonged and continue to death in immunodeficient individuals.

7.5.1.4 HELMINTHS

The Genus Anisakis

Anisakiasis is an infection of the human intestinal tract caused by the ingestion of raw or undercooked fish containing larval stages of the nematodes *Anisakis simplex* or *Pseudoterranova decipiens*. Infections caused by the latter roundworm are not a serious threat to human health, but those caused by *A. simplex* are more problematic because this agent penetrates the gastrointestinal tissue and causes disease that is difficult to diagnose. The primary hosts are warm-blooded marine mammals such as seals, walruses, and porpoises. Their larvae pass via krill to fish such as cod, pollack, halibut, rockfish, salmon, and herring.

7.5.1.5 UNCONVENTIONAL AGENTS

Prions

Transmissible spongiform encephalopathies in animals and humans are caused by prions. These conditions include scrapie in sheep, bovine spongiform encephalopathy (BSE or mad cow disease) in cattle, and Creutzfeldt Jacob disease (CJD) in humans. It is commonly accepted that BSE was first caused in the United Kingdom when cattle were fed carcass meal from scrapie-infected sheep. It is also accepted that humans contracted the non-classic form of CJD after consuming cattle meat, in particular nerve tissue.

Mycotoxins

Mycotoxins are the toxic products of certain microscopic fungi which, in some circumstances, develop on or in foodstuffs of plant or animal origin.

They are ubiquitous and widespread at all levels of the food chain. Hundreds of mycotoxins have been identified and are produced by some 200 varieties of fungi. In terms of their implications for human health and the economy, mycotoxins are by far the most important contaminants of the food chain. Of particular importance in current toxicological studies are the combined and possible synergistic effects that some of the mycotoxins may have on human and animal life. Some of the important mycotoxins include fumonisins, zearalenone, trichothecenes, and ochratoxins.

Pesticide Residues

Restrictions are now being placed on some of the older organochlorine pesticides because of their environmental persistence and their potential accumulation in fatty tissues. Although exposure to these pesticides is usually below acceptable daily intake (ADI) levels, breastmilk in both developed and developing countries has occasionally been found to contain relatively high levels of organo-chlorine pesticides. It has also been found that, although there may be a high variability in residue levels of pesticides in individual units of commodities, this is unlikely to cause any direct adverse health effects.

Drug Residues

The intake of veterinary drug residues in food at levels below the ADI is also considered to be safe. In recent years, however, growing concern has been expressed about the development of antimicrobial drug resistance. Some important contributing factors to the development of this resistance are the widespread use of veterinary drugs, the misuse of such drugs and the feeding of low doses to animals in order to promote weight gain and improve feed efficiency. Resistant microorganisms may be passed on to humans via food originating from the animals that harboured them. In addition, the development of resistance may also lead to the application of larger and larger therapeutic doses to food producing animals.

Environmental Contaminants

Chemicals such as dioxins, chlorinated biphenyls, furans, and heavy metals may contaminate the environment as a result of industrial activities. From the environment, these chemicals may enter the food chain via plants or animals and cause a variety of health problems. These are considered as emerging problems in countries that are in the early stages of industrialization.

Genetically Modified Organisms

The production of genetically modified foodstuffs offers tremendous opportunities and benefits for future food production. However, the emergence of this new technology has also given rise to a number of problems, although such problems are often regarded as potential or perceived, rather than real. The concerns that have been expressed relate mainly to changes in the nutritional quality of food, an increase in toxicity or hazards with respect to food intolerances or food allergies, and the development of antimicrobial resistance.

7.6 APPLICATION OF NANOTECHNOLOGY FOR QUALITY CONTROL OF FOOD PRODUCTS

Nanotechnology may be defined as the creation and utilization of materials, devices, and systems through the control of matter on the nanometer scale (1 billionth of a meter). Nanotechnology has been widely applied for the rapid detection of microbes in various samples including food. These technologies extend the limits of current molecular diagnostics and enable point-of-care diagnosis as well as the development of personalized medicine. Although the potential applications of nanotechnology are unlimited, most important current applications are foreseen in the areas of biomarker research, cancer diagnosis, and sensitive and specific detection of infectious microorganisms.

Nanotechnology has indeed become a field of its own within the biotechnology world, in just a short period of time. Having grown well beyond its origins in semiconductor industry, the application of nanotechnology to biological problems has matured enough not only to deserve its own name, nano-biotechnology, but also to have spawned institutes devoted to its further development. A decade ago, nanoparticles were studied because of their size-dependent physical and chemical properties, now they have entered a commercial exploration period. Living organisms are built of cells that are typically 10 μm across. However, the cell parts are much smaller and are in the sub-micron size domain. Even smaller are the proteins with a typical size of just 5 nm, which is comparable with the dimensions of smallest manmade nanoparticles. This simple size comparison gives an idea of using nanoparticles as very small probes that would allow us to spy at the cellular machinery without introducing too much interference (Tatan, 2002).

The properties of materials can be different at the nanoscale for two main reasons. First, nanomaterials have a relatively larger surface area when compared to the same mass of material produced in a larger form. This can make materials more chemically reactive (in some cases materials that are inert in their larger form are reactive when produced in their nanoscale form), and affect their strength or electrical properties. Second, quantum effects can begin to dominate the behavior of matter at the nanoscale—particularly at the lower end—affecting the optical, electrical, and magnetic behavior of materials. Materials can be produced that are nanoscale in one dimension (e.g., very thin surface coatings), in two dimensions (e.g., nanowires and nanotubes) or in all three dimensions (e.g., nanoparticles). Common examples of nanomaterials found in scientific literature are fullerenes, nanotubes, buckyballs, quatum dots, dendrimers, and nanoshells.

Fullerenes are pure carbon molecules composed of at least 60 atoms of carbon. Because a fullerene takes a shape similar to a soccer ball or a geodesic dome, it is sometimes referred as a buckyball after the inventor, Buckminster Fuller. Buckyballs are perfectly smooth, round, inert, nontoxic, and because of their size, they can interact easily with cells, proteins, and viruses. Since they are hollow inside they can be used for delivering medicine more efficiently to the inside of cells, and also in diagnostic imaging by enclosing radioactive agent inside them.

Nanotubes are a sequence of nanoscale C60 atoms arranged in a long thin cylindrical structure. Carbon nanotubes were discovered in 1991 by S. Iijima. Besides having a single cylindrical wall (SWNTs), nanotubes can have multiple walls (MWNTs) cylinders inside the other cylinders. Nanotubes have a very broad range of electronic, thermal, and structural properties that change depending on the different kinds of nanotube (defined by its diameter, length, and twist). They are among the stiffest and strongest fibers known and have remarkable electronic properties and many other unique characteristics.

Quantum dots are a nano-scale crystalline structure made from cadmium selenide that absorbs white light and then re-emits it a couple of nanoseconds later in a specific color The color of the emitted light depends on the size of the dots. A rainbow of colors can be emitted from a single material simply by changing the dot size

Dendrimer is a regularly branched synthetic polymer, three-dimensional molecule which resembles the branches of a tree. The name comes from the Greek word-dendron, meaning "tree." The first few dendrimers were described by Vogtle in 1978.

Nanoshells are concentric sphere nanoparticles consisting of a dielectric (typically gold sulfide or silica) core and a metal (gold) shell. They are considered a very special kind of nanoparticle because they combine infrared optical activity with the uniquely biocompatible properties of gold colloid.

7.7 APPLICATION OF NANOMATERIALS

7.7.1 DIAGNOSTICS

7.7.1.1 DETECTION OF MICROORGANISMS

Detection of Bacteria

The rapid and sensitive detection of pathogenic bacteria is extremely important in medical diagnosis and for implementing measures against bioterrorism. Limitations of most of the conventional diagnostic methods are lack of ultrasensitivity or delay in getting results. A bioconjugated nanoparticle-based bioassay for in situ pathogen quantification can detect a single bacterium within 20 min. The quantum dot nanoparticles provide an extremely high fluorescent signal for bioanalysis and can be easily incorporated in a biorecognition molecule such as an antibody. The antibody-conjugated nanoparticles can readily and specifically identify a variety of bacteria such as *Escherichia coli* O157:H7 through antibody-antigen interaction and recognition. The nanoparticle-based colorimetric assay, in comparison to a previously reported absorbance-based method, increases detection sensitivity by over four orders of magnitude and has been applied to the rapid detection of mecA in methicillin-resistant *Staphylococcus aureus* genomic DNA samples. Detection of a small number of *Salmonella* is achieved due to a change in the surface stress on the silicon nitride cantilever surface in situ upon binding of bacteria. Scanning electron micrographs indicate that adsorption of less than 25 organisms may suffice for detection. A nanotechnology-based technique, Sensing of Phage-Triggered Ion Cascade (SEPTIC), uses a nanowell device with two antenna-like electrodes to detect the electric-field fluctuations that result when a bacteriophage infects a specific bacterium and then identifies the bacterium. This method had a 100% success rate in detecting and identifying strains of *E. coli* quickly and accurately.

Detection of Viruses

Rapid, selective, and sensitive detection of viruses is crucial for implementing an effective response to viral infection, such as through medication or quarantine. Established methods for viral analysis include plaque assays, immunological assays, transmission electron microscopy, and PCR-based testing of viral nucleic acids. These methods have not achieved rapid detection at a single virus level and often require a relatively high level of sample manipulation that is inconvenient for infectious materials. Direct, real-time electrical detection of single virus particles can be achieved with high selectivity by using nanowire field effect transistors. Measurements made with nanowire arrays modified with antibodies for influenza A showed discrete conductance changes characteristic of binding and unbinding in the presence of influenza A but not paramyxovirus or adenovirus. Simultaneous electrical and optical measurements using fluorescently labeled influenza A were used to demonstrate conclusively that the conductance changes correspond to binding/ unbinding of single viruses at the surface of nanowire devices. Larger arrays of reproducible nanowire devices might simultaneously screen for the presence of 100 or more different viruses. The possibility of largescale integration of these nanowire devices suggests potential for simultaneous detection of a large number of distinct viral threats at the single virus level.

7.7.2 THERAPEUTICS

Metals and metal oxide nanoparticles were found to have antimicrobial property. Arsenical compounds were used for syphilis and Zn, Cd and tin salts date back to 1960s. Proteins, small molecules, nucleic acids loaded in nanoparticle are not recognized by immune system and can be targeted to particular type of cell. Drug carriers can be nanoparticles like nanotubes, dendrimers, fullerenes, liposomes, nanopolymers of poly (butyl cyanoacrylate) (PBCA) or magnetic nanoparticles depending on the requirement. Drug targeting can be done by using biomarkers. Blood vessels express molecular markers that distinguish the vasculature of individual organs, tissue, and tumor. Unique blood vessel biomarkers include CGFECVRQCPERC peptide (denoted as GFE) which binds to membrane dipeptidase on the endothelial cells in lung blood vessels, KDEPQRRSARLSAKPAPPKPEPKPKKAPAKK (F3) preferentially binds to blood vessels and tumor cells in various tumors, CGNKRTRGC (LyP-1) recognizes lymphatic vessels and tumor cells in certain tumors. These blood vessel biomarkers can be used for drug targeting using intravenous route.

Nanoparticle-encapsulated drug has been found to cross blood–brain barrier and blood–retina barrier, therefore can be used for various neurological diseases like Alzheimer's and Parkinson's disease, as well as for retinal diseases like diabetic retinopathy, retinitis pigmentosa, and macular degeneration. A nanoparticle-based drug delivery system for the treatment of brain tumors using Doxorubicin is enclosed in a nanopolymer PBCA particle and coated with polysorbate 80. Polysorbate 80 camouflage LDL cholesterol, which attract apolipoproteins to cross the blood–brain barrier. Sensor nanochips implanted under the skin are also being developed to continuously monitor key body parameters including pulse, temperature, and blood glucose. Drug dispensing can also be done as per requirement. Simple closed-loop delivery systems that measure glucose levels and release insulin to treat diabetes are currently under clinical trials.

Tissue engineering is one more therapeutic aspect to make bone and dental implants. Natural bone surface is quite often contains features that are about 100 nm across. If the surface of an artificial bone implant were left smooth, the body would try to reject it. Because of that smooth surface is likely to cause production of a fibrous tissue covering the surface of the implant. This layer reduces the bone-implant contact, which may result in loosening of the implant and further inflammation. It was demonstrated that by creating nano-sized features on the surface of the hip or knee prosthesis, one could reduce the chances of rejection as well as to stimulate the production of osteoblasts. The osteoblasts are the cells responsible for the growth of the bone matrix and are found on the advancing surface of the developing bone.

Photodynamic cancer therapy is based on the destruction of the cancer cells by laser generated atomic oxygen, which is cytotoxic. A greater quantity of a special dye that is used to generate the atomic oxygen is taken in by the cancer cells when compared with a healthy tissue. Hence, only the cancer cells are destroyed then exposed to a laser radiation. Unfortunately, the remaining dye molecules migrate to the skin and the eyes and make the patient very sensitive to the daylight exposure. This effect can last for up to six weeks. To avoid this side effect, the hydrophobic version of the dye molecule was enclosed inside a porous nanoparticle. The dye stayed trapped inside the Ormosil nanoparticle and did not spread to the other parts of the body. At the same time, its oxygen generating ability has not been affected and the pore size of about 1 nm freely allowed for the oxygen to diffuse out.

One of the simplest medical nanomaterials is a surface perforated with holes, or nanopores. These pores are large enough to allow small molecules such as oxygen, glucose, and insulin to pass but are small enough to impede the passage of much larger immune system particles. Supplying encapsulated new cells to the body could also be a valuable way to treat other enzyme- or hormone-deficiency diseases, including encapsulated neurons that could be implanted in the brain and then be electrically stimulated to release neurotransmitters, possibly as part of a future treatment for Alzheimer's or Parkinson's diseases. Micro-capsules can also be used for immuno-isolation, drug delivery and cell-based sensing.

7.7.3 RESEARCH

The ever increasing research in proteomics and genomic generates escalating number of sequence data and requires development of high throughput screening technologies. Single quantum dots of compound semiconductors were successfully used as a replacement of organic dyes in various bio-tagging applications. This idea has been taken one step further by combining differently sized and hence having different fluorescent colors quantum dots, and combining them in polymeric microbeads. A precise control of quantum dot ratios has been achieved. The selection of nanoparticles used in those experiments had 6 different colors as well as 10 intensities. It is enough to encode over 1 million combinations. The uniformity and reproducibility of beads was high letting for the bead identification accuracies of 99.99%. Photoluminescence life is long and there is stability against photo bleaching, therefore long term tracking of biological process can be done.

Functionalized magnetic nanoparticles have found many applications including cell separation and probing, it is possible to produce magnetic nanowires with spatially segregated fluorescent parts. In addition, because of the large aspect ratios, the residual magnetization of these nanowires can be high. Hence, weaker magnetic field can be used to drive them. It has been shown that a self-assembly of magnetic nanowires in suspension can be controlled by weak external magnetic fields. This would potentially allow controlling cell assembly in different shapes and forms. Moreover, an external magnetic field can be combined with a lithographically defined magnetic pattern ("magnetic trapping").

7.8 FUTURE PROSPECTS OF NANOBIOTECHNOLOGY

Advances in nanobiotechnology are providing nanofabricated devices that are small, sensitive, and inexpensive enough to facilitate direct observation, manipulation, and analysis of single biological molecule from single cell. This opens new opportunities and provides powerful tools in the fields such as genomics, proteomics, molecular diagnostics, and high-throughput screening. Nanoparticles are the most versatile material for developing diagnostics. In vitro diagnostics has no safety issues of fate of nanoparticles in the human body. Currently, QD technology is the most widely employed nanotechnology for diagnostic developments. Among the recently emerging technologies, cantilevers are the most promising. This technology complements and extends current DNA and protein microarray methods, because nanomechanical detection requires no labels, optical excitation, or external probes and is rapid, highly specific, sensitive, and portable. This will have applications in genomic analysis, proteomics, and molecular diagnostics. Nanosensors are promising for detection of bioterrorism agents that are not detectable with current molecular diagnostic technologies and some have already been developed. Nanotechnology has potential advantages in applications in POC diagnosis: on patient's bedside, self-diagnostics for use in the home, integration of diagnostics with therapeutics and for the development of personalized medicines. In the near future, nanodiagnostics would drastically reduce the waiting time for the test results. For example, the patients with sexually transmitted diseases could give the urine sample when they first arrive at the outpatient clinic or physician's practice; the results could then be ready by the time they go in to see the doctor. They could then be given the prescription immediately, reducing the length of time worrying for the patient and making the whole process cheaper.

Future trends in diagnostics will continue in miniaturization of biochip technology to nano-range. There is also a trend to build the diagnostic devices from bottom up starting with the smallest building blocks. Whether interest and application of nanomechanical detection will hold in the long range remains to be seen. Another trend is to move away from fluorescent labeling as miniaturization reduces the signal intensity but there have been some improvements making fluorescence viable with nanoparticles.

Freitas in his book has hypothesized many nanodevices, which can perform the function of the natural body cells, like microbivores which is an artificial mechanical white blood cell, respirocyte is the artificial mechanical red blood cell and clottoctes an artificial mechanical blood.

7.9 SAFETY AND TOXICOLOGICAL ISSUES

Unwanted nanoparticles may enter body via lungs, intestines, and skin causing health-related problems. Since nanoparticles have a tendency to aggregate and such aggregated particles behave differently from individual nanoparticle. Carbon black nanoparticles have been found to interfere with cell signaling. Nanoparticles injected deliberately during medical procedures or released from implants can cross blood–brain barrier. Since matters behave differently at nano scale, separate investigations and a database of health risks associated with different nanoparticles should be made. Although much research are necessary before nanotechnology is common place in Veterinary and human medicine as well as Food Science, there are numerous glimpses of the future in applications for drug delivery, disease diagnosis and treatment, breeding and identity preservation (IP). Nanotechnology, although still in the early stages of its development, is beginning to equip scientists, engineers and biologists to work at the cellular and molecular levels for significant benefits in healthcare and animal medicine. It is reasonable to presume over the next couple of decades that unique developments in nanobiotechnology will be revolutionizing animal health and medicine. Although at present various nanodevices proposed by nanobiotechnology seem to be a fantasy, hopefully they will soon emerge as a reality.

KEYWORDS

- **foodborne pathogens**
- **nanoparticles**
- **quality assurance**
- **health applications**
- **food safety**
- **food microbiology**
- **toxicology**

REFERENCES

1. Motarjemi, Y.; Käferstein, F. K. Global Estimation of Foodborne Diseases. *World Health Statistics Quarterly* **1997,** *50* (1–2)**,** 5–11.
2. Käferstein, F. K. Food Safety: A Commonly Underestimated Public Health Issue. Introduction. *World Health Statistics Quarterly* **1997,** *50* (1–2), 3–4.
3. Patyal, A. Rathore, R. S.; Mohan, H. V.; Dhama, K.; Kumar, A. Prevalence of Arcobacter spp. in Humans, Animals and Foods of Animal Origin Including Sea Food from India. *Transbound Emerg. Dis.* **2011,** *58* (5), 402–410.
4. Nandy, S.; Dutta, S.; Ghosh, S.; Ganai, A.; Jyothi, R. Ramani Bai, J. T., et al. Foodborne-Associated *Shigella sonnei*, India, 2009 and 2010 [letter]. *Emerg. Infect. Dis.* 2011.
5. Nelapati, S.; Nelapati, K.; Chinnam, B. K. *Vibrio parahaemolyticus*—An Emerging Foodborne Pathogen–A Review *Vet. World* **2012,** *5* (1), 48–62.

CHAPTER 8

Citrus: Origin, Evolution, Postharvest, Heat, pH, Filtration, and Transgenic

FRANCISCO TORRENS[1*] and GLORIA CASTELLANO[2]

[1]*Institut Universitari de Ciència Molecular, Universitat de València, Edifici d'Instituts de Paterna, P. O. Box 22085, E-46071 València, Spain*

[2]*Departamento de Ciencias Experimentales y Matemáticas, Facultad de Veterinaria y Ciencias Experimentales, Universidad Católica de Valencia San Vicente Mártir, Guillem de Castro-94, E-46001 València, Spain*

[*]*Corresponding author. E-mail: torrens@uv.es*

ABSTRACT

Findings draw an evolutionary framework for citrus, a scenario that challenges current taxonomic and phylogenetic thoughts, and points to a reformulation of the genus *Citrus*. The effects of postharvest time of fruits, heat treatment, pH, and filtration of juice on limonin content in Newhall navel orange *Citrus sinensis* Osbeck cv. Newhall juice was investigated. The revealed results might not only provide useful information for the development of novel debittering methods for the manufacture of juice from navel oranges, but also might shed light on the delayed bitterness mechanism of navel orange juice, suggesting that the formation of delayed bitterness might mainly occur, because acid-promoted rather than enzyme-catalyzed lactonization of limonoate A-ring lactone produces limonin in navel orange juice. In general, when people observe the plants that today serve them as food, they should be conscious that each one of them contains an evolutionary history, based on the principles of variation by mutation and selection, in part because of its permanence as wild species and, in part, because of the action of man, which contributed to select the genetic changes produced sometimes by reasons of searching for appropriate food, and, other times, directed by cultural or religious values.

8.1 INTRODUCTION

Setting the scene: The origin and evolution of *Citrus* spp., the effects of postharvest time, heat treatment, pH, and filtration on the limonin content in Newhall navel orange juice, and cultivations of genetically modified organisms (GMOs) of *Citrus* spp. Humanity history is characterized by a constant search for food. At present, we are capable for feeding 7,700,000,000 persons, although 800,000,000 are hungry or undernourished; by 2050, there will be 9,000,000,000 inhabitants to be fed. The techniques of traditional genetic improvement seem sufficient for reaching food safety, so that knowing the role that GMOs cultivation play in genetic improvement strategies of plants with which people produces food for them and animals seems indispensable. Findings draw an evolutionary framework for citrus, a scenario that challenges current taxonomic and phylogenetic thoughts, and points to a reformulation of the genus *Citrus*. The effects of postharvest time of fruits, heat treatment, pH, and filtration of juice on limonin content in Newhall navel orange *Citrus sinensis* Osbeck cv. Newhall juice were investigated. The results might provide useful information for the development of novel debittering methods for the manufacture of juice from navel oranges. In general, when people observe the plants that today serve them as food, they should be conscious that each one of them contains an evolutionary history, based on the principles of variation by mutation and selection, in part because of its permanence as wild species and, in part, because of the action of man, which contributed to select the genetic changes produced sometimes by reasons of searching for appropriate food, and, other times, directed by cultural or religious values.

Monoterpene β-myrcene in gastric and duodenal ulcers, and its mechanisms of action was evaluated.[1] *Citrus aurantium* and its flavonoids regulated TNBS-induced inflammatory bowel disease (IBD) via anti-inflammation and suppressing isolated jejunum contraction.[2] Polymethoxylated flavones were separated from *Dahongpao* tangerine *C. tangerina* via macroporous adsorptive resins combined with preparative high-performance liquid chromatography (prep-HPLC).[3] The effect of effluent recirculation on biogas production via two-stage anaerobic digestion of citrus waste was informed.[4] Nutraceutical essential oils (EOs) produced by olives and *Citrus* peel of Tuscany varieties were reported as sources of functional ingredients.[5] The Botanic Garden of the Universitat de València includes a collection with ca. 50 species of citrus, which range from the most well known and commonly grown (e.g., oranges, mandarins, lemons, and grapefruit) to others used as rootstock, widely in perfumery, for example, bergamot, or some others with important religious significance, for example, citron.[6]

In earlier publications in *Nereis*, etc., the molecular classification of yams,[7] soya bean, Spanish legumes, commercial soya bean,[8,9] fruits proximate and mineral content,[10] and food spices proximate content[11] by principal component, cluster and meta-analyses. Polyphenols,[12] flavonoids,[13] stilbenoids,[14] triterpenoids, steroids,[15] isoflavonoids,[16] natural sesquiterpene lactones,[17,18] artemisinin and their derivatives[19–22] were analyzed. The classification of *Citrus*, principal components, cluster and meta-analyses,[23] the extraction of natural products found in vegetal species clove and citrus[24] and world of biological activities and safety of *Citrus* spp. Eos[25] was informed. The main aim of the present report is to review the origin and evolution of *Citrus* spp., the effects of postharvest time, heat treatment, pH, and filtration on the limonin content in Newhall navel orange juice, *Citrus* spp. GMO cultivations, Cathie Martin's purple tomatoes and orange fruits, citrus as the paradigm of València agriculture, culture, how environment and agronomic practice condition pigmentation, what happens when people graft a plant on other that is a GMO and whether plants tolerant to weedkillers damage the environment. The aim of this work is to initiate a debate by suggesting a number of questions, which can arise when addressing subjects of the origin and evolution of *Citrus* spp., the effects of postharvest time, heat treatment, pH, and filtration on the limonin content in Newhall navel orange juice and *Citrus* spp. GMO cultivations, and their points of view about how environment and agronomic practice condition pigmentation, what happens when people graft a plant on other that is a GMO and whether plants tolerant to weedkillers damage the environment, etc.

8.2 ORIGIN AND EVOLUTION OF *CITRUS*: FROM HESPERIDES' GARDEN TO HIMALAYAS

The genus *Citrus*, comprising some most widely cultivated fruit crops worldwide, includes an uncertain number of species. Talon group described 10 natural *Citrus* spp., via genomic, phylogenetic, and biogeographic analyses of 60 accessions representing diverse citrus germ plasms, and proposed that citrus diversified during the late Miocene epoch, via a rapid southeast Asian radiation that correlates with a marked weakening of the monsoons.[26] A second radiation enabled by migration via Wallace line gave rise to the Australian limes in the early Pliocene. Identification and analyses of hybrids and admixed genomes provided insights into the genealogy of major commercial cultivars of citrus. Among mandarins and sweet orange, they found an extensive net of relatedness that illuminates the domestication of the groups. Widespread pummelo admixture among the mandarins, and its correlation with fruit size and acidity, suggested a plausible role of pummelo introgression in the selection of palatable mandarins. Their work provided an evolutionary framework for *Citrus*.

8.3 POSTHARVEST/HEAT/pH/FILTRATION EFFECTS ON LIMONIN CONTENT IN ORANGE

Because of its pleasant sensory and healthy beneficial properties, orange juice attracts the largest number of juice consumers worldwide. Delayed bitterness is one of the major problems facing the citrus juice industry, which is characterized by gradual formation of bitter compounds in juice on standing after its extraction from the fruit, which is usually non-bitter when eaten fresh. Limonin, an intensely bitter triterpenoid dilactone (*cf.* Fig. 8.1), was identified as the main compound responsible for the delayed bitterness in navel orange juice.[27]

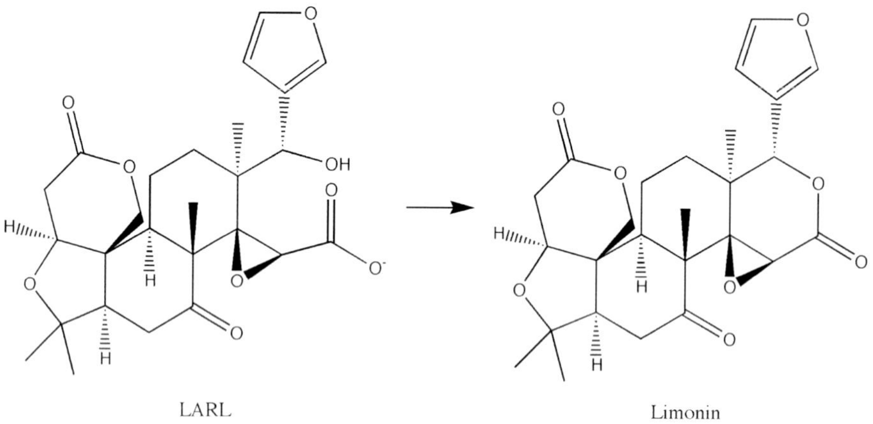

FIGURE 8.1 Conversion of non-bitter precursor LARL to bitter compound limonin.

8.4 CATHIE MARTIN'S PURPLE TOMATOES AND ORANGE FRUITS

Martin's group, famous for developing purple tomatoes, proved that in blood orange (a variety of sweet orange *Citrus* × *sinensis*) fruits, which are grown mainly in Sicily Island, the deep reddish-blue color that fruits present is because of anthocyanins-biosynthesis activation.[28] The oranges are demanded and are called *blood* because its juice presents an intense coloration that remembers blood. In València, it would be easy to modify via genetic engineering citriculture elite varieties to add the character, that is, that they produce anthocyanins with their antioxidant (AO) effect so beneficial for health. In indigenous populations of South Africa, it would be a taboo as blood orange juice presents a blood-like coloration and tribal culture rejects blood consumption. Both cultural factor and benefit for health should be taken into account when it comes to designing products for concrete populations.

8.5 CITRUS, THE PARADIGM OF VALÈNCIA AGRICULTURE

Orange-trees domestication origin is found in China (*C. sinensis*) subtropical zones.[28] However, people frequently forget that plants travel with man, and València persons consider that oranges fruits are from València, forgetting that citrus cultivation in València is recent and produced as substitution of mulberry *Morus alba* trees, which were grown massively because their leaves served to feed silkworms *Bombyx mori*, which trees were planted, in turn, to replace previous elms. When València-silk production and commercialization reached crisis point ending 18th and early 19th centuries, because of problems derived from a textile-mill insufficient modernization and Bourbon-administration politics, the progressive substitution of mulberry trees with citrus was produced. Citrus domestication explains some phenotypic characteristics that the most common commercial varieties present: sweet orange, bitter orange, clementine, lemon, lime, and grapefruit. All are mandarin (*C. reticulata*)–citron (*C. medica*)–*pummelo* (*C. maxima*) trees interspecific hybrids, which result important species of genus *Citrus*; for example, citron tree produces anthocyanins (reddish-blue pigments with great AO capacity) in its young leaves and flowers, but mandarin or *pummelo* trees that produce them were never described. A regulating gene that codes a myeloblastosis (MYB)-type transcription factor (*Ruby*) controls anthocyanins biosynthesis. All natural variations in pigmentation, because of anthocyanins in different *Citrus* spp., can be explained by differences in *Ruby* activity caused by point mutations, nucleotides loses, or transposable elements (Barbara McClintock's *jumping genes*) insertions. Likely to that occurred with the grasses that originated wheat *Triticum* spp., hybridizations occur naturally between *Citrus* spp. and with species of nearby genera, which sexual promiscuity, together with its culture by man via millennia, produced the citrus diversity that nowadays are commercialized, which origin is difficult to establish attending to only morphological characters or geographic data. *Ruby* isolation and characterization allowed unravelling the molecular bases of the deep red color of the flesh of blood orange fruits, which is unique among commercialized sweet oranges that synthesizes a great amount of anthocyanins. On characterizing *Ruby* transcription start in hybrid lemon tree, a mutation was identified in an allele of sweet orange tree derived of hybrid parental *pummelo*, as a consequence of which, sweet orange tree lost synthesizing-anthocyanins capacity. However, in blood sweet orange tree, a retrotransposon [transposon via ribonucleic acid (RNA) intermediates, *jumping gene*] insertion in region promoting mutated transcription, in sweet orange tree, leads to the fact that *Ruby* transcription could be reactivated via other transcription start, which allows anthocyanins production and their cold activation. Man took part, which knew to preserve blood sweet orange tree, especially in Sicilian geographic areas adequate for its cultivation, and that, nowadays, when people know the healthy potential of anthocyanin AO pigments, turns a desirable and economically valuable character in citrus fruits.

8.6 CULTURE, ENVIRONMENT, AND AGRONOMIC PRACTICE CONDITION PIGMENTATION

Anthocyanin-pigmentation generalized loss could start as a consequence of climate conditions proper of where different *Citrus* spp. started.[28] Citron tree, the only parental that continues producing anthocyanins, started in dry highlands, where anthocyanins play a role protecting versus high bright intensity and ultraviolet (UV) radiation. On the contrary, both mandarin and grapefruit trees emerged in subtropical lowlands, with high humidity and lengthy rainfall, in which conditions, radiation stress decays a lot and protecting pigments can be dispensable, so that *Ruby* mutants selection could be favored on avoiding the energy cost that anthocyanins synthesis supposes. However, citron trees were grown for millennia in wide Asian zones as sacred plants, and their fruits presented great religious symbolism. Present in Northeast India zones, citron trees joined Jewish tradition and continue used in religious rituals. In Chinese tradition, early citrus use was strictly religious. Kumquat flowers represented luck and prosperity, and mandarin- and sweet-orange-trees white flowers were symbols of purity and innocence, whose symbolism was maintained during Christendom. Golden fruits–white flowers combination continues a central element of the symbolism of Japanese sacred tree *C. tachibana*. When people observe the plants that today serve them as food, they should be conscious that every one of them contains an evolutionary history based on mutation and selection variation principles, in part because of its permanence as wild species and, in part, because of man action, which contributed to select genetic changes produced, sometimes, by reasons of searching for appropriate food and, some other times, directed by cultural or religious values.

8.7 WHAT HAPPENS WHEN PEOPLE GRAFT A PLANT ON OTHER THAT IS A GMO?

Plants present a great regeneration and growth capacity.[28] People talk about plant cells are totipotent, so that from an isolated cell, people can obtain a complete specimen. Plants are able to fuse their parts coming from different specimens to create a new chimera plant, which technique (*graft*) is used in fruticulture, where a variety aerial part, the graft, fuses with the radicular part of other, the foot. It is really surprising that when people join tissues with sectioned phloem and xylem vessels, cells be able to regenerate both functional vessels that allow the normal life of resulting chimera plant, which incorporates properties of both cultivars, for example, in València citriculture, sweet orange tree was grafted on bitter orange tree foot, which provided trees with high-quality tasty fruits for fresh consumption, with a foot that better tolerated the conditions of the cultivation soils. Unfortunately, in orange-trees case, the combination of sweet orange trees grafted on bitter orange tree conditions that grafted plants be sensitive to *quick-decline* virus, precisely because the phloem vessels of graft–foot connection zone are plugged as a response to viral infection. Citrus quick decline supposed a challenge for València citriculture, which could be solved thanks to Luis Navarro (València Institute for Agrarian Research), which developed a method of production of viruses-free plants from the in vitro cultivation of orange-tree caulinar apices, which method is used by all València tree nurseries to guarantee that viruses-free citrus plants sell. The EU authorized field experiments of cultivations of dwarf orange trees grafted on feet resistant to attack by fungi.

8.8 DO PLANTS TOLERANT TO WEEDKILLERS DAMAGE THE ENVIRONMENT?

In València, it is well known the damage for orange and mandarin fruit crops that occurs by cross-ferilization, between sexually compatible varieties of neighbouring smallholdings.[28] Farmers grow parthenocarpic varieties, that is, their flowers are sexually self-incompatible even though they are able to develop their fruits without fertilization, which are seedless. Smallholdings are small and every farmer grows the variety that he enjoys the most. Because of dominant winds or beehives installation in zones nearby orchards, in order that they manufacture orange-blossom honey, pollen transmission is favored from some orchards to others. Cross-fertilization occurs and real damages are derived from it, since consumers do not accept citric fruit crops that contain seeds. València farmers prevent the damages making use of agricultural insurances, which could be taken as an example in the case that a conflict emerge between GMO and ecological maize farmers.

8.9 DISCUSSION

Challenges exist on previous proposals for citrus taxonomy, for example, several genera (*Fortunella*, *Eremocitrus*, *Microcitrus*) are nested within the citrus clade. These and other distinct clades, which were identified, are more appropriately considered *Citrus* spp., on a par with those that formerly were referred to as the three *true* or *biological* species (*C. reticulata*, *C. maxima*, *C. medica*). The related genus *Poncirus*, a subject of continuous controversy since it was originally proposed to be within the genus *Citrus*, is clearly a distinct clade that is separate from *Citrus*, based on sequence divergence and whole-genome phylogeny. Insights exist in the origin, evolution, and domestication of citrus, and the genealogy of the most important wild and cultivated varieties. Findings draw an evolutionary framework for citrus, a scenario that challenges current taxonomic and phylogenetic thoughts, and points to a reformulation of the genus *Citrus*.

Delayed bitterness causes severe economic loss in citrus juice industry worldwide, which is mostly because of limonoid compounds formation, especially limonin, in juice. Effects of postharvest time of fruits, heat treatment, pH, and filtration of juice on limonin content in Newhall navel orange *C. sinensis* Osbeck cv. Newhall juice were investigated. The research indicated that: (1) limonin content in juice would gradually increase to a maximal level and then remained almost constant thereafter as storage time going on, whereas the maximum constant value (MCV) of limonin content in juice significantly decayed with the rise in postharvest time of fruits being juiced; (2) heat treatment and acidification of juice only speeded up the formation of limonin to the maximal level while without changing limonin-content MCV; (3) the juice after filtration exhibited much lower limonin-content MCV compared with the unfiltered one. The experimental observations might not only provide useful information for the development of new debitterness method for navel orange juice, but also strongly support acid-promoted delayed bitterness mechanism, suggesting that the formation of delayed bitterness might primary occur, because acid-promoted rather than enzyme-catalyzed lactonization of limonoate A-ring lactone (LARL) produces limonin in navel orange juice.

8.10 FINAL REMARKS

From the present results and discussion, the following final remarks can be drawn.

1. Findings draw an evolutionary framework for citrus, a scenario that challenges current taxonomic and phylogenetic thoughts, and points to a reformulation of the genus *Citrus*.
2. The effects of postharvest time of fruits, heat treatment, pH, and filtration of juice on limonin content in Newhall navel orange *Citrus sinensis* Osbeck cv. Newhall juice were investigated. The revealed results might not only provide useful information for the development of novel debittering methods for the manufacture of juice from navel oranges, for example, by juicing the fruits stored for an appropriate time instead of freshly collected ones and immediate removal of some insoluble tissues from the obtained juice, but also might shed light on the delayed bitterness mechanism of navel orange juice, suggesting that the formation of delayed bitterness might mainly occur, because acid-promoted rather than enzyme-catalyzed lactonization of limonoate A-ring lactone produces limonin in navel orange juice.
3. In general, when people observe the plants that today serve them as food, they should be conscious that each one of them contains an evolutionary history, based on the principles of variation by mutation and selection, in part because of its permanence as wild species and, in part, because of the action of man, which contributed to select the genetic changes produced sometimes by reasons of searching for appropriate food, and, other times, directed by cultural or religious values.

ACKNOWLEDGMENTS

The authors thank support from Generalitat Valenciana (Project No. PROMETEO/2016/094) and Universidad Católica de Valencia *San Vicente Mártir* (Project No. 2019-217-001).

KEYWORDS

- **Hesperides' garden**
- **Himalayas**
- **delayed bitterness**
- **limonin**
- **navel orange**
- **newhall**
- **purple orange fruit**

REFERENCES

1. Bonamin, F.; Costa, C. A. R. A.; da Rocha, L. R. M.; Hiruma-Lima, C. A. Avaliacão do monoterpeno β-mirceno nas úlceras gástrica e duodenal e seus mecanismos de ação. *Rev. Fitoterapia* **2010,** *10* (S1), 68.
2. He, W.; Li, Y.; Liu, M.; Yu, H.; Chen, Q.; Chen, Y.; Ruan, J.; Ding, Z.; Zhang. Y.; Wang, T. *Citrus aurantium* L. and Its Flavonoids Regulate TNBS-Induced Inflammatory Bowel Disease Through Anti-Inflammation and Suppressing Isolated Jejunum Contraction. *Int. J. Mol. Sci.* **2018,** *19*, 3057–1-14.
3. Li, Z.; Zhao, Z.; Zhou, Z. Simultaneous Separation and Purification of Five Polymethoxylated Flavones From *Dahongpao* Tangerine (*Citrus tangerina* Tanaka) Using Macroporous Adsorptive Resins Combined With Prep-HPLC. *Molecules* **2018,** *23*, 2660–1-17.
4. Lukitawesa; Wikandari, R.; Millati, R.; Taherzadeh, M. J.; Niklasson, C. Effect of Effluent Recirculation on Biogas Production Using Two-Stage Anaerobic Digestion of Citrus Waste. *Molecules* **2018,** *23*, 3380–1-11.
5. Ascrizzi, R.; Taglieri, I.; Sgherri, C.; Flamini, G.; Macaluso, M.; Sanmartin, C.; Venturi, F.; Quartacci, M.F.; Pistelli, L.; Zinnai, A. Nutraceutical Oils Produced by Olives and *Citrus* Peel of Tuscany Varieties As Sources of Functional Ingredients. *Molecules* **2019,** *24*, 65–1-16.
6. Ancillo, G.; Medina, A. *Citrus*; Botanical Monographs The Botanic Garden of the Universitat de València No. 2, Universitat de València: València, Spain, 2015.
7. Torrens-Zaragozá, F. Molecular Categorization of Yams by Principal Component and Cluster Analyses. *Nereis* **2013,** *2013* (5), 41–51.

8. Torrens, F.; Castellano, G. From Asia to Mediterranean: Soya Bean, Spanish Legumes and Commercial *Soya Bean* Principal Component, Cluster and Meta-Analyses. *J. Nutr. Food Sci.* **2014,** *4* (5), 98.
9. Torrens, F.; Castellano, G. Principal Component, Cluster and Meta-Analyses of Soya Bean, Spanish Legumes and Commercial Soya Bean. In *High-Performance Materials and Engineered Chemistry*; Torrens, F., Balköse, D., Thomas, S., Eds.; Apple Academic–CRC: Waretown, NJ, 2018; pp 267–294.
10. Torrens-Zaragozá, F. Classification of Fruits Proximate and Mineral Content: Principal Component, Cluster, Meta-Analyses. *Nereis* **2015,** *7*, 39–50.
11. Torrens-Zaragozá, F. Classification of Food Spices by Proximate Content: Principal Component, Cluster, Meta-Analyses. *Nereis* **2016,** *8*, 23–33.
12. Castellano, G.; Tena, J.; Torrens, F. Classification of Polyphenolic Compounds by Chemical Structural Indicators and Its Relation to Antioxidant Properties of *Posidonia oceanica* (L.) Delile. *MATCH Commun. Math. Comput. Chem.* **2012,** *67*, 231–250.
13. Castellano, G.; González-Santander, J. L.; Lara, A.; Torrens, F. Classification of Flavonoid Compounds by Using Entropy of Information Theory. *Phytochemistry* **2013,** *93*, 182–191.
14. Castellano, G.; Lara, A.; Torrens, F. Classification of Stilbenoid Compounds by Entropy of Artificial Intelligence. *Phytochemistry* **2014,** *97*, 62–69.
15. Castellano, G.; Torrens, F. Information Entropy-Based Classification of Triterpenoids and Steroids From *Ganoderma*. *Phytochemistry* **2015,** *116*, 305–313.
16. Castellano, G.; Torrens, F. Quantitative Structure–Antioxidant Activity Models of Isoflavonoids: A Theoretical Study. *Int. J. Mol. Sci.* **2015,** *16*, 12891–12906.
17. Castellano, G.; Redondo, L.; Torrens, F. QSAR of Natural Sesquiterpene Lactones as Inhibitors of Myb-Dependent Gene Expression. *Curr. Top. Med. Chem.* **2017,** *17*, 3256–3268.
18. Torrens, F.; Redondo, L.; León, A.; Castellano, G. Structure–Activity Relationships of Cytotoxic Lactones as Inhibitors and Mechanisms of Action. *Curr. Drug Discov. Technol.*, In press.
19. Torrens, F.; Redondo, L.; Castellano, G. Artemisinin: Tentative Mechanism of Action and Resistance. *Pharmaceuticals* **2017,** *10*, 20–4-4.
20. Torrens, F., Redondo, L.; Castellano, G. Reflections on Artemisinin, Proposed Molecular Mechanism of Bioactivity and Resistance. In *Applied Physical Chemistry with Multidisciplinary Approaches*; Haghi, A.K., Balköse, D., Thomas, S., Eds.; Apple Academic–CRC: Waretown, NJ, 2018; pp 189–215.
21. Torrens, F.; Castellano, G. Chemical/Biological Screening Approaches to Phytopharmaceuticals. In *Research Methods and Applications in Chemical and Biological Engineering*; Pourhashemi, A., Deka, S. C., Haghi, A. K., Eds.; Apple Academic–CRC, Waretown, NJ, In press.
22. Torrens, F.; Castellano, F. Chemical Components from Artemisia austro-yunnanensis, Anti-inflammatory Effects and Lactones. In *Molecular Chemistry and Biomolecular Engineering: Integrating Theory and Research with Practice*; Pogliani, L., Torrens, F., Haghi, A. K., Eds.; Apple Academic–CRC: Waretown, NJ, In press.
23. Torrens, F.; Castellano, G. Classification of Citrus: Principal Components, Cluster, and Meta-analyses. In *Applied Physical Chemistry with Multidisciplinary Approaches*; Haghi, A.K., Balköse, D., Thomas, S., Eds.; Apple Academic–CRC: Waretown, NJ, 2018, pp 217–234.

24. Torrens, F.; Castellano, G. Extraction of Natural Products Found in Vegetal Species: Clove/Citrus. In: *Applied Food Science and Engineering with Industrial Applications*; Aguilar, C.N., Carvajal-Millan, E., Eds.; Apple Academic–CRC: Waretown, NJ, In press.
25. Torrens, F.; Castellano, G. *World of Biological Activities and Safety of Citrus spp. Essential Oils*. In *Molecular Chemistry and Biomolecular Engineering: Integrating Theory and Research with Practice*; Pogliani, L., Torrens, F., Haghi, A. K., Eds.; Apple Academic–CRC: Waretown, NJ, In press.
26. Wu, G. A.; Terol, J.; Ibanez, V.; López-García, A.; Pérez-Román, E.; Borredá, C.; Domingo, C.; Tadeo, F. R.; Carbonell-Caballero, J.; Alonso, R.; Curk, F.; Du, D.; Ollitrault, P.; Roose, M. L.; Dopazo, J.; Gmitter, Jr., F. G.; Rokhsar, D. S.; Talon, M. Genomics of the Origin and Evolution of *Citrus*, *Nature (London)*, **2018,** *554*, 311–316.
27. Zhang, J.; Yang, Z.; Liang, Y.; Zhang, L.; Ling, W.; Guo, C.; Liang, G.; Luo, G.; Ye, Q.; Zhong, B. Effects of Postharvest Time, Heat Treatment, Ph and Filtration on the Limonin Content in Newhall Navel Orange (*Citrus sinensis* Osbeck cv. Newhall) Juice. *Molecules* **2018,** *23*, 2691–1-10.
28. Beltrán, J.P. *Cultivos Transgénicos*; ¿Qué Sabemos de? No. 89, CSIC–Catarata: Madrid, Spain, 2018.

CHAPTER 9

Disruptive Big Data, Intelligence, and Physics: Process Understanding

FRANCISCO TORRENS[1,*] and GLORIA CASTELLANO[2]

[1]Institut Universitari de Ciència Molecular, Universitat de València, Edifici d'Instituts de Paterna, P. O. Box 22085, E-46071 València, Spain

[2]Departamento de Ciencias Experimentales y Matemáticas, Facultad de Veterinaria y Ciencias Experimentales, Universidad Católica de Valencia San Vicente Mártir, Guillem de Castro-94, E-46001 València, Spain

[*]Corresponding author. E-mail: Francisco Torrens (torrens@uv.es)

ABSTRACT

Connectivity is a key competitiveness component. Technology and *big data* are important for health, health care, and precision personalized medicine (PPM). Using conventional deep learning (DL) algorithms of artificial intelligence adds little extra knowledge, compared to more powerful random forest and other common algorithms. The present health care system is untenable. The economic risk of passiveness is the notable loss of competitiveness in the next years. The contemporary society is a digital one. A threat to work, freedom and privacy exists. An effect of democratic systems is real. Visualization transforms data into knowledge. One should not confuse: information, knowledge, and wisdom.

9.1 INTRODUCTION

Setting the scene: the economic and legal view of disruptive tools *big data* (BD) and *artificial knowledge* (*intelligence*, AI), data scientists, turning data into gold, data science (DS), the challenge of knowledge, and physics, AI deep learning (DL), and process understanding for data-driven Earth system science and advancing climate phenomena. Connectivity turns out to be a key competitiveness component. Technology and BD result important for health, health care, and precision personalized medicine (PPM). Relevant information vs. data becomes significant in health, health care, and PPM. The BD show meaning in health, health care, and PPM. Using conventional AI DL algorithms, to build quantitative structure–activity relationships (QSARs), adds little extra knowledge when comparing to more powerful random forest and other common algorithms. The present health care system is untenable. People must innovate and transform the present sanatorium into the hospital of the future: innovation in care processes, innovation in purchase processes, innovation in the way of paying and technological innovation.

The economic risk of passiveness results the notable loss of competitiveness in the next years. The contemporary society ensures a digital society. There is a threat to work, freedom, and privacy. An effect of democratic systems exists. Visualization can transform data into knowledge. It should not be confused the concepts of information, knowledge, and wisdom.

In earlier publications, it was informed the empirical didactics of molecular shape,[1] the phylogenesis of anthropoid apes,[2] the fractal analysis of proteins' tertiary structure (P3S),[3] fractal hybrid orbitals (FHOs) in biopolymer chains,[4] FHOs in protein models,[5] FHOs P3S analysis,[6] resonance in interacting induced-dipole polarizing force fields (FFs), application to FF derivatives,[7] the modeling of complex multicellular systems, tumor–immune cells competition,[8] molecular diversity classification *via* information theory,[9] a tool for the interrogation of macromolecular structure,[10] a new tool for the study of resonance in chemical education,[11] dialectic walk on science,[12] Brownian motion, random trajectory, diffusion, fractals, chaos theory, dialectics,[13] the work with nanomaterials, reductionism and positivism philosophical, and ethical considerations,[14] BD in AI, and models for health and health care.[15] The present report reviews the economic and legal view of disruptive tools BD and AI, data scientists, turning data into gold, DS, the challenge of knowledge, and physics, DL and process understanding for data-driven Earth system science. The aim of this work is to initiate a debate by suggesting a number of questions (Q), which can arise when addressing subjects of physics, technology, BD, and AI in different fields, and providing, when possible, answers (A) and hypotheses (H).

9.2 DISRUPTIVE BD AND AI: ECONIMIC/LEGAL VIEW

Pedreño Muñoz and Plaza Penadés presented BD/AI book raising the following questions.[16,17]

Q1. Which is in these moments the present normative framework of BD and AI?
Q2. Which are the aspects of BD and AI that should be regulated?

Pedreño Muñoz proposed the following hypotheses and Q on the economic impact of AI.

H1. (Ng, 2017). *AI is the new electricity.*
Q3. Does Ng exaggerate?
Q4. Are rulers conscious/proactive enough to face changes/challenges associated with BD/AI?
Q5. Are chosen strategies right to face all potentialities/risks/benefits that AI could provide?
Q6. How are the countries preparing themselves faced with the new changes that approach?
Q7. In addition, what does it happen in the rest of the world?
Q8. Sectorial impacts of AI: either transformation or disruption?
A8. AI is a technological disruption almost without precedents but it is not the only one.
Q9. How will robotics or automation derived from AI affect employment?
Q10. (McKinsey Global Institute, 2017). *What's now and next in analytics/AI/automation?*
H2. Academic failure vs. working failure.
Q11. However, where do small and medium-sized enterprises start?
He provided the following conclusion (C).
C1. The economic risk of passiveness: The notable lose of competitiveness in the next years.

Plaza Penadés proposed questions and hypotheses on the legal aspects of AI and BD.

Q12. Which is the use and purpose of personal data?
Q13. How will identity be protected?
Q14. How to protect the valuable data that one has?
H3. Text mining.
H4. Data mining.
Q15. I have an algorithm, how do I protect it?

9.3 DATA SCIENTISTS OR HOW TO TURN THEM INTO GOLD

Moreno proposed questions and Hs on data scientists or how to turn them into gold.[18]

Q1. How to turn them into gold?
H1. (Clive Humby). Data are the new oil (comparison of BD to *Big Oil*).
H2. Data are 21st century oil and companies know it.
Q2. Do we really want to turn into data analytics professionals?
Q3. However, which is the best option to go into the world of BD?

9.4 DATA SCIENCE: THE CHALLENGE OF KNOWLEDGE

Ariño moderated a debate on DS/knowledge challenge with Soria and de la Barra proposing H/Q/A (*cf.* Fig. 9.1).[19]

H1. (Ariño). The contemporary society is a digital society.
Q1. (Ariño). What is DS nowadays?
Q2. (Ariño). Which is your personal experience?
Q3. (Ariño). Which are the present and future applications?
Q4. (Ariño). How will they improve life?
Q5. (Ariño). Which are the risks?
Q6. (Ariño). Which are the threats?
Q7. (Ariño). What is DS?
A7. (De la Barra). That data analysis matters in business units.
Q8. (Soria). Why are we here?
Q9. (Ariño). Is DS a more masculine university course?
Q10. (Ariño). Are there women?
A10. (De la Barra). Yes.
Q11. (Ariño). In your post, is there symmetry?
A11. (Soria). No, 30% women.
H2. (De la Barra). Phases: Data→(1) Scalable data storage→(2) Data ingestion and processing→(3) Transform Q to algorithm→(4) Asking the right Q→C-suit members→Value (1.2., engineering; 3.4., semantics).
H3. (De la Barra). The most important is the strategic document for data.
Q12. (De la Barra). The key is in knowing: Which data should one compile?
Q13. (Ariño). Is this business model going to improve people's life?

A13. (De la Barra). It is already but ambits exist where it is seen clearer: medicine, food, etc.
H4. (De la Barra). The *illusion of choice*: seven or eight companies choose what people consume.
H5. (Susskind and Susskind, 2015). *The Future of the Professions*.[20]
H6. (Ariño). There is a threat to work, freedom and privacy.
H7. (Ariño). There is an effect of democratic systems.
H8. (Soria). Data Visualization→Knowledge.
H9. (Moreso, 2017). *Are We Becoming Idiots or Have We Been Always?*[21]
H10. (O'Neil, 2016). *Weapons of Math Destruction*.[22]
H11. (Ariño). It should not be confused: information, knowledge and wisdom.

FIGURE 9.1 Data science: Servers of the Universitat de València Information-Technology Facility.

9.5 PHYSICS AND AI TO ADVANCE CLIMATE PHENOMENA/EARTH UNDERSTANDING

Machine learning (ML) approaches are increasingly used to extract patterns and insights from the ever-increasing stream of geospatial data, but current approaches are not optimal when system behavior is dominated by spatial or temporal context. Rather than amending classical ML, Reichstein group argued that the contextual cues should be used as part of DL (an approach that is able to extract spatio-temporal features automatically) to gain process understanding, for example, of Earth system science problems, improving the predictability of seasonal forecasting and modeling of long-range spatial connections via multiple timescales.[23] Next step will be a hybrid modeling approach, coupling physical process models–data-driven ML versatility.

9.6 FINAL REMARKS

From the present results and discussion, the following final remarks can be drawn:

1. Connectivity is one of the key components of competitiveness.
2. On BD and health: technology and BD; relevant information vs. data; BD in health and PPM.
3. Using conventional DL algorithms, to build quantitative structure–activity relationship models, adds little extra knowledge compared to random forest and other commonly used algorithms.
4. The present health care system is untenable. People should innovate and transform into the hospital of the future: innovation in care processes; innovation in purchase processes; innovation in the way of paying; technological innovation.
5. The economic risk of passiveness is the notable loss of competitiveness in the next years.
6. The contemporary society is a digital society.
7. There is a threat to work, freedom and privacy.
8. There is an effect of democratic systems.
9. Visualization transforms data into knowledge.
10. It should not be confused: information, knowledge, and wisdom.

ACKNOWLEDGMENTS

The authors thank support from Generalitat Valenciana (Project No. PROMETEO/2016/094) and Universidad Católica de Valencia *San Vicente Mártir* (Project No. 2019-217-001).

KEYWORDS

- **artificial knowledge**
- **artificial intelligence**
- **disruptive tool**
- **economic view**
- **legal view**
- **data scientist**
- **turning data into gold**

REFERENCES

1. Torrens, F.; Sánchez-Pérez, E.; Sánchez-Marín, J. Didáctica empírica de la forma molecular. *Enseñanza de las Ciencias* **1989,** *Número Extra (III Congreso)* (1), 267-268.
2. Torrens, F. Filogénesis de los simios antropoides. *Encuentros en la Biología* **2000,** *8* (60), 3–5.
3. Torrens, F. Análisis fractal de la estructura terciaria de las proteínas. *Encuentros en la Biología* **2000,** *8* (64), 4–6.
4. Torrens, F. Fractal Hybrid Orbitals in Biopolymer Chains. *Russ. J. Phys. Chem.* (Engl. Transl.) **2000,** *74*, 115–120.
5. Torrens, F. Fractals for Hybrid Orbitals in Protein Models. *Complexity Int.* **2001,** *8*, torren01–1–13.
6. Torrens, F. Fractal Hybrid Orbitals Analysis of the Tertiary Structure of Protein Molecules. *Molecules* **2002,** *7*, 26–37.
7. Torrens, F.; Castellano, G. Resonance in Interacting Induced-Dipole Polarizing Force Fields: Application to Force-Field Derivatives. *Algorithms* **2009,** *2*, 437–447.
8. Torrens, F.; Castellano, G. Modelling of Complex Multicellular Systems: Tumour–Immune Cells Competition. *Chem. Central J.* **2009,** *3* (Suppl. I), 75–1-1.
9. Torrens, F; Castellano, G. Molecular Diversity Classification *via* Information Theory: A Review. *ICST Trans. Complex Syst.* **2012,** *12* (10–12), e4–1-8.
10. Torrens, F.; Castellano, G. A Tool for Interrogation of Macromolecular Structure. *J. Mater. Sci. Eng. B* **2014,** *4* (2), 55–63.
11. Torrens, F.; Castellano, G. Una nueva herramienta para el estudio de la resonancia en docencia química. *Avances en Ciencias e Ingeniería* **2014,** *5* (1), 81–91.
12. Torrens, F.; Castellano, G. Dialectic Walk on Science. In *Sensors and Molecular Recognition*; Laguarda Miro, N., Masot Peris, R., Brun Sánchez, E., Eds.; Universidad Politécnica de Valencia: València, Spain, in press; Vol. 11, pp 271–275.
13. Torrens, F.; Castellano, G. Brownian Motion, Random Trajectory, Diffusion, Fractals, Theory of Chaos, and Dialectics. In *Modern Physical Chemistry: Engineering Models, Materials, and Methods with Applications*; Haghi, R., Besalú, E., Jaroszewski, M., Thomas, S., Praveen, K. M., Eds.; Apple Academic/CRC: Waretown, NJ, in press.

14. Torrens, F.; Castellano, G. El Trabajo con Nanomateriales: Consideraciones Filosóficas: Reduccionismo/Positivismo y Éticas. In *Tecnología e Innovación Social: Hacia un Desarrollo Inclusivo y Sostenible*; Feltrero, R., Ed.; Desafíos Intelectuales del Siglo XXI No. 1, Global Knowledge Academics: Cantoblanco, Madrid, Spain, 2018; pp 11–35.
15. Torrens, F.; Castellano, G. Big Data in Artificial Knowledge: Models for Health and Healthcare. In *Physical Biochemistry, Biophysics, and Molecular Chemistry: Applied Research and Interactions*; Torrens, F., Mahapatra, D. K., Haghi, A. K., Eds.; Apple Academic/CRC: Waretown, NJ, in press.
16. Pedreño Muñoz, A.; Moreno Izquierdo, L.; Plaza Penadés, J. *Big Data e Inteligencia Artificial: Una Visión Económica y Legal de Estas Herramientas Disruptivas*; Universitat de València: València, Spain, 2018.
17. Pedreño Muñoz, A.; Plaza Penadés, J. *Libro Blanco sobre Big Data e Inteligencia Artificial*; Gobierno de España: Madrid, Spain, in press.
18. Moreno, J. M. Científico de datos o cómo convertirlos en oro. *El Economista* **2019,** *14* (3952), 31–31.
19. Ariño, A.; Soria, E.; de la Barra, M. Personal Communication.
20. Susskind, R.; Susskind, D. *The Future of the Professions: How Technology Will Transform the Work of Human Experts*; Oxford University: Oxford, UK, 2015.
21. Moreso, S. *¿Nos Estamos Volviendo Idiotas o lo Hemos Sido Siempre?*; Bang: Barcelona, Spain, 2017.
22. O'Neil, C. *Weapons of Math Destruction: How Big Data Increases Inequality and Threatens Democracy*; Crown: Largo, MD, 2016.
23. Reichstein, M.; Camps-Valls, G.; Stevens, B.; Jung, M.; Denzler, J.; Carvalhais, N.; Prabhat, Deep Learning and Process Understanding for Data-Driven Earth System Science. *Nature (London)* **2019,** *566*, 195–204.

CHAPTER 10

Agro-Ecological Conditions of Farming Intensification in the Russian Ural Federal District

RAFAIL A. AFANAS'EV* and MICHAEL O. SMIRNOV*

Pryanishnikov All-Russian Scientific Research Institute of Agrochemistry, d. 31A, Pryanishnikova St., Moscow 127550, Russia

**Corresponding author. E-mail: rafail-afanasev@mail.ru; User53530@yandex.ru*

ABSTRACT

The chapter provides a detailed analysis of agro-climatic conditions of the Ural Federal districts well as agrochemical properties of soils in this region. It is shown that this region has great potential for the development of agricultural production: scientifically based application of fertilizers could dramatically increase the productivity regional agriculture.

10.1 INTRODUCTION

For the further development of agriculture and increase of agricultural production, a detailed study of the agro-climatic conditions of the areas of cultivation of crops is necessary. Only this will significantly increase the productivity of agriculture through the development of science-based agricultural practices, including the rational use of fertilizers and other agrochemical means.[1–3] In the present study, the possibilities of increase in agricultural productivity in the Ural Federal district were considered in this respect. Geographically, the Ural Federal district includes four administrative regions—Kurgan, Sverdlovsk, Chelyabinsk, and Tyumen. The name of the district is formed from the word Ural. Ural is a geographical region in Russia, stretching between the East European and West Siberian plains. The main part of this region is the Ural mountain system.

10.2 METHODS

To achieve this goal, studies of soils were carried out to identify their types and subtypes in the region. The studies of soil granulometric composition, as well as the topography of arable lands and the identification of the risk of water erosion in this regard were also carried out. The acidity of arable soils was also determined. The methods adopted in agrochemistry[4] were used to determine the acidity of arable soils and their availability with mobile phosphorus and exchange potassium. The study of the main agro-climatic indicators of the studied region was also carried out: the annual amount of precipitation, evaporation per year, the height of snow cover, the reserve of productive moisture in the layer 0–100 cm. The temperature regime in the subjects of the Ural region was also considered. The data on the yield of the main cultures (cereal crops, potatoes) cultivated in the Ural district (the area sometimes called the Trans-Urals) were summarized.

10.3 DISCUSSION

Although the subjects of the Ural district differ in climatic and economic conditions of agricultural production, but in general they are characterized by the presence of potentially fertile soils, such as chernozem and gray forest (Fig. 10.1).

At the same time, the most widespread in the Urals were leached chernozems, typical, common, meadow, occupying more than 50% of the total area of arable land. More than 30% of arable land belongs to sod, sod-carbonate, gray forest soils, and chernozem podzolized.[5] Chernozems are the richest in Kurgan and Tyumen regions, where they occupy up to 80% of arable land. Gray forest soils are more common in the Sverdlovsk and Chelyabinsk regions, rarely—in the Tyumen region. According to the granulometric composition, the arable soils of the Trans-Urals mainly belong to clays and clay loams, the area of which is about 60% of the total area of arable land. Up to 38% of arable land is represented by medium and sandy loams; sands and sabulous clays (loamy sands) occupy a little more than 2% (Fig. 10.2).

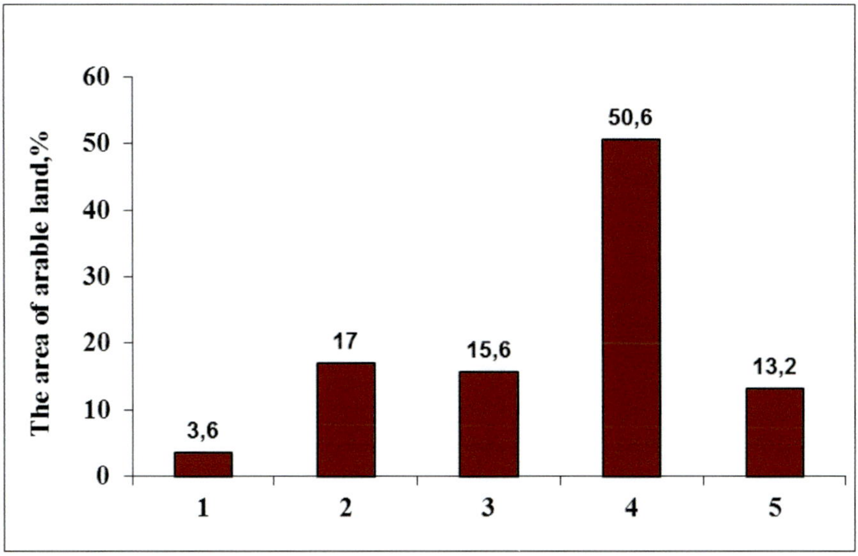

FIGURE 10.1 Soil types and subtypes in the Ural region (% of the total arable area). Legend: 1—podzolic and sod-podzolic; 2—sod, sod-calcareous, light gray, and gray forest; 3—dark gray forest, chernozem podzolized; 4—leached chernozem, typical, common, meadow, and 5—a number of other.

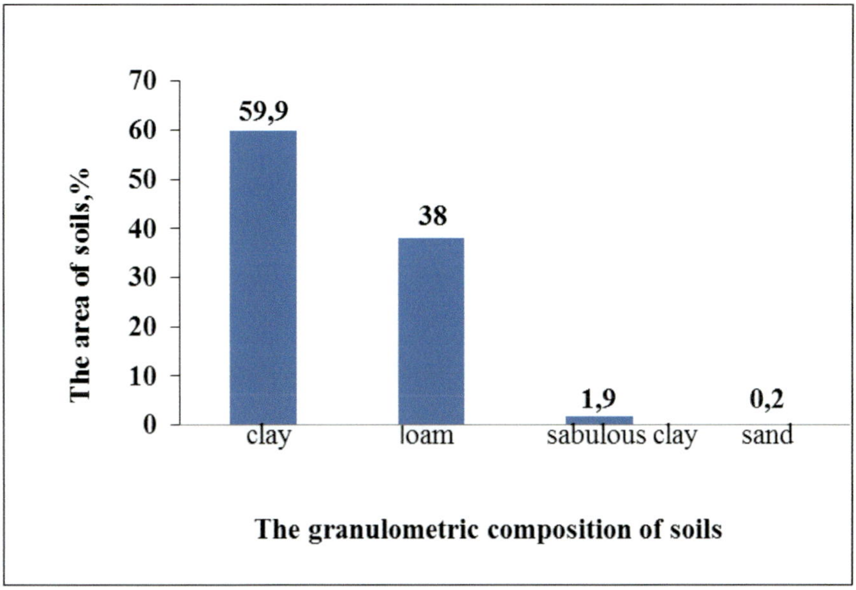

FIGURE 10.2 The area of soils with different granulometric composition in Ural region

Soils of heavy granulometric composition are typical for Kurgan, Sverdlovsk, and Chelyabinsk regions, where they occupy from 64% to 70% of arable land. Arable soils of light granulometric composition, that is, sandy and loamy sands, are more common in Sverdlovsk (14.2%) and Tyumen (7%) regions.

According to the relief of arable lands, the Trans-Urals refer mainly to low-lying locations. In Figure 10.3 it can be seen that the largest area of arable land—almost 90%—has slopes of less than 2°, which causes a weak expression of water erosion. At the same time, the Sverdlovsk region, despite the proximity of the Ural mountains, has the largest area of farmland, the slopes of which do not exceed 2°. However, the proximity to the Ural mountain range has led to the presence of more than 120,000 hectares of farmland (14.2%) with high rates of steepness of slopes—from 5° to 10° or more. In the Tyumen region, the upland location a little more than one third of the total area of agricultural land, at the same time as this area has a greater area of land with slope of the terrain from 2° to 5°.

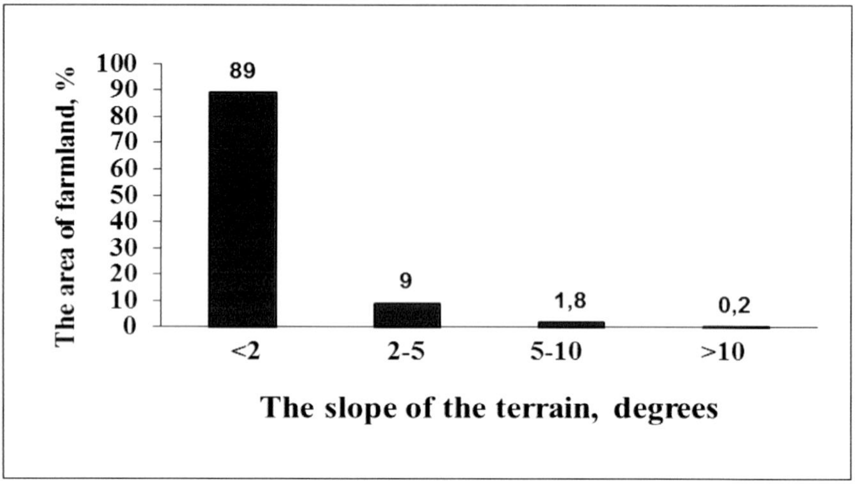

FIGURE 10.3 Area of agricultural land with different terrain slopes in the Ural region.

Because of the relatively flat terrain, the risk of water erosion of arable soils in the Ural region is considered to be quite small no more than 2%, although in the Sverdlovsk region it increases 7–8% (Table 10.1).

Of the negative properties of the soils of the Ural region, affecting the efficiency of farming, it is necessary to note the presence of salt licks. Thus, the area of solonetz complexes in the Kurgan Region reaches 21% of arable land, in the Chelyabinsk Region—15%, in the Tyumen Region—almost 7%, and in the region in total—13% with their practical absence in the Sverdlovsk Region. Soil fertility in the Ural region is also limited by the increased acidity of the arable layer (Fig. 10.4). In the Sverdlovsk region, the area of acid soils is, according to the agro-chemical service for 2010, more than 730,000 hectares, or about 63% of the arable land surveyed.[6] Significant areas of acid soils are also found in the Kurgan and Tyumen regions.

TABLE 10.1 Quantitative Characteristics of Arable Soils of the Ural Region, as a Percentage of the Total Area of Arable Land.

Administrative region	Salt complexes	Exposed to wind erosion	Washed
Kurgan	21.29	1.68	0.04
Sverdlovsk	0.18	–	7.51
Tyumen	6.73	–	2.0
Chelyabinsk	14.88	24.12	2.19
In total	13.0	8.6	3.13

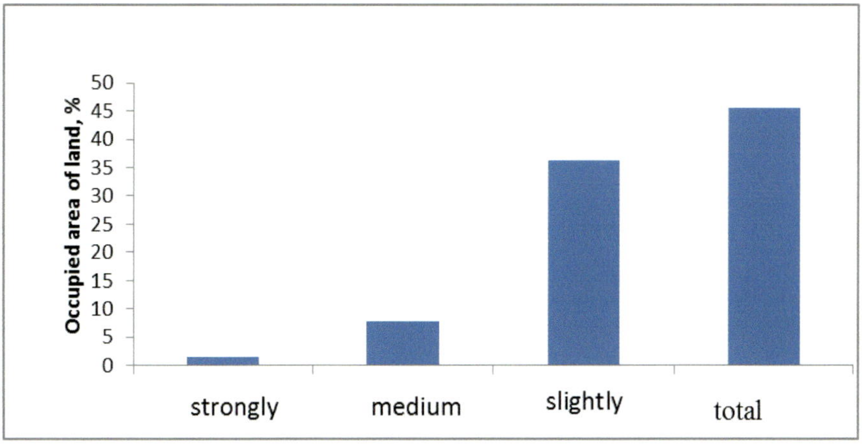

FIGURE 10.4 The area of arable land of gradual acidity in the Ural region: from strongly to slightly acid and total.

In general, in the Urals area of arable land in need improvement by amelioration, reaches many hundreds of thousands of hectares. At the same time, due to the presence of a large area of chernozem soils, the Ural region is a well-endowed soil organic matter-humus (Fig. 10.5), the mineralization of which can largely provide crops with mineral nitrogen in the Ural region.

In the Ural Federal district, soils with humus content of more than 4% occupy 70% of the area, of which the largest part is located in the Kurgan and Chelyabinsk regions. But unfortunately, the soils of the region are characterized by a relatively low content of mobile phosphorus in the area, and most importantly, in an area exceeding half of the arable wedge (Fig. 10.6).

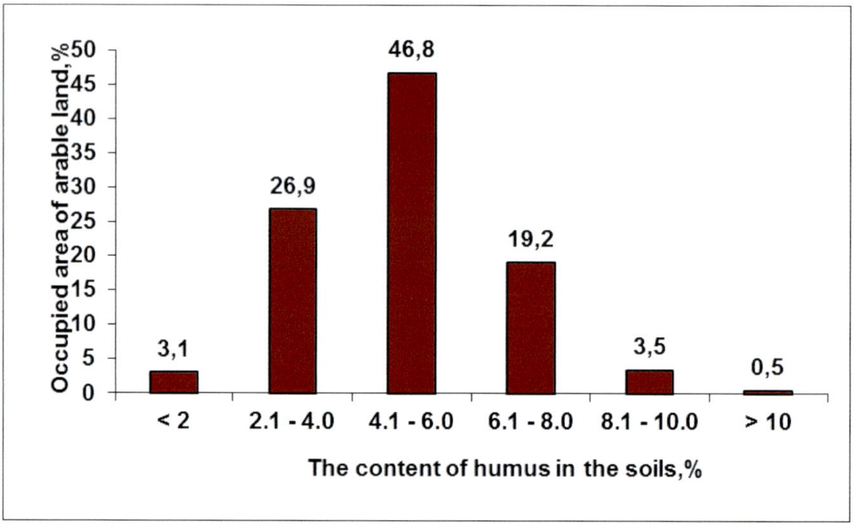

FIGURE 10.5 The area of arable land with different humus content.

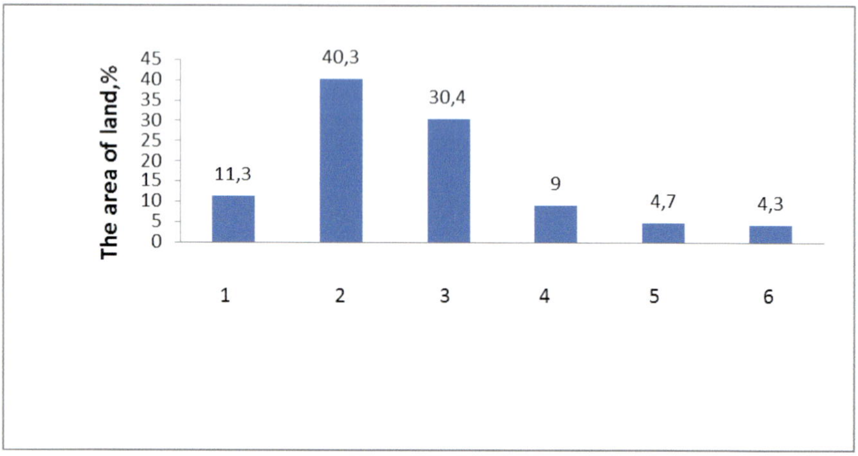

FIGURE 10.6 Areas of arable soils of the Ural region with different content of mobile phosphorus: 1—very low, 2—low, 3—medium, 4—elevated, 5—high, and 6—very high.

In general, the negative situation with the provision of arable soils of the Ural region with available phosphorus for plants necessitates the need to adjust agriculture toward a more balanced use of chemicals, including, first of all, phosphorization of soils and other ways to increase the provision of crops with phosphorus nutrition.

The situation in the region with the provision of agricultural crops with exchangeable potassium has been more successful. Soils with average and high values of this indicator occupy a total of about 98% of the arable area (Fig. 10.7). According to the highest content of mobile potassium, as well as the content of humus in the soil, again allocated Kurgan and Chelyabinsk region.

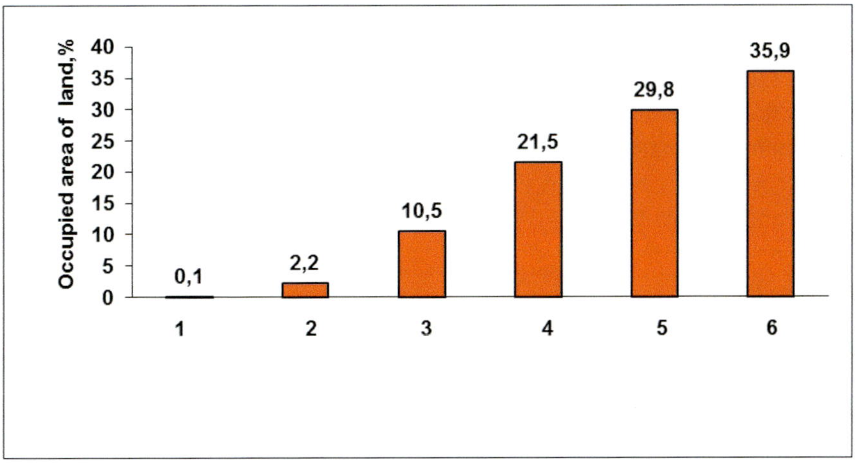

FIGURE 10.7 Areas of arable soils of the Ural region with different content of exchangeable potassium: 1—very low, 2—low, 3—medium, 4—elevated, 5—high, and 6—very high.

This means that in the foreseeable future, the demand for potassic nutrition in crops can be largely satisfied by the use of soil reserves of this element. In other words, to ensure a balanced nutrition of crops in the Ural region, phosphorus fertilizers are of paramount importance, while the need for potassic nutrition of plants, especially in the Kurgan and Chelyabinsk regions, can be mainly covered by natural soil reserves.

According to the main agro-climatic indicators, the Ural region, with the exception of the Sverdlovsk region, belongs to the zone of insufficient moisture (Table 10.2).

According to calculations, the hydrothermal indicator of weather conditions in these areas, as a rule, is less than one, and only in the Sverdlovsk region of the hydrothermic coefficient can exceed this value. The greatest aridity of the climate is inherent in the Kurgan region, in which the amount of precipitation for the year on average does not exceed 300–400 mm with evaporation of 450–485 mm. The area is also characterized by the lowest soil moisture reserves in the early spring period, not exceeding 120 mm in the first meter layer of soil before the sowing campaign. According to the agrochemical service,[7] most of the precipitation in the Kurgan region during the growing season falls on summer and autumn. Spring precipitation from their annual amount is a very small proportion. That is why crops suffer greatly from drought in the spring. A more significant spring soil moisture reserve, estimated at 157 mm, is typical only for the Sverdlovsk region and Northern Tyumen regions. According to the temperature conditions, the subjects of the Ural region also differ markedly in the provision of agrocenoses with heat (Table 10.3). The difference in the minimum duration of the frost-free period in the subjects of the Trans-Urals reaches a month or more, the sum of temperatures above 10°Celsius-890–900°. The pronounced continental climate of the Ural region also causes low temperatures of the winter period, reaching in the coldest month on average −20, and in the Tyumen region −29°.

TABLE 10.2 Indicators of Atmospheric Moisture Regime in the Subjects of Ural Region.

Administrative region	Annual rainfall, mm	Evaporation per year, mm	Snow depth, cm	Reserve of productive moisture in the layer 0–100 cm in the spring, mm
Kurgan	300–400	450–485	30–40	120
Sverdlovsk	400–700	375–440	50–80	157
Chelyabinsk	300–500	450–520	40–70	128
Tyumen	350–500	230–420	30–90	136

TABLE 10.3 Temperature Regime in the Subjects of the Ural Region.

Administrative region	The sum of temperatures >10°C	The duration of frost-free period, days	The temperature of the warmest month, degree Celsius	Temperature of the coldest month, degree Celsius
Kurgan	1900–2200	115–120	18.5–21.6	−16.5–20.1
Sverdlovsk	1200–1800	80–120	14.6–18.5	−15.5–20.0
Chelyabinsk	1800–2300	108–125	15.6–20.2	−16.0–20.0
Tyumen	1000–2000	70–120	15.0–18.9	−17.5–29.0

Of technical and food crops here can cultivate sunflower, mustard, flax, potatoes, vegetable crops, and to sow fodder crops. In the Sverdlovsk and Chelyabinsk regions, along with the listed crops, it is allowed to cultivate winter wheat.[8]

The yield of agricultural crops in the Urals, as in other regions of the country, largely depends not only on natural but also anthropogenic factors, that is, on the applied agricultural technologies. As shown in Tables 10.4 and 10.5, the highest yield of grain crops on average in the region was observed in the 5 years 1976–1980, potatoes in the 4 years 1981–1984.

TABLE 10.4 Grain Yield in the Ural Region.

Administrative region	All categories of farms, over the years			SVS 1979–1993
	1971–1975	1976–1980	1981–1984	
Kurgan	12.8	17.3	13.3	28.9
Sverdlovsk	14.0	18.9	16.9	36.0
Chelyabinsk	11.1	14.5	10.8	26.4
Tyumen	13.4	15.4	16.0	36.4
In total	12.8	16.5	14.2	31.9

SVS, state variety sections.

TABLE 10.5 The Yield of Potatoes in the Ural Region.

Administrative region	All categories of farms, over the years			SVS 1979–1993
	1971–1975	1976–1980	1981–1984	
Kurgan	66	69	75	190
Sverdlovsk	100	119	110	231
Chelyabinsk	75	93	96	264
Tyumen	74	69	85	221
In total	78.8	87.5	91.5	226.5

SVS, state variety sections.

In these years, that is, in the period from 1976 to 1984, the grain yield was significantly higher than in the preceding 5-year period 1971–1975. This, apparently, was associated with an increase of fertilizer application during the years of intensive chemicalization. Chiefly the influence of agricultural technologies on yield of agricultural crops, particularly cereals and potatoes, is shown on the State variety sections (SVS) here they are, as a rule, was cultivated with respect for the relevant evidence-based recommendations. Thus, the yield of grain at SVS—31.9 t/ha is almost twice the highest yield—16.5 t/ha, obtained in production conditions, where these recommendations for various reasons were not fully maintained. Even more remarkably, the yield of potatoes: SVS—226 kg/ha, production—96 kg/ha. A significant effect on the yield of crops in the region was affected by level of fertilizer application. In the years of intensive chemicalization, the use of mineral and organic fertilizers in the conversion to the active substance reached in the Sverdlovsk region—184 kg/ha per year, Tyumen—103 kg/ha, Kurgan—66 kg/ha per year, Chelyabinsk—64 kg/ha. It is the intensity of application of fertilizers to a certain extent, along with climatic conditions, that can be explained by the difference in the yield of grain and potatoes in the Sverdlovsk region compared to the yield of these crops in other parts of the region. These data show that the soil and climatic potential of the Ural region allows even for the former agricultural technologies to double the productivity of agriculture through the development of scientifically based agricultural practices, including the rational use of fertilizers and other agrochemical means.

Analysis of natural and anthropogenic resources of the Ural Federal district shows that the region has significant opportunities for further development of agriculture and agricultural production. The most important features of the region include the presence of large areas of potentially fertile soils, mainly black soil and gray forest. Despite certain restrictions on agro-climatic conditions, these resources make it possible to successfully cultivate a wide range of crops in the region. According to agrochemical properties, most of the arable lands of the region belong to the well-provided organic matter and, accordingly, mineralized nitrogen, as well as exchangeable potassium. More problematic is the provision of crops with mobile phosphorus due to the low content of mobile phosphates on most arable soils, as well as a number of other negative qualitative indicators in agricultural land. In this regard, a significant part of arable lands in the region requires improving the basic agrochemical, agrophysical, and biological properties of soils by liming acidic soils, phosphorization of soils that are poorly provided with mobile phosphates, improvement by amelioration of solonetz complexes, and a number of other agrotechnical measures. At the same time, agricultural crops cultivated in the region need regular scientifically justified use of mineral and organic fertilizers, which, as the experience of local SVS has shown, can almost, double the level of productivity of regional agriculture. It should also be noted that an important role in the development of science-based methods of agriculture of the Trans-Urals, including the use of agrochemical means, played by regional agricultural research institutes and agrochemical units, which are responsible for the further development of modern agricultural technologies, assisting in their development of agricultural production.

10.4 CONCLUSIONS

1. A detailed analysis of the data allows us to conclude that a significant area of the Ural Federal district has potentially fertile soils.
2. The resources of these areas make it possible to successfully cultivate a large number of crops in this region.
3. Most of the land area of the region is characterized by a good supply of organic matter, namely nitrogen and potassium.
4. Due to the low soil availability of mobile phosphorus in this region, a significant part of the land in this region requires various amelioration improvements.
5. Regular scientifically justified use of mineral and organic fertilizers, can double the level of productivity of regional agriculture.

KEYWORDS

- **soil**
- **climate**
- **fertilizers**
- **agrochemical indicators**
- **yield**
- **potential fertility**

REFERENCES

1. Timiryazev, K. A. The Life of Plants. “Novosti” (“News” in Rus.) Publishing House: Moscow, 2006; p 320 (in Russian).
2. Pryanishnikov D. N. Selected Works, Vol. 1; Publishing House: Moscow: “Kolos” (“Ear” in Rus.), 1965; p 768 (in Russian).
3. Mineev, V. G.; Sychev, V. G.; Gamzikov, G. P., et al. *Agrochemistry*; Mineev, V. G, Ed.; Publishing House of Pryanishnikov All-Russian Scientific Research Institute of Agrochemistry: Moscow, 2017; p 854c. (in Russian).

4. Agrochemicals Methods of Investigations of Soils. "Nauka" ("Science" in Rus.) 1975; p 656 (in Russian).
5. Natural and Economic Conditions of Agricultural Production in the Russian Federation. Part I. Publishing House of Russian Academy of Agricultural Sciences: Moscow, 1986; p 300 (in Russian).
6. Agrochemical Characteristics of Soil Quality of the Russian Federation. Register of Soil Fertility. Publishing House of Pryanishnikov All-Russian Scientific Research Institute of Agrochemistry: Moscow, 2013; p 208 (in Russian).
7. Consultancy in Agrochemical Service of the Russian Federation. Results and Prospects. (40 years of Agrochemical Service; Sychev, V. G., Ed.; Publishing House of Pryanishnikov All-Russian Scientific Research Institute of Agrochemistry: Moscow, 2005; p 569 (in Russian).
8. Natural and Economic Conditions of Agricultural Production in the Russian Federation. Part III. Publishing House of Russian Academy of Agricultural Sciences: Moscow, 1986, p 259 (in Russian).

PART II

Public Health, Ethical Problems, and Social Issues

CHAPTER 11

Primary Health, Smoking Habit, and Immunotherapy in Cancer

FRANCISCO TORRENS[1*] and GLORIA CASTELLANO[2]

[1]*Institut Universitari de Ciència Molecular, Universitat de València, Edifici d'Instituts de Paterna, P. O. Box 22085, E-46071 València, Spain*

[2]*Departamento de Ciencias Experimentales y Matemáticas, Facultad de Veterinaria y Ciencias Experimentales, Universidad Católica de Valencia San Vicente Mártir, Guillem de Castro-94, E-46001 València, Spain*

**Corresponding author. E-mail: torrens@uv.es*

ABSTRACT

The work of primary health physicians is *essential* in health care. *The Ideal Healer*. *Argumentum ad verecundiam*. Head and neck, and larynx cancers are related to smoking habit. Reinke's edema, peptic laryngitis, and Whartin's tumor are related to smoking habit. Tobacco kills more than 50% of those that consume it. To stop smoking: addiction to nicotine, undo a social behavior, and changes in surroundings. Personalized medicine: The aim is to identify new objectives of treatment. Self-deception justified in that it is my will, I wish to reach there whatever happens. *On a shadow in X-ray*: *for a man, think in cancer; for a woman, think in typhus*. *Smoking Habit Prevention and Care Strategy in Valencia Region*. Theory of tumoral immonoediting: elimination, equilibrium, and escape. Immunotherapy of melanoma. *The Emperor of All Maladies*: *A Biography of Cancer*. Alcoholic liver disease is the main cause of disease, death etc. *Worldwide lack of early referral of patients with alcoholic liver disease*: GLADIS *Survey*. *Disparities between research attention and burden in liver diseases: implications on uneven advances in pharmacological therapies in Europe and* USA. *Histology and alcohol abstinence determine long-term prognosis*. The lack of vitamin D causes addiction.

Normal liver → Steatosis → Fibrosis → Cirrhosis → Hepatocellular carcinoma Alcoholic steatohepatitis –↑———↑

Don't ask, don't tell. Invalid paradigm: Alcoholic hepatopathy is because of polymorphonuclear neutrophil.

11.1 INTRODUCTION

Setting the scene: primary health for medical students, smoking habit, breathing, infectious diseases, immunotherapy in cancer, alcoholic hepatopathy (AH), and translational research to identify new therapeutic targets. The work of primary health physicians is *essential* in health care. Hippocrates' *The Ideal Healer. Argumentum ad verecundiam.* Head and neck, and larynx (90% by smoking habit) cancers are related to smoking habit. Reinke's edema and peptic laryngitis are related to smoking habit. Whartin's tumor is related to smoking habit. Tobacco kills more than 50% of those that consume it. To stop smoking: addiction to nicotine, undo a social behavior, and changes in surroundings. Personalized medicine (PM): The aim is to identify new objectives of treatment. Self-deception justified in that it is my will, I wish to reach there whatever happens. *On a shadow in X-ray*: *for a man, think in cancer*; *for a woman, think in typhus*. *Strategy of Prevention and Care of Smoking Habit in the Valencia Region.* Theory of tumoral immonoediting: elimination, equilibrium, and escape (3E). Immunotherapy of melanoma. *The Emperor of All Maladies*: *A Biography of Cancer*. Alcoholic liver disease (ALD) is the main cause of disease, death etc. *Worldwide lack of early referral of patients with* ALD: *Final results of the Global Alcoholic Liver Disease Survey* (GLADIS). *Disparities between research attention and burden in liver diseases. Histological parameters and alcohol abstinence determine long-term prognosis in patients with* ALD. The lack of vitamin D causes addiction.

Normal liver → Steatosis → Fibrosis → Cirrhosis → Hepatocellular carcinoma (HCC) Alcoholic steatohepatitis (ASH) –↑———↑

Don't ask, don't tell. Invalid paradigm: AH is because of polymorphonuclear neutrophil (PMN).

Neither chronicity nor cancer is like that of early human immunodeficiency virus (HIV)/acquired immunodeficiency syndrome (AIDS).

Earlier studies provided information about the modeling of complex multicellular systems and tumor–immune cells competition[1]; theoretic entropy for molecular classification of oxadiazolamines as potential therapeutic agents[2]; molecular classification of 5-amino-2-aroylquinolines and 4-aroyl-6,7,8-trimethoxyquinolines as highly potent tubulin polymerization inhibitors[3]; polyphenolic phytochemicals in cancer prevention, therapy, bioavailability versus bioefficacy[4]; molecular classification of antitubulin agents with indole ring binding at colchicine-binding site[5]; molecular classification of 2-phenylindole-3-carbaldehydes as potential antimitotic agents in human breast cancer cells[6]; cancer and its hypotheses[7]; precision PM from theory to practice and cancer[8]; and clinical translational research, cancer, diabetes, and cardiovascular disease.[9] It was reported how HIV/AIDS destroy immune defenses, hypothesis,[10] 2014 emergence, spread, uncontrolled Ebola outbreak,[11,12] Ebola virus disease, questions, ideas, hypotheses and models.[13] The present report reviews primary health for medical students, the smoking habit, breathing, infectious diseases, the beginning or end of immunotherapy in cancer, AH and translational research to identify new therapeutic targets. The aim of this work is to initiate a debate by suggesting a number of questions (Q), which can arise when addressing subjects of primary health for medical students, the smoking habit and infectious diseases, immunotherapy in cancer, AH, and hypotheses on smoking habit, immunotherapy, cancer and AH, in different fields, and providing, when possible, answers (A) and hypotheses (H).

11.2 PRIMARY HEALTH FOR MEDICAL STUDENTS

In *Primary Health* (PH) *for Medical Students* Day, Arbaizar Marínez raised following questions[14]:

Q1. Is the specialty in PH a good choice?
Q2. Is PH the cornerstone of the system?
Q3. How to make your group work useful?
Q4. Dr: They told me that I should be operated on, what do you think?
Q5. Are PH physicians needed?

She provided the following conclusions (Cs):

C1. The work of PH physicians is essential in health care.
C2. (Hippocrates, *ca.* 320 BC) *The Ideal Healer*.

Climent presented a documentary film raising Q on listening art from patient to person.[15]

Q6. What are patient's needs/preferences/expectations on relation/communication with physicians?

11.3 SMOKING HABIT AND BREATHING: SMOKING HABIT AND INFECTIOUS DISEASES

Carrión Valero organized Smoking Habit (SH) and Infectious Diseases Day.[16] In a social gathering on resident–physician–end SH, Marco Alacid proposed the following Q/H from endocrinology[17]:

Q1. What could we do to be a part of the solution?
H1. *Argumentum ad verecundiam.*

García Zamora proposed following hypotheses and question from otorhinolaryngology[18]:

H2. Head and neck (HNC), and larynx (LC, 90% by SH) cancers are related to SH.
H3. Reinke's edema and peptic laryngitis are related to SH.
Q2. (Veile, Zimmermann, Lorenz and Becher, 2018) *Is smoking a tinnitus risk factor* (RF)?[19]
H4. Whartin's tumor is related to SH.

Salamea Ávila proposed the following hypothesis from pneumology[20]:

H5. (World Health Organization, WHO) Tobacco kills more than 50% of those that consume it.

Arbaizar Marínez proposed the following H/Q on SH at the end of resident physician[21]:

H6. To stop smoking: addiction to nicotine, undo a social behavior and changes in the surroundings.
Q3. Primary health (PH) resources are limited; is it the time to carry out intensive interventions?

In social gathering on SH chair view, Benlloch García proposed the following Q/H from pneumology[22]:

Q4. Where are we?
H7. PM: The aim is to identify new objectives of treatment.

Fernández Laso proposed the following hypothesis from pneumology[23]:

H8. Self-deception justified in that it is my will, I wish to reach there whatever happens.

Ruano Marco proposed following hypothesis, question and answer from intensive care medicine[24]:

H9. (Old saying) On a shadow in X-ray: for a man, think in cancer; for a woman, think in typhus.

Q5. If nicotine is responsible for many ills of tobacco, what should be used to treat SH?

A5. I think that the solution is the use of the nicotinic partial agonists.

In social gathering on SH in health professions, Mascarós Balaguer raised the following Q from PH[25]:

Q6. What is a brief intervention?

Sanchis Chafer proposed following questions and answer on SH from highly competitive sport[26]:

Q7. What is nicotine?

A7. It is one of the greatest psychostimulants in the world.

Q8. Why is it so important in sport?

Q9. Is snus the substitute of cigarettes in sport?

Q10. Is there any improvement of 17% in efficiency?

A10. It seems yes, but only in hot climates.

Q11. Is tobacco common for sportspersons?

A11. Yes, 15% élite sportspersons are consumers of tobacco.

Q12. Which tobacco do sportspersons use?

A12. Snus

Q13. Which sport has the most habitual use of tobacco?

Q14. What is greater or lower consumption in sportspersons?

A14. Sportspersons consume more alcohol and less tobacco.

Quiles Izquierdo proposed the following question and H on a challenge for Valencia Region 2030[27]:

Q15. Is region without tobacco smoke possible?

H10. (Generalitat Valenciana, 2018) SH *Prevention and Care Strategy in Valencia Region*.[28]

11.4 IMMUNOTHERAPY IN CANCER: BEGINNING OR END

Antón group reviewed tumoral immunology and immunotherapy of cancer.[29] Hubbard–Lucey group revised a comprehensive analysis of the clinical immuno-oncology (IO) landscape.[30] Cañete proposed the following Q/A/H on immunotherapy in cancer as the beginning or end[31]:

Q1. Can immune response (IR) versus a tumor be possible?
A1. Yes
Q2. Do tumoral antigens exist?
A2. Yes
H1. Theory of tumoral immonoediting: 3E.
Q3. What can people do?
A3. Immunotherapy and different types of immunotherapy: active and passive.
Q4. How does chimeric antigen receptor T-cell (CAR-T) therapy work?
Q5. How can people use immunotherapy in cancer?
H2. (Allison, 1945) Immunotherapy of melanoma.
Q6. Till when should physicians continue administering the memory?
A6. Possibly, always some memory should be administered.
H3. (Mukherjee, 2011) *The Emperor of All Maladies*: *A Biography of Cancer*.[32]

11.5 ALCOHOLIC HEPATOPATHY: TRANSLATIONAL RESEARCH

Bataller Alberola proposed the following H/Q/A on AH and translational research to identify targets[33]:

H1. Alcoholic liver disease (ALD) is the main cause of disease, death etc.
H2. (Shah, 2017) *Worldwide lack of early referral of patients with* ALD: GLADIS *Survey*.[34]
H3. (Bataller, 2017) *Disparities between research attention and burden in liver diseases*.[35]
H4. (Lackner, 2017) *Histology and alcohol abstinence determine long-term prognosis*.[36]
H5. The lack of vitamin (Vit) D causes addiction.
H6. Normal liver → Steatosis → Fibrosis → Cirrhosis → HCC
ASH ————————————↑————↑
Q1. Does alcohol relapse influence on long-term survival?
Q2. How can I help the patient?
H7. *Don't ask, don't tell* (if you are a gay, then do not tell it to the army).
Q3. What are the specific treatments?
A3. Prednisolone since 1971!
Q4. What are the main cellular and molecular effects of alcohol?
A4. In mouse: tumor necrosis factor (TNF)α.
Q5. What did it go wrong with anti-TNFα therapy?

A5. The TNFα is not increased in humans.
H8. (Invalid paradigm) AH is because of PMN.
Q6. What are the main molecular drivers of hepatocellular failure?

11.6 FINAL REMARKS

From the present results and discussion, the following final remarks can be drawn.

1. The work of primary health physicians is essential in health care.
2. Hippocrates' *The Ideal Healer.*
3. *Argumentum ad verecundiam.*
4. Head and neck, and larynx (90% by smoking habit) cancers are related to smoking habit.
5. Reinke's edema and peptic laryngitis are related to smoking habit.
6. Whartin's tumor is related to smoking habit.
7. Tobacco kills more than 50% of those that consume it.
8. To stop smoking: addiction to nicotine, undo a social behavior, and changes in surroundings.
9. Personalized medicine: The aim is to identify new objectives of treatment.
10. Self-deception justified in that it is my will, I wish to reach there whatever happens.
11. *On a shadow in X-ray: for a man, think in cancer; for a woman, think in typhus.*
12. *Smoking Habit Prevention and Care Strategy in Valencia Region.*
13. Theory of tumoral immonoediting: elimination, equilibrium, and escape.
14. Immunotherapy of melanoma.
15. *The Emperor of All Maladies: A Biography of Cancer.*
16. Alcoholic liver disease is the main cause of disease, death etc.
17. *Worldwide lack of early referral of patients with alcoholic liver disease*: GLADIS.
18. *Disparities between research attention and burden in liver diseases.*
19. *Histology and alcohol abstinence determine long-term prognosis.*
20. Lack of vitamin D causes addiction.
21. Normal liver → Steatosis → Fibrosis → Cirrhosis → Hepatocellular carcinoma
    Alcoholic steatohepatitis ————↑————↑
22. *Don't ask, don't tell.*

23. Invalid paradigm: Alcoholic hepatopathy is because of polymorphonuclear neutrophil.

ACKNOWLEDGMENTS

The authors thank support from Generalitat Valenciana (Project No. PROMETEO/2016/094) and Universidad Católica de Valencia *San Vicente Mártir* (Project No. UCV.PRO.17-18.AIV.03).

KEYWORDS

- **immuno-oncology landscape**
- **cancer immunotherapy**
- **clinical trial**
- **tumor immunology**
- **paradigm**
- **protocol**
- **cancer treatment**

REFERENCES

1. Torrens, F.; Castellano, G. Modelling of Complex Multicellular Systems: Tumour–Immune Cells Competition. *Chem. Central J.* **2009,** *3*(1), 75–1-1.
2. Torrens, F.; Castellano, G. Information Theoretic Entropy for Molecular Classification: Oxadiazolamines as Potential Therapeutic Agents. *Curr. Comput.-Aided Drug Des.* **2013,** *9*, 241–253.
3. Torrens, F.; Castellano, G. Molecular Classification of 5-amino-2-aroylquinolines and 4-aroyl-6,7,8-trimethoxyquinolines as Highly Potent Tubulin Polymerization Inhibitors. *Int. J. Chemoinf. Chem. Eng.* **2013,** *3*(2), 1–26.
4. Estrela, J. M.; Mena, S.; Obrador, E.; Benlloch, M.; Castellano, G.; Salvador, R.; Dellinger, R. W. Polyphenolic Phytochemicals in Cancer Prevention and Therapy: Bioavailability *versus* Bioefficacy. *J. Med. Chem.* **2017,** 60, 9413–9436.
5. Torrens, F.; Castellano, G. Molecular Classification of Antitubulin Agents with Indole Ring Binding at Colchicine-Binding Site. In *Molecular Insight of Drug Design*; Parikesit, A. A., Ed.; InTechOpen: Vienna, 2018; pp 47–67.
6. Torrens, F.; Castellano, G. Molecular Classification of 2-Phenylindole-3-carbaldehydes as Potential Antimitotic Agents in Human Breast Cancer Cells. In *Theoretical Models*

and Experimental Approaches in Physical Chemistry: Research Methodology and Practical Methods; Haghi, A. K., Thomas, S., Praveen, K. M., Pai, A. R., Eds.; Apple Academic–CRC: Waretown, NJ, in press.
7. Torrens, F.; Castellano, G. Cancer and Hypotheses on Cancer. In *Molecular Chemistry and Biomolecular Engineering: Integrating Theory and Research with Practice*; Pogliani, L., Torrens, F., Haghi, A. K., Eds.; Apple Academic–CRC: Waretown, NJ, in press.
8. Torrens, F.; Castellano, G. Precision Personalized Medicine from Theory to Practice: Cancer. In *Green Chemistry and Biodiversity: Principles, Techniques, and Correlations*; Aguilar, C. N., Ameta, S. C., Haghi, A. K., Eds.; Apple Academic–CRC: Waretown. NJ, in press.
9. Torrens, F.; Castellano, G. Clinical Translational Research: Cancer, Diabetes, and Cardiovascular Disease. In *Physical Biochemistry, Biophysics, and Molecular Chemistry: Applied Research and Interactions*; Torrens, F., Mahapatra, D. K., Haghi, A. K., Eds.; Apple Academic–CRC: Waretown, NJ, in press.
10. Torrens, F.; Castellano, G. AIDS Destroys Immune Defences: Hypothesis. *New Front. Chem.* **2014,** *23*, 11–20.
11. Torrens-Zaragozá, F.; Castellano-Estornell, G. Emergence, Spread and Uncontrolled Ebola Outbreak. *Basic Clin. Pharmacol. Toxicol.* **2015,** *117*(2), 38-38.
12. Torrens, F.; Castellano, G. 2014 Spread/Uncontrolled Ebola Outbreak. *New Front. Chem.* **2015,** *24*, 81–91.
13. Torrens, F.; Castellano, G. Ebola Virus Disease: Questions, Ideas, Hypotheses and Models. *Pharmaceuticals* **2016,** *9*, 14–6-6.
14. Arbaizar Marínez, A. M. Book of Abstracts, V Jornada sobre Atenció Primària per a Estudiants de Medicina, València, Spain, November 27, 2018; Universitat de València–Sociedad Valenciana de Medicina Familiar y Comunitaria: València, Spain, 2018; O-1.
15. Climent, M. *El Arte de Escuchar: De Paciente a Persona*; Sociedad Valenciana de Medicina Familiar y Comunitaria: València, Spain, 2018.
16. Carrión Valero, F., Ed., Book of Abstracts, 10ª Valencia Day *Smoking Habit and Breathing: Smoking Habit and Infectious Diseases*, València, Spain, November 16, 2018; Instituto de Investigación Sanitaria: València, Spain, 2018.
17. Marco Alacid, C. Book of Abstracts, 10ª Valencia Day *Smoking Habit and Breathing: Smoking Habit and Infectious Diseases*, València, Spain, November 16, 2018; Instituto de Investigación Sanitaria: València, Spain, 2018; RT-2.
18. García Zamora, E. Book of Abstracts, 10ª Valencia Day *Smoking Habit and Breathing: Smoking Habit and Infectious Diseases*, València, Spain, November 16, 2018; Instituto de Investigación Sanitaria: València, Spain, 2018; RT-2.
19. Veile, A.; Zimmermann, H.; Lorenz, E.; Becher, H. Is Smoking a Risk Factor for Tinnitus? A Systematic Review, Meta-Analysis and Estimation of the Population Attributable Risk in Germany. *BMJ Open* **2018,** *8*, e016589–1-12.
20. Salamea Ávila, O. B. Book of Abstracts, 10ª Valencia Day *Smoking Habit and Breathing: Smoking Habit and Infectious Diseases*, València, Spain, November 16, 2018; Instituto de Investigación Sanitaria: València, Spain, 2018; RT-2.
21. Arbaizar Marínez, A. M. Book of Abstracts, 10ª Valencia Day *Smoking Habit and Breathing: Smoking Habit and Infectious Diseases*, València, Spain, November 16, 2018; Instituto de Investigación Sanitaria: València, Spain, 2018; O-1.
22. Benlloch García, E. Book of Abstracts, 10ª Valencia Day *Smoking Habit and Breathing: Smoking Habit and Infectious Diseases*, València, Spain, November 16, 2018; Instituto de Investigación Sanitaria: València, Spain, 2018; RT-3.

23. Fernández Laso, E. Book of Abstracts, 10ª Valencia Day *Smoking Habit and Breathing: Smoking Habit and Infectious Diseases*, València, Spain, November 16, 2018; Instituto de Investigación Sanitaria: València, Spain, 2018; RT-3.
24. Ruano Marco, M. Book of Abstracts, 10ª Valencia Day *Smoking Habit and Breathing: Smoking Habit and Infectious Diseases*, València, Spain, November 16, 2018; Instituto de Investigación Sanitaria: València, Spain, 2018; RT-3.
25. Mascarós Balaguer, E. Book of Abstracts, 10ª Valencia Day *Smoking Habit and Breathing: Smoking Habit and Infectious Diseases*, València, Spain, November 16, 2018; Instituto de Investigación Sanitaria: València, Spain, 2018; RT-4.
26. Sanchis Chafer, A. Book of Abstracts, 10ª Valencia Day *Smoking Habit and Breathing: Smoking Habit and Infectious Diseases*, València, Spain, November 16, 2018; Instituto de Investigación Sanitaria: València, Spain, 2018; RT-4.
27. Quiles Izquierdo, J. Book of Abstracts, 10ª Valencia Day *Smoking Habit and Breathing: Smoking Habit and Infectious Diseases*, València, Spain, November 16, 2018; Instituto de Investigación Sanitaria: València, Spain, 2018; O-2.
28. Generalitat Valenciana (GVA). *Estrategia de Prevención y Atención al Tabaquismo en la Comunitat Valenciana*; GVA: València, Spain, 2018.
29. Antón, A., Ed., *Inmunología Tumoral e Inmunoterapia del Cáncer*; Amazing Books: Zaragoza, Spain, 2018.
30. Tang, J.; Shalabi, A.; Hubbard-Lucey, V. M. Comprehensive Analysis of the Clinical Immuno-Oncology Landscape. *Ann. Oncol.* **2018,** *29*, 8491.
31. Cañete, A., Personal Communication.
32. Mukherjee, S. *The Emperor of All Maladies: A Biography of Cancer*; Public Broadcasting Service: Arlington, VA, 2011.
33. Bataller Alberola, R., Personal Communication.
34. Shah, N. D.; Ventura-Cots, M.; Zhang, C.; Zahiragic, N.; Yu, Y.; Yacoub, M. A.; Wu, P.; Wandera, A.; Vorobioff, J. D.; Thurairajah, P. H.; Tan, S.; Spreckic, S.; Siow, W.; Scheurich, C.; Sáez-Royuela, F.; Rodil, A.; Reis, D.; Ono, S. K.; Nabeshima, M. A.; Teo, E. K.; Karoney, M. J.; Castellanos Fernandez, M. I.; Farias, A. Q.; Domech, C. R.; da Costa, P. M.; Alfadhli, A.; Yang, L.; Some, F.; Kochhar, R.; Kluwe, J.; Kim, W.; Isakov, V.; Husic-Selimovic, A.; Hsiang, J. C.; George, J.; El Kassas, M.; Guridi, Z. D.; Carrilho, F. J.; Bessone, F.; Badia, E.; Alboraie, M.; Cortez-Pinto, H.; Bataller, R. Worldwide lack of early referral of patients with alcoholic liver disease: Final results of the Global Alcoholic Liver Disease Survey (GLADIS), *Hepatology* **2017,** *66*(1), 700A-701A.
35. Ndugga, N.; Lightbourne, T. G.; Javaherian, K.; Cabezas, J.; Verma, N.; Barritt, IV, A. S.; Bataller, R. Disparities between Research Attention and Burden in Liver Diseases: Implications on Uneven Advances in Pharmacological Therapies in Europe and the USA. *BMJ Open* **2017,** *7*, e013620–1-9.
36. Lackner, C.; Spindelwoeck, W.; Haybaeck, J.; Douschan, P.; Rainer, F.; Terracciano, L.; Haas, J.; Berghold, A.; Bataller, R.; Stauber, R. E. Histological Parameters and Alcohol Abstinence Determine Long-Term Prognosis in Patients with Alcoholic Liver Disease. *J. Hepatol.* **2017,** *66*, 610–618.

CHAPTER 12

Reflections on (Palaeo) Climate/Global Changes: Past/Present/Future

FRANCISCO TORRENS[1*] and GLORIA CASTELLANO[2]

[1]*Institut Universitari de Ciència Molecular, Universitat de València, Edifici d'Instituts de Paterna, P. O. Box 22085, E-46071 València, Spain*

[2]*Departamento de Ciencias Experimentales y Matemáticas, Facultad de Veterinaria y Ciencias Experimentales, Universidad Católica de Valencia San Vicente Mártir, Guillem de Castro-94, E-46001 València, Spain*

[*]*Corresponding author. E-mail: torrens@uv.es*

ABSTRACT

From the past viewpoint, to climate factors outside people's existence and, in action for millions of years, a new one has joined since capitalism emerged: burning fossil fuels. From the present point of view, the coal companies are making Europe sick: They are responsible for €22,000,000,000 in health costs in Europe. From the future viewpoint, new computers and simulators will be needed to understand environmental processes. Without knowing the past, it will not be possible to live the present or build the future. To know the past has no sense if we do not know as a species what we want for us and *Homo sapiens* in the future. To prospect species is essential to feel us recognized as humans that want to learn from ourselves and our environment. To challenge the natural evolutionary processes that transform us into absolute protagonist of our history. The collective acquisition of a planetary conscience is another important step in our history. All projects that develop have no sense if we do not organize them as species parameters. Sectors such as communications, computing, data processing and energy are increasingly inter-related. Universal knowledge characteristics are: intrinsicalness, generality, rationality, programmability, systematicness, integrality, transformativeness and unity.

12.1 INTRODUCTION

Setting the scene: some reflections on the environment, environmental problems, climate change, global change, palaeoclimate change etc., and their points of view from the past, the present, and the future. From the past point of view, to climate factors outside people's existence and, in action for million years, a new one has joined since capitalism emerged. It was about a condition generated by human action because of recent activities, for example, the burning of fossil fuels since 18th century, which contribute to the continued rise in CO_2 and CH_4 levels in the atmosphere. From the present point of view, companies owner of coal power plants are making Europe sick because they are responsible for €22,000,000,000 in health costs in Europe. From the future point of view, new computers and simulators will be needed to understand fully global environmental processes, for example, a two-quantum bit (*qubit*) molecular architecture for electron-mediated nuclear quantum simulation.

The trap of diversity was discussed as how neoliberalism fragmented the identity of the *working class*.[1] The societies of the persons without value were analyzed as Fourth Industrial Revolution, de-substantiation of capital and generalized de-valorization.[2] Tarín Sanz and Rivas Otero reviewed the working class as a possible subject of change in 21st century.[3] The growing accumulation of CO_2 in the atmosphere constitutes a large-scale experiment on the Earth's environment with unknown long-term consequences. The radiational balance between the Earth and space may be affected significantly by CO_2 accumulations, i.e. the so-called *greenhouse effect* (GHE), which may produce significant increases in the Earth's atmospheric temperature. The increase in temperature, in turn, may produce major changes in the environment, for example melting of polar ice, severe weather, and adverse effects on agriculture. Without knowing the past, it will be possible to neither live the present nor build the future. To know that the past has no sense if we do not know as a species what we want for us and the *Homo sapiens* in the future. To prospect that species is essential to feel us recognized as humans that want to learn from ourselves and our environment. To challenge the natural evolutionary processes that transform us into absolute protagonist of our history. The collective acquisition of a planetary conscience is another important step in our history. All projects that develop have no sense if we do not organize them as species parameters. Sectors such as communications, computing, data processing, and energy are increasingly inter-related. Universal knowledge characteristics follow: intrinsicalness, generality, rationality, programmability, systematicness, integrality, transformativeness, and unity.

Earlier studies were reported the periodic table of the elements (PTE)[4–6], quantum simulators[7–15], science, ethics of developing sustainability via nanosystems, devices[16], *green nanotechnology* as an approach toward environment safety[17], molecular devices, machines as hybrid organic–inorganic structures[18], quantum biting its tail, sustainable chemistry[19], quantum molecular *spintronics*, nanoscience and graphenes[20]: It was informed about cancer, its hypotheses[21], precision personalized medicine from theory to practice, cancer[22], how human immunodeficiency virus/acquired immunodeficiency syndrome (HIV/AIDS) destroy immune defenses, hypothesis[23], 2014 emergence, spread, uncontrolled Ebola outbreak[24,25], Ebola virus disease, questions, ideas, hypotheses, models[26], metaphors that made history, reflections on philosophy, science and deoxyribonucleic acid (DNA)[27], scientific integrity, ethics, science communication, psychology[28], capital versus. nature contradiction, and inclusive spreading of science.[29] In the present report, it is reviewed some reflections on climate and palaeoclimate via fauna and flora, climate on Earth, unexpected changes, the future of *H. sapiens* to 21st century, last gap, the coal companies making Europe sick, and a two-*qubit* molecular architecture for electron-mediated nuclear quantum simulation. The aim of this work is to compare the past, present, and future points of view of environmental problems. This study deals with initiating a debate by suggesting a number of questions (Q), which can arise when addressing subjects of the environment, environmental problems, climate, palaeoclimate and global changes etc., and their points of view from the past, the present, and the future of *H. sapiens* to 21st century, and providing, when possible, hypotheses (H) on the key concepts of how, without knowing the past, it will be possible to neither live the present nor build the future.

12.2 CLIMATE VIA FAUNA AND FLORA

It is enough to look around to check how the urban destroyed the natural environment.[30] Nowadays, to enjoy nature, a great number of persons must move away from the place where they live and, frequently, people make parks to preserve behind bars what natural selection (NS) took a long time to build. People humanized the landscape till such a point that now it becomes necessary to have spaces that protect what remains from the selected past. In addition, people lose biodiversity. Since the triumph of Industrial Revolution, at the end of 19th century, most inhabitants of this planet lost the ancestral knowledge of vegetation and its properties.

12.3 CLIMATE ON EARTH: UNEXPECTED CHANGES

To climate factors outside people's existence and, in action for million years, a new one has joined since capitalism emerged. It is about a condition generated by human action because of recent activities (e.g. exponential burning of fossil fuels since 18th century and meat production), which contribute to the continued rise in CO_2 and CH_4 levels in the atmosphere. With them, GHE accelerated and is contributing to global heating. Scientific proofs confirmed it by different ways. In addition, air bubbles trapped in polar ice in early Holocene, 10,000 years ago, showed that the proportion of pollutant particles (parts per million in volume, ppmv) in fossil O was high (275 ppmv), but since at the end of 19th century and with early Industrial-Revolution socialization, they began to rise exponentially till reaching in the last decades to 400 ppmv. Although humans not be responsible of Earth's cyclic changes, but they contribute negatively to rise in temperature that accelerating processes and, as a consequence, increasing energy in the environment. All those affect directly the growth of plants that are in the trophic chain of almost all animals. The light and heat that they receive are the same in neither all latitudes nor all seasons so that the growth and reproduction cycles of the species that depend on them associate with the moments of greater energy in the environment, which is usually when more humidity, light, and heat exist. Suffice the general questions to understand Earth climate variation, which present impact on how plants and animals adapt themselves to the changes, till the point that they can modify species composition in a trophic chain and, even, they contribute to their extinction. The scientific knowledge of phenomena series results essential to understand people's environment since, only passing from general to particular knowledge, people can turn into conscious beings, which is the basic way of behavior for a future humanity, more formed and informed and, of course, more critical with its actions.

12.4 FUTURE OF *HOMO SAPIENS*: TO 21ST CENTURY

Carbonell and Tristán proposed the following questions and hypotheses on the future of *H. sapiens*:

Q1. How did the different species of our genus that preceded us live?
Q2. Why does it happen with us, humans, that now live in 21st century?
Q3. For what is to revive the past good if we are not able to govern our present and plan in future?

Q4. What do we want in the future?

Q5. However, with all knowledge of human history in the case, and now, what?

Q6. What were the objectives of our work?

Q7. How should they develop in order that they have a great impact in scientific community/society?

H1. Without knowing the past, it will be possible to neither live the present nor build the future.

Q8. What do we want for us and *H. sapiens* of the planet in the future?

H2. To know that the past has no sense if unknowing as a species what we want for us/*H. sapiens* in future.

H3. Prospect species is essential to feel us recognized as humans that want to learn ourselves.

Q9. Which relationship has the prospective and retrospective that I have analyzed till then?

Q10. Who are we?

Q11. Where are we going?

Q12. What do we want for the future?

Q13. What is to be *human*?

Q14. Do we want to be thousand millions more in the planet?

Q15. Do we want that inequalities continue in the species?

Q16. Do we want to destroy the ship with our action in which we navigate?

Q17. Perhaps to know whether conscious life exists in cosmos?

Q18. In addition, how many universes do they exist?

Q19. Do we want to be changed into bionic beings?

H4. To challenge natural evolutionary processes that transform us into absolute protagonist of history.

Q20. Which objectives have we as a species?

H5. The collective acquisition of a planetary conscience is another important step in our history.

H6. All projects that develop have no sense if we do not organize them as species parameters.

Q21. What are we doing now?

Q22. What do we want from us?

Q23. How could the history of humans be interpreted?

Q24. Who does know how the history of humans could be interpreted?

Q25. Which view will the humans of our history have?

Q26. Which mutation will allow us to pass from one to other state?

Q27. Which effects will the substitution of technical and cultural for NS, etc. have?

Q28. How will the new scientists analyze the study of their past that is our present?
Q29. How will they interpret already amortized acts, objects, or information?
Q30. How will knowledge generated with so much effort influence the posthunanity?
Q31. How many questions that were made could we answer?

12.5 LAST GAP: THE COAL COMPANIES MAKING EUROPE SICK

The company owners of C power plants are responsible for €22,000,000,000 in health costs in Europe. The platform *Spain beyond Coal*, integrated by the main Spanish environmental organizations, forms part of the European alliance *Europe beyond Coal*, which has an objective of the end of C in Europe and published the report *Last Gap: the Coal Companies Making Europe Sick* (cf. Fig. 12.1). They ask the government that more subsidies to C should not be granted and that it should establish year 2025 as deadline to Spanish C thermal power plants to close, and electrics for giving up investing in their C power plants and officially seek closures as soon as possible, in order to guarantee that they must be ordered and planed. The Spanish platform launched an action campaign, because of 24th Conference of the Parties to the United Nations Framework Convention on Climate Change (COP_{24}, Katowice, Poland, 2018). Communication activities were carried out and on December 1, 2018, specific actions versus C were developed in all Spain.

12.6 TWO QUBITS FOR ELECTRON-MEDIATED NUCLEAR QUANTUM SIMULATION

A switchable interaction between pairs of highly coherent *qubits* is a crucial ingredient for the physical realization of quantum information processing. One promising route to enable quantum logic operations involves the use of nuclear spins as protected elementary units of information, qubits. Sessoli group proposed a simple way to use fast electronic spin excitations to switch the effective interaction between nuclear spin qubits and the realization of a two-qubit molecular architecture based on highly coherent vanadyl moieties to implement quantum logic operations.[31] Controlled generation of entanglement between qubits is possible via chemically tuned magnetic coupling between electronic spins, which is clearly evidenced by the splitting of the

V^{IV} hyperfine lines in the continuous-wave electron paramagnetic resonance spectrum. The system was characterized by pulsed electron paramagnetic resonance (EPR) spectroscopy, evidencing remarkably long coherence times. The experimentally derived spin Hamiltonian parameters were used to simulate the system dynamics under the sequence of pulses required to implement quantum gates in a realistic description, which includes the harmful effect of *decoherence*, which shows the possibility of using the molecular complex to implement a controlled-Z (cZ) gate and simple quantum simulations. They proposed a proof-of-principle experiment based on the simulation of the quantum tunneling of the magnetization in an $S = 1$ spin system.

FIGURE 12.1 Action in Tarragona (Spain) of the platform *Tarragona beyond Coal.*

12.7 FINAL REMARKS

From the previous discussion, the following final remarks can be drawn:

1. Without knowing the past, it will not be possible to live the present or build the future.
2. To know past has no sense if unknowing as a species what we want for present and future.

3. Prospect species essential to feel us recognized as humans that want to learn ourselves.
4. Challenging natural evolutionary processes transform us into absolute history protagonist.
5. The collective acquisition of a planetary conscience is another important step in history.
6. All projects that develop have no sense if we do not organize them as species parameters.
7. Both advanced development informatic modeling and atmospheric actual measurements exactly agree: In computer modeling, scientifics are allowed to relate the atmosphere to the seas and to other essential elements of the climatic structure of the planet, in order to plan the question of what can be the effects of the added carbon dioxide. Answers of computers increasingly corroborate direct observation.
8. During the next decades, humankind must face severe restrictions in energy production. Worldly, most part of the energy is produced on the basis of hydrocarbons combustion and the trend will be to the substitution of the use of fossil fuels with the so-called *renewable energies* (solar, eolic etc.). While other forms of energy production will not be feasible in large scale, man should be accustomed to optimize the existing processes and reduce the use of hydrocarbons.
9. Governments must encourage the public financing of research on technologies of renewable energies that finally will cause the solution that the climatic change obliges.
10. Sectors such as communications, computing, data processing, and energy are increasingly interrelated. Universal knowledge characteristics follow: intrinsicalness, generality, rationality, programmability, systematicness, integrality, transformativeness, and unity.

Future works will deal with a large global CO_2 model and its application to different landscapes.

ACKNOWLEDGMENTS

The authors thank support from Generalitat Valenciana (Project No. PROMETEO/2016/094) and Universidad Católica de Valencia *San Vicente Mártir* (Project No. 2019-217-001).

KEYWORDS

- **fauna**
- **flora**
- **palaeoclimate change**
- ***Homo sapiens***
- **coal power plant**
- **health policy**
- **sustainabiliy**

REFERENCES

1. Bernabé D. *La Trampa de la Diversidad: Cómo el Neoliberalismo Fragmentó la Identidad de la Clase Trabajadora*; Akal: Tres Cantos, Madrid, Spain, 2018.
2. Piqueras, A. *Las Sociedades de las Personas sin Valor: Cuarta Revolución Industrial, Des-substanciación del Capital, Desvalorización Generalizada*; El Viejo Topo: Vilassar de Dalt, Barcelona, Spain, 2018.
3. Tarín Sanz, A.; Rivas Otero, J. M., Eds., *La Clase Trabajadora: ¿Sujeto de Cambio en el Siglo XXI?*; Siglo XXI: Tres Cantos, Madrid, Spain, 2018.
4. Torrens, F.; Castellano, G. Reflections on the Nature of the Periodic Table of the Elements: Implications in Chemical Education. In *Synthetic Organic Chemistry*; Seijas, J. A., Vázquez Tato, M. P., Lin, S. K., Eds.; MDPI: Basel, Switherland, 2015; Vol. 18; pp 1–15.
5. Torrens, F.; Castellano, G. Nanoscience: From a Two-Dimensional to a Three-Dimensional Periodic Table of the Elements. In *Methodologies and Applications for Analytical and Physical Chemistry*; Haghi, A. K., Thomas, S., Palit, S., Main, P., Eds.; Apple Academic–CRC: Waretown, NJ, 2018; pp 3–26.
6. Torrens, F.; Castellano, G. Periodic Table. In *New Frontiers in Nanochemistry: Concepts, Theories, and Trends*; Putz, M. V., Ed.; Apple Academic–CRC: Waretown, NJ, in press.
7. Torrens, F.; Castellano, G. Ideas in the History of Nano/Miniaturization and (Quantum) Simulators: Feynman, Education and Research Reorientation in Translational Science. In *Synthetic Organic Chemistry*; Seijas, J. A., Vázquez Tato, M. P., Lin, S. K., Eds.; MDPI: Basel, Switzerland, 2015; Vol. 19; pp 1–16.
8. Torrens, F.; Castellano, G. Reflections on the Cultural History of Nanominiaturization and Quantum Simulators (Computers). In *Sensors and Molecular Recognition*; Laguarda Miró, N., Masot Peris, R., Brun Sánchez, E., Eds.; Universidad Politécnica de Valencia: València, Spain, 2015; Vol. 9; pp 1–7.
9. Torrens, F.; Castellano, G. Nanominiaturization and Quantum Computing. In *Sensors and Molecular Recognition*; Costero Nieto, A. M., Parra Álvarez, M., Gaviña Costero, P., Gil Grau, S., Eds.; Universitat de València: València, Spain, 2016; Vol. 10 pp 31-1-5.
10. Torrens, F.; Castellano, G. Nanominiaturization, Classical/Quantum Computers/Simulators, Superconductivity, and Universe. In *Methodologies and Applications for Analytical*

and Physical Chemistry; Haghi, A.K., Thomas, S., Palit, S., Main, P., Eds.; Apple Academic–CRC: Waretown, NJ, 2018; pp 27–44.

11. Torrens, F.; Castellano, G. Superconductors, Superconductivity, BCS Theory and Entangled Photons for Quantum Computing. In *Physical Chemistry for Engineering and Applied Sciences: Theoretical and Methodological Implication*; Haghi, A. K., Aguilar, C. N., Thomas, S., Praveen, K. M., Eds.; Apple Academic–CRC: Waretown, NJ, 2018; pp 379–387.
12. Torrens, F.; Castellano, G. EPR Paradox, Quantum Decoherence, Qubits, Goals and Opportunities in Quantum Simulation. In *Theoretical Models and Experimental Approaches in Physical Chemistry: Research Methodology and Practical Methods*; Haghi, A. K., Ed.; Apple Academic–CRC: Waretown, NJ, 2018; Vol. 5, pp 317–334.
13. Torrens, F.; Castellano, G. Nanomaterials, Molecular Ion Magnets, Ultrastrong and Spin–Orbit Couplings in Quantum Materials. In *Physical Chemistry for Chemists and Chemical Engineers: Multidisciplinary Research Perspectives*; Vakhrushev, A.V., Haghi, R., de Julián-Ortiz, J.V., Allahyari, E., Eds.; Apple Academic–CRC: Waretown, NJ, in press.
14. Torrens, F.; Castellano, G. Nanodevices and Organization of Single Ion Magnets and Spin Qubits. In *Chemical Science and Engineering Technology: Perspectives on Interdisciplinary Research*; Balköse, D., Ribeiro, A. C. F., Haghi, A. K., Ameta, S. C., Chakraborty, T., Eds.; Apple Academic–CRC: Waretown, NJ, in press.
15. Torrens, F.; Castellano, G. Superconductivity and Quantum Computing via Magnetic Molecules. In *New Insights in Chemical Engineering and Computational Chemistry*; Haghi, A. K., Ed.; Apple Academic–CRC: Waretown, NJ, in press.
16. Torrens, F.; Castellano, G. Developing Sustainability via Nanosystems and Devices: Science–Ethics. In *Chemical Science and Engineering Technology: Perspectives on Interdisciplinary Research*; Balköse, D., Ribeiro, A.C.F., Haghi, A.K., Ameta, S.C., Chakraborty, T., Eds.; Apple Academic–CRC: Waretown, NJ, in press.
17. Torrens, F.; Castellano, G. Green Nanotechnology: An Approach towards Environment Safety. In *Advances in Nanotechnology and the Environmental Sciences: Applications, Innovations, and Visions for the Future*; Vakhrushev, A. V.; Ameta, S. C.; Susanto, H., Haghi, A. K., Eds.; Apple Academic–CRC: Waretown, NJ, in press.
18. Torrens, F.; Castellano, G. Molecular Devices/Machines: Hybrid Organic–Inorganic Structures. In *Research Methods and Applications in Chemical and Biological Engineering*; Pourhashemi, A., Deka, S. C., Haghi, A. K., Eds.; Apple Academic–CRC: Waretown, NJ, in press.
19. Torrens, F.; Castellano, G. The Periodic Table, Quantum Biting its Tail, and Sustainable Chemistry. In *Chemical Nanoscience and Nanotechnology: New Materials and Modern Techniques*; Torrens, F., Haghi, A. K., Chakraborty, T., Eds.; Apple Academic–CRC: Waretown, NJ, in press.
20. Torrens, F.; Castellano, G. Quantum Molecular Spintronics, Nanoscience and Graphenes. In *Molecular Physical Chemistry*; Haghi, A. K., Ed.; Apple Academic–CRC: Waretown, NJ, in press.
21. Torrens, F.; Castellano, G. Cancer and Hypotheses on Cancer. In *Molecular Chemistry and Biomolecular Engineering: Integrating Theory and Research with Practice*; Pogliani, L., Torrens, F., Haghi, A. K., Eds.; Apple Academic–CRC: Waretown, NJ, in press.
22. Torrens, F.; Castellano, G. Precision Personalized Medicine from Theory to Practice: Cancer. In *Molecular Physical Chemistry*; Haghi, A. K., Ed.; Apple Academic–CRC: Waretown, NJ, in press.

23. Torrens, F.; Castellano, G. AIDS destroys immune defences: Hypothesis. *New Front. Chem.* **2014**, *23*, 11–20.
24. Torrens-Zaragozá, F.; Castellano-Estornell, G. Emergence, spread and uncontrolled Ebola outbreak. *Basic Clin. Pharmacol. Toxicol.* **2015**, *117*(2), 38-38.
25. Torrens, F.; Castellano, G. 2014 spread/uncontrolled Ebola outbreak. *New Front. Chem.* **2015**, *24*, 81–91.
26. Torrens, F.; Castellano, G. Ebola Virus Disease: Questions, Ideas, Hypotheses and Models. *Pharmaceuticals* **2016**, *9*, 14–6-6.
27. Torrens, F.; Castellano, G. Metaphors That Made History: Reflections on Philosophy/Science/DNA. In *Molecular Physical Chemistry*; Haghi, A. K., Ed.; Apple Academic–CRC: Waretown, NJ, in press.
28. Torrens, F.; Castellano, G. Scientific Integrity/Ethics: Science Communication and Psychology. In *Molecular Physical Chemistry*; Haghi, A. K., Ed., Apple Academic–CRC: Waretown, NJ, in press.
29. Torrens, F.; Castellano, G. Capital vs. Nature Contradiction and Inclusive Spreading of Science. In *Concepts, Theories, and Trends in Physical Chemistry*; Haghi, A. K., Ed., Apple Academic–CRC: Waretown, NJ, in press.
30. Carbonell, E.; Tristán, R. M. *Atapuerca: 40 Años Inmersos en el Pasado*; RBA: Barcelona, Spain, 2017.
31. Atzori, M.; Chiesa, A.; Morra, E.; Chiesa, M.; Sorace, L.; Carretta, S.; Sessoli, R. A Two-Qubit Molecular Architecture for Electron-Mediated Nuclear Quantum Simulation. *Chem. Sci.* **2018**, *9*, 6183–6192.

CHAPTER 13

Ethical Issues in Clinical Research

FRANCISCO TORRENS[1,*] and GLORIA CASTELLANO[2]

[1]*Institut Universitari de Ciència Molecular, Universitat de València, Edifici d'Instituts de Paterna, P. O. Box 22085, E-46071 València, Spain*

[2]*Departamento de Ciencias Experimentales y Matemáticas, Facultad de Veterinaria y Ciencias Experimentales, Universidad Católica de Valencia San Vicente Mártir, Guillem de Castro-94, E-46001 València, Spain*

[*]*Corresponding author. E-mail: torrens@uv.es*

ABSTRACT

About 40% of researchers are aware of misconduct but they do not report it. About 17% of authors of clinical drug trials know of fabrication in research. It is critical that *culture of research* be developed within the system, which should be based on basic fundamentals of integrity, openness, and honest work. Misconduct reasons are: prevalent in good manufacturing practice laboratories, widespread in research laboratories, less-rigorous controls in research laboratories, research laboratories are less likely to be inspected, and research laboratories have similar motives. The shift in focus to inspecting original data directly in the *electronic system* exposed an *industry-wide* issue that will require several more years of efforts to *improve*. Open peer review does not compromise process at least when referees protect their anonymity. Publishing reports do not significantly compromise referees' willingness to review recommendations or turn-around times. Younger and nonacademic scholars are more willing to accept to review and provide more positive and objective recommendations. Male referees tend to write more constructive reports during a pilot. An 8.1% of referees agree to reveal their identity in the published report.

13.1 INTRODUCTION

Setting the scene: fraud, conflict of interest, other enforcement issues and misconduct in clinical research, repairing research integrity, and the effect of publishing peer review reports on referee behavior.

Fraud appears to be widespread. Scientific fraud reappears with alarming consistency from paleontology to nanotechnology. Indirect evidence in research was obtained from statistics. Several studies found that more than 40% of surveyed researchers were aware of research misconduct. However, they did not report it. About 17% of surveyed authors of clinical drug trials reported that they personally knew of fabrication in research occurring over the previous 10 years. It results critical that a *culture of research* be developed within the system, which should be based on basic fundamentals of integrity, openness, and honest work. Misconduct reasons follow: prevalent in good manufacturing practice (GMP) laboratories, widespread in research laboratories (RLs), less-rigorous controls in RLs, RLs result less likely to be inspected, and RLs present similar motives. The shift in focus to inspecting original data directly in the *electronic system* exposed an *industry-wide* issue that will require several more years of efforts to *improve*. Open peer review does not compromise process at least when referees protect their anonymity. Publishing reports did not significantly compromise referees' willingness to review, recommendations or turn-around times. Younger and nonacademic scholars were more willing to accept to review and provided more positive and objective recommendations. Male referees tended to write more constructive reports during a pilot. About 8.1% of referees agreed to reveal their identity in the published report.

In earlier publications, it was reported the periodic table of the elements (PTE)[1–3], quantum simulators,[4–12] science, ethics of developing sustainability via nanosystems, devices,[13] *green nanotechnology* as an approach toward environment safety,[14] molecular devices, machines as hybrid organic–inorganic structures,[15] PTE, quantum biting its tail, sustainable chemistry,[16] quantum molecular *spintronics*, nanoscience and graphenes.[17] It was informed cancer, its hypotheses,[18] precision personalized medicine from theory to practice, cancer,[19] how human immunodeficiency virus/acquired immunodeficiency syndrome (HIV/AIDS) destroy immune defenses, hypothesis,[20] 2014 emergence, spread, uncontrolled Ebola outbreak,[21,22] Ebola virus disease, questions, ideas, hypotheses, models,[23] metaphors that made history, reflections on philosophy, science, deoxyribonucleic acid (DNA),[24] scientific integrity, ethics, science communication, psychology[25] scientific creativity, social stability, and attention to diversity.[26] In the present report, it is reviewed some reflections on fraud, conflict of interest, and other enforcement issues in clinical research, a concern on fraud and misconduct in clinical research, research misconduct, repairing research integrity, some metaphors that have made history in philosophy, and the effect of publishing peer review reports on referee behavior. The aim of this work is to initiate a debate by suggesting a number of questions (Q), which can arise when addressing subjects of fraud, conflict of interest, enforcement, misconduct, and open peer review. It was provided, when possible, answers (A), facts (F), and hypotheses (H) on fraud, misconduct, and open peer review.

13.2 FRAUD, CONFLICT OF INTEREST, AND ENFORCEMENT IN CLINICAL RESEARCH

Sheehan proposed the following questions, answer, facts, and hypothesis on fraud in research.[27]

- Q1. How is fraud defined in the courts?
- Q2. How does fraud frequently manifest itself in scientific research?
- Q3. How is fraud defined?
- A3. The knowing breach of the standard of good faith and fair dealing as understood in the community, involving deception or breach of trust, for money.
- Q4. How does fraud apply to conflict-of-interest questions in medical research?
- Q5. Was the bad conduct intentional?
- Q6. How to determine whether to prosecute a case?
- Q7. Does a case go to trial?
- Q8. What if I signed up for a clinical trial?
- Q9. What would I expect?
- Q10. What would I rely upon?
- Q11. What is the standard of good faith and fair dealing with respect to my family or me?
- Q12. Are community standards met?
- Q13. Who is also a treating physician?
- Q14. What is responsibility to research participants of main investigator who is treating physician?
- Q15. What did it motivate scientists to commit fraud?
- Q16. Was *good faith and fair dealing* violated in a research project?
- Q17. Did the investigator ignore warning signs?
- Q18. Did the investigator decide not to consult guidance?

Q19. Did the investigator seek advice and not follow it?
Q20. Was researcher's conduct consistent with community standard of good faith and fair dealing?
Q21. Why was the decision made to ignore or contradict the guidance?
Q22. Who do they obtain grants and use the funds to meet other departmental goals?
Q23. How extensive is the problem of fraud in medical research?
F1. (Ranstam, 2000). 40% of researchers were aware of misconduct but did not report it.[28]
F2. (Gardner, 2005). 17% of authors of clinical drug trials knew of fabrication in research.[29]
Q24. Who are caught cheating?
Q25. Who have much less power than the researchers and institutions have?
Q26. How best to protect research participants and patients?
Q27. How to protect the interests of research subjects?
Q28. What might potential jurors expect in terms of conduct on the part of a researcher?
H1. Lawyers' maxim: *The guilty fleeth where no man pursueth.*
Q29. Has anyone been told to lie or has been threatened?

13.3 FRAUD AND MISCONDUCT IN CLINICAL RESEARCH: A CONCERN

Gupta proposed facts, questions, and answers on fraud and misconduct in clinical research.[30]

F1. (Ranstam, 2000). 40% of surveyed researchers were aware of misconduct but did not report it.
F2. (Gardner, 2005). 17% of authors of clinical drug trials knew of fabrication in research.
Q1. Is fraud and misconduct the same?
A1. Fraud should have an element of deliberate action, which is not the case with misconduct.
Q2. Why does anyone commit fraud/misconduct?
A2. Reasons for fraud/misconduct in clinical research could vary from personal to professional.
Q3. Are there different types of fraud/misconduct?

A3. Fraud can be fabrication, falsification and plagiarism of data, or even deception in conduct.
Q4. Can research fraud be picked up early?
A4. Red flags during the conduct of a clinical trial should prompt the monitor to be more vigilant.
Q5. What could be the impact of fraud?
A5. Costs to sponsor in terms of additional resource for investigating fraud and cost of repeating.
Q6. How can people strengthen research misconduct and fraud detection?
A6. They should have internal control/review mechanisms for monitoring ethical/quality aspects.
Q7. Who are involved in clinical research?
Q8. What are various countries doing to manage research fraud?
A8. Despite fraud being recognized as a criminal act by all nations there are no international rules.
Q9. Can people prevent fraud from ever happening?
A9. It might not but definitely measures can be taken to reduce its incidence (*cf.* Table 13.1).[31]
Q10. How to report fraud?

TABLE 13.1 Measures that Can be Taken to Reduce Fraud Incidence to a Great Extent.

Adopt zero tolerance (all suspected misconduct must be reported and all allegations must be thoroughly and fairly investigated).
Protect whistleblowers (careful attention must be paid to the creation and dissemination of measures to protect whistleblowers).
Clarify how to report (establish clear policies, procedures, and guidelines related to misconduct and responsible conduct).
Train the mentors (researchers must be educated to pay more attention to how they work with their junior team members).
Use alternative mechanisms (institutions need continuing mechanisms to review and evaluate the research and training environment of their institution, for example, internal auditing of research records).
Model ethical behavior (institutions successfully stop cheating when they have leaders who communicate what is acceptable behavior, develop fair and appropriate procedures for handling misconduct cases, develop and promote ethical behavior, and provide clear deterrents that are communicated).

Source: Ref. [31].

Q11. How do researchers work with their junior team members?
Q12. What is acceptable behavior?
Q13. Can people build the *culture* of research?
A13. A *culture of research* should be developed based on integrity, openness, and honest work.

He provided the following conclusion (C).

C1. *Culture of research* developed in system based on integrity/openness/honest work is critical.

13.4 RESEARCH MISCONDUCT

Newton proposed questions, answers, hypotheses, and facts on research misconduct.[32]

Q1. Who does he want to continue to publish original research?
H1. Misconduct reasons: prevalent in GMP laboratories; widespread in RLs; less-rigorous controls; RLs are less likely to be inspected; RLs have similar motives.
Q2. What did it cause the spike in these infractions?
A2. Regulators learned to do data forensic auditing.
Q3. How to look for data discrepancies/ways that data could be manipulated to create outcome?
H2. Shift to inspecting original data directly in the *electronic system* will require years to *improve*.
F1. (Ranstam, 2000). 40% of researchers were aware of misconduct and they did not report it.
F2. (Gardner, 2005). 17% authors of clinical drug trials personally knew of fabrication in research.
Q4. Who have no conflict of interest in the work they perform?
Q5. How could RLs be the same as quality control (QC) laboratories?
A5. People will manipulate data when they are rewarded (or, not punished) for doing it.
H3. Challenge of research: *Publish or perish.*
Q6. What will cause the issue to be exposed?
Q7. How will the improvements in data integrity be pushed into research?
Q8. Will it come from governments, NIH, publishers, or the universities themselves?
Q9. Will it be voluntary, tied to standards or external accreditations, or codified as law?

H4. Contradictory studies: *Coffee is bad for you; it is good for you*, etc.
Q10. Is problem lack of data statistical power or data selection trying to support unsupportable H?

13.5 SOME METAPHORS THAT HAVE MADE HISTORY IN PHILOSOPHY

Strauss published an introduction to philosophy.[33] Table 13.2 gives some metaphors that have made history in philosophy.[34]

TABLE 13.2 Some Metaphors that Have Made History in Philosophy.

Metaphor
Humans are born to die.
We all die in the end.
(Cervantes). As much true, Sancho, as someday we will be dead.
(Keynes). In the long run, we are all dead.
(Louis L'Amour). The more one learns the more he understands his ignorance.
(Albert Einstein). The more I learn, the more I realize how much I don't know.
(Javier Sábada). The delimitation that supposes the beach between the ocean of the unattainable and the rationality of the land.
Value: a monetary metaphor used for intellectual or philosophical value.
Economic force: the one that is governed by the benefit.

13.6 THE EFFECT OF PUBLISHING PEER REVIEW REPORTS ON REFEREE BEHAVIOR

Squazzoni group raised a Q on effect of publishing peer review reports on referee behavior.[35]

Q1. How the practice of publishing peer review reports affects the peer review process?

They provided the following conclusions (Cs).

C1. Publishing reports did not compromise referees' willingness to review/recommendations/times.
C2. Younger/nonacademic scholars accepted to review/gave positive/objective recommendations.
C3. Male referees tended to write more constructive reports during the pilot.

C4. Only 8.1% of referees agreed to reveal their identity in the published report.

They proposed the following hypotheses (H), questions, and facts (F).

H1. Open peer review does not compromise process at least when referees protect their anonymity.

Q2. Did knowing that report would be published affect referees' willingness, recommendations, time, and tone?

Q3. (Rodríguez-Bravo, 2017; Tennant, 2017). Were innovations perceived differently by categories?

F1. More senior academic professors agreed less to review whereas younger scholars were keener.

Q4. Had the open review condition a different effect on specific subgroups of referees?

H2. The apparent decline of review invitation acceptance simply reflected a time trend.

F2. Referees without professor/Ph.D., younger/nonacademic, were actually more keen to review.

H3. The differences in recommendations could reflect a self-selection process.

F3. Referees who wrote more positive reviews were keener to reveal their identity later.

F4. Only a small minority of referees (8.1%) accepted to have their names published together.

F5. Referee status effect: Younger/nonacademic referees submitted more positive recommendations.

F6. They did not find any significant effect on turn-round time.

Q5. Did the linguistic style of reports change during the pilot?

Q6. Was the tone of the report mainly negative or positive?

Q7. Was the style used in the reports predominantly objective?

F7. They found a positive and significant interaction effect with gender.

F8. They found that younger and nonacademic referees were more objective.

H4. Open peer review does not compromise the inner workings of the peer review system.

F9. Only younger and nonacademic referees were slightly sensitive to the pilot.

F10. They were keener to accept more objective and less demanding when under open peer review.

F11. The tone of the report was less negative and subjective when referees were male and younger.
H5. Referees opting to reveal their identity is a reputational signal for future cooperation.
F12. Only 8.1% of referees agreed to reveal their identity.
H6. The veil of anonymity is the key also for open peer review.
Q8. (Fang and Casadevall, 2015). Is competition ruining science?
H7. (Casadevall, 2015; Siddhartha, 2017). Anonymity reflects need for protection from retaliation.
Q9. (Squazzoni, Bravo, and Takacs, 2013). Does incentive provision increase peer-review quality?
H8. (Heavlin, 2017; Takacs, 2013). Peer review reflects epistemic differences in evaluation standards/disciplinary traditions.
H9. Open peer review influences author behavior/publication strategies making journals attractive.
Q10. (Ross-Hellauer, 2017). What is open peer review?
H10. (Ross-Hellauer, 2017; Tennant, 2017). Open-peer-review feasibility/sustainability is context specific.
H11. (Ross-Hellauer, 2017; Tennant, 2017). Current experiments diversity reflects awareness.

13.7 FINAL REMARKS

From the present result and discussion, the following final remarks can be drawn.

1. A 40% of surveyed researchers were aware of misconduct. However, they did not report it.
2. A 17% of surveyed authors of clinical drug trials reported that they personally knew of fabrication in research, occurring over the previous 10 years.
3. It results critical that *culture of research* be developed within the system, which should be based on basic fundamentals of integrity, openness, and honest work.
4. Misconduct reasons follow: prevalent in good manufacturing practice laboratories, widespread in research laboratories, less-rigorous controls in research laboratories, research laboratories result less likely to be inspected, and research laboratories present similar motives.

5. The shift in focus to inspecting original data directly in the *electronic system* exposed an *industry-wide* issue that will require several more years of efforts to *improve*.
6. Open peer review does not compromise process at least when referees protect their anonymity.
7. Publishing reports did not significantly compromise referees' willingness to review, recommendations or turn-around times.
8. Younger and nonacademic scholars were more willing to accept to review and provided more positive and objective recommendations.
9. Male referees tended to write more constructive reports during the pilot.
10. Only 8.1% of referees agreed to reveal their identity in the published report.

ACKNOWLEDGMENTS

The authors thank support from Generalitat Valenciana (Project No. PROMETEO/2016/094) and Universidad Católica de Valencia *San Vicente Mártir* (Project No. 2019-217-001).

KEYWORDS

- **clinical research**
- **conflict of interest**
- **concern**
- **research misconduct**
- **metaphor**
- **misconduct awareness**
- **misconduct reporting**

REFERENCES

1. Torrens, F.; Castellano, G. Reflections on the Nature of the Periodic Table of the Elements: Implications in Chemical Education. In *Synthetic Organic Chemistry*; Seijas, J. A., Vázquez Tato, M. P., Lin, S. K., Eds.; MDPI: Basel, Switzerland, 2015; Vol. 18, pp 1–15.

2. Torrens, F.; Castellano, G. Nanoscience: From a Two-Dimensional to a Three-Dimensional Periodic Table of the Elements. In *Methodologies and Applications for Analytical and Physical Chemistry*; Haghi, A. K., Thomas, S., Palit, S., Main, P., Eds.; Apple Academic/CRC: Waretown, NJ, 2018; pp 3–26.
3. Torrens, F.; Castellano, G. Periodic Table. In *New Frontiers in Nanochemistry: Concepts, Theories, and Trends*; Putz, M. V., Ed.; Apple Academic/CRC: Waretown, NJ, in press.
4. Torrens, F.; Castellano, G. Ideas in the History of Nano/Miniaturization and (Quantum) Simulators: Feynman, Education and Research Reorientation in Translational Science. In *Synthetic Organic Chemistry*; Seijas, J. A., Vázquez Tato, M. P., Lin, S. K., Eds.; MDPI: Basel, Switzerland, 2015; Vol. 19, pp 1–16.
5. Torrens, F.; Castellano, G. Reflections on the Cultural History of Nanominiaturization and Quantum Simulators (Computers). In *Sensors and Molecular Recognition*; Laguarda Miró, N., Masot Peris, R., Brun Sánchez, E., Eds.; Universidad Politécnica de Valencia: València, Spain, 2015; Vol. 9, pp 1–7.
6. Torrens, F.; Castellano, G. Nanominiaturization and Quantum Computing. In *Sensors and Molecular Recognition*; Costero Nieto, A. M., Parra Álvarez, M., Gaviña Costero, P., Gil Grau, S., Eds.; Universitat de València: València, Spain, 2016; Vol. 10, pp 1–5.
7. Torrens, F.; Castellano, G. Nanominiaturization, Classical/Quantum Computers/Simulators, Superconductivity, and Universe. In *Methodologies and Applications for Analytical and Physical Chemistry*; Haghi, A. K., Thomas, S., Palit, S., Main, P., Eds.; Apple Academic/CRC: Waretown, NJ, 2018; pp 27–44.
8. Torrens, F.; Castellano, G. Superconductors, Superconductivity, BCS Theory and Entangled Photons for Quantum Computing. In *Physical Chemistry for Engineering and Applied Sciences: Theoretical and Methodological Implication*; Haghi, A. K., Aguilar, C. N., Thomas, S., Praveen, K. M., Eds.; Apple Academic/CRC: Waretown, NJ, 2018; pp 379–387.
9. Torrens, F.; Castellano, G. EPR Paradox, Quantum Decoherence, Qubits, Goals and Opportunities in Quantum Simulation. In *Theoretical Models and Experimental Approaches in Physical Chemistry: Research Methodology and Practical Methods*; Haghi, A. K., Ed.; Apple Academic/CRC: Waretown, NJ, 2018; Vol. 5, pp 317–334.
10. Torrens, F.; Castellano, G. Nanomaterials, Molecular Ion Magnets, Ultrastrong and Spin–Orbit Couplings in Quantum Materials. In *Physical Chemistry for Chemists and Chemical Engineers: Multidisciplinary Research Perspectives*; Vakhrushev, A. V., Haghi, R., de Julián-Ortiz, J. V., Allahyari, E., Eds.; Apple Academic/CRC: Waretown, NJ, in press.
11. Torrens, F.; Castellano, G. Nanodevices and Organization of Single Ion Magnets and Spin Qubits. In *Chemical Science and Engineering Technology: Perspectives on Interdisciplinary Research*; Balköse, D., Ribeiro, A. C. F., Haghi, A. K., Ameta, S. C., Chakraborty, T., Eds.; Apple Academic/CRC: Waretown, NJ, in press.
12. Torrens, F.; Castellano, G. Superconductivity and Quantum Computing via Magnetic Molecules. In *New Insights in Chemical Engineering and Computational Chemistry*; Haghi, A. K., Ed.; Apple Academic/CRC: Waretown, NJ, in press.
13. Torrens, F.; Castellano, G. Developing Sustainability via Nanosystems and Devices: Science–Ethics. In *Chemical Science and Engineering Technology: Perspectives on Interdisciplinary Research*; Balköse, D., Ribeiro, A. C. F., Haghi, A. K., Ameta, S. C., Chakraborty, T., Eds.; Apple Academic/CRC: Waretown, NJ, in press.
14. Torrens, F.; Castellano, G. Green Nanotechnology: An Approach towards Environment Safety. In *Advances in Nanotechnology and the Environmental Sciences: Applications,*

Innovations, and Visions for the Future; Vakhrushev, A. V., Ameta, S. C., Susanto, H., Haghi, A. K., Eds.; Apple Academic/CRC: Waretown, NJ, in press.

15. Torrens, F.; Castellano, G. Molecular Devices/Machines: Hybrid Organic–Inorganic Structures. In *Research Methods and Applications in Chemical and Biological Engineering*; Pourhashemi, A., Deka, S. C., Haghi, A. K., Eds.; Apple Academic/CRC: Waretown, NJ, in press.
16. Torrens, F.; Castellano, G. The Periodic Table, Quantum Biting its Tail, and Sustainable Chemistry. In *Chemical Nanoscience and Nanotechnology: New Materials and Modern Techniques*; Torrens, F., Haghi, A. K., Chakraborty, T., Eds.; Apple Academic/CRC: Waretown, NJ, in press.
17. Torrens, F.; Castellano, G. Quantum Molecular Spintronics, Nanoscience and Graphenes. In *Molecular Physical Chemistry*; Haghi, A. K., Ed.; Apple Academic/CRC: Waretown, NJ, in press.
18. Torrens, F.; Castellano, G. Cancer and Hypotheses on Cancer. In *Molecular Chemistry and Biomolecular Engineering: Integrating Theory and Research with Practice*; Pogliani, L., Torrens, F., Haghi, A. K., Eds.; Apple Academic/CRC: Waretown, NJ, in press.
19. Torrens, F.; Castellano, G. Precision Personalized Medicine from Theory to Practice: Cancer. In *Molecular Physical Chemistry*; Haghi, A. K., Ed.; Apple Academic/CRC: Waretown, NJ, in press.
20. Torrens, F.; Castellano, G. AIDS Destroys Immune Defences: Hypothesis. *New Front. Chem.* **2014,** *23*, 11–20.
21. Torrens-Zaragozá, F.; Castellano-Estornell, G. Emergence, Spread and Uncontrolled Ebola Outbreak. *Basic Clin. Pharmacol. Toxicol.* **2015,** *117* (Suppl. 2), 38–38.
22. Torrens, F.; Castellano, G. 2014 Spread/Uncontrolled Ebola Outbreak. *New Front. Chem.* **2015,** *24*, 81–91.
23. Torrens, F.; Castellano, G. Ebola Virus Disease: Questions, Ideas, Hypotheses and Models. *Pharmaceuticals* **2016,** *9*, 14–6-6.
24. Torrens, F.; Castellano, G. Metaphors That Made History: Reflections on Philosophy/Science/DNA. In *Molecular Physical Chemistry*; Haghi, A. K., Ed.; Apple Academic/CRC: Waretown, NJ, in press.
25. Torrens, F.; Castellano, G. Scientific Integrity and Ethics: Science Communication and Psychology. In *Biochemistry, Biophysics, and Molecular Chemistry: Applied Research and Interactions*; Torrens, F., Mahapatra, D. K., Haghi, A. K., Eds.; Apple Academic/CRC: Waretown, NJ, in press.
26. Torrens, F.; Castellano, G. Scientific Creativity, Social Stability and Attention to Diversity. In *Theoretical Chemistry*; Haghi, A. K., Ed.; Apple Academic/CRC: Waretown, NJ, in press.
27. Sheehan, J. G. Fraud, Conflict of Interest, and Other Enforcement Issues in Clinical Research. *Cleveland Clin. J. Med.* **2007,** *74*, S63–S67.
28. Ranstam, J.; Buyse, M.; George, S. L.; Evans, S.; Geller, N. L.; Scherrer, B.; Lesaffre, E.; Murray, G.; Edler, L.; Hutton, J. L.; Colton, T.; Lachenbruch, P. For the ISCB Subcommittee on Fraud. Fraud in Medical Research: An International Survey of Biostatisticians. *Control. Clin. Trials* **2000,** *21*, 415–427.
29. Gardner, W.; Lidz, C. W.; Hartwig, K. C. Authors' Reports About Research Integrity Problems in Clinical Trials. *Contemp. Clin. Trials* **2005,** *26*, 244–251.
30. Gupta, A. Fraud and Misconduct in Clinical Research: A Concern. *Perspect. Clin. Res.* **2013,** *4*, 144–147.

31. Titus, S. L.; Wells, J. A.; Rhoades, L. J. Repairing Research Integrity. *Nature (London)* **2008,** *453*, 980–982.
32. Newton, M. Research Misconduct. *Sci. Comput. World* **2019,** *2019* (Laboratory Information Guide), 24–26.
33. Strauss, L. *El Gusto de Jenofonte: Una Introducción a la Filosofía*; Biblioteca Nueva: Madrid, Spain, 2018.
34. Pascual, S., Ed.; *Per què Filosofia?* Neopàtria: Alzira, València, Spain, 2017.
35. Bravo, G.; Grimaldo, F.; López-Iñesta, E.; Mehmani, B.; Squazzoni, F. The Effect of Publishing Peer Review Reports on Referee Behavior in Five Scholarly Journals. *Nat. Commun.* **2019,** *10*, 322–1-8.

CHAPTER 14

Deadly Canine Distemper: A Global Multispecies Disease

V. NAVEENKUMAR[1*], K. PORTEEN[2], M. VIJAYA BHARATHI[3], and B. S. PRADEEP NAG[4]

[1]*Department of Veterinary Preventive Medicine, Madras Veterinary College, Chennai, Tamil Nadu, India*

[2]*Department of Veterinary Public Health and Epidemiology, Madras Veterinary College, Chennai, Tamil Nadu, India*

[3]*Cattle and Buffalo Breeding Unit, Post Graduate Research Institute in Animal Science, Kattupakkam, Tamil Nadu Veterinary and Animal Sciences University, Chennai, Tamil Nadu, India*

[4]*Department of Veterinary Obstetrics and Gynaecology, National Dairy Research Institute, Karnal, India*

[*]*Corresponding author. E-mail: naviviswanathan300@gmail.com*

ABSTRACT

Canine distemper (CD) is a multispecies, deadly, and contagious viral disease; mainly it affects canines, characterized by multisystemic involvement such as respiratory, nervous, gastrointestinal, immune, and cutaneous system. Canine distemper virus (CDV) belongs to Morbillivirus genus and this virus is closely related to measles, rinderpest, Peste-des-petits-ruminant virus, and recently added species namely, Cetacean morbillivirus, phocine distemper virus, and Feline morbillivirus. Due to its high mutation and host variation, CDV is considered to be a deadly and dangerous disease among all carnivores. After the epidemics of Serengeti Lion population with CD, proved the role of CD in felines too. Genetic relatedness between CDV and measles enlighten scientific community to better understand the pathogenesis of measles in human population. Speculative information on multiple sclerosis in human and CDV connection signifies the role of CD with zoonotic threat in near future. Understanding the CDV in multiple species along with its associated epidemiological factors is of prime importance to counteract the future challenges of Morbillivirus diseases.

14.1 GENERAL INFORMATION

In 1905, Henri Care described the first incidence of Canine Distemper (CD) in France (Sykes, 2013). CD is the worldwide contagious disease of domestic dogs *(Canis familiaris)* due to its high case fatality rate next to rabies (Swango, 1995). Domestic ferrets *(Mustela putorius furo)* with CDV possessing 100% case fatality rate (Davidson, 1986) whereas 50–70% of domestic dogs with CDV infection may remain subclinical and or asymptomatic carriers (Greene and Appel, 1990). CDV is closely related to human measles virus which suggested using the dog as sentinel model to study the pathogenesis of measles virus. The nondomestic animals for instance raccoon infects the domestic dogs which may cause severe interspecies transmission of CDV. Among the wildlife population, CDV may highly be influenced by catastrophic outbreaks with high mortality (Sykes, 2013).

After the CDV outbreak documentation from Baikal and Caspian seals and large felids in the Serengeti Park, dogs and wolves were identified as a potential vector for this infectious disease transmission (Beineke et al., 2015). Due to the expanding host range of CDV and its survival and maintenance within wildlife reservoir hosts considerably hampers the CDV disease control programs.

14.2 VIRUS

Morbilliviruses belong to the family Paramyxoviridae includes various pathogenic viruses, that is, measles virus, rinderpest virus, CDV, and Peste-des-petits-ruminants virus which causes devastating disease in both humans and animals (Beineke et al., 2015). CDV is a short-lived virus in the environment and can survive at a lower temperature (e.g., 48 h at 25°C and 14 days at 5°C) (Shen and Gorham, 1980; Deem et al., 2000). CDV is an enveloped virus and can also easily be destroyed by heat, drying, and disinfectants (Sykes, 2013). The phylogenetic association of phocine distemper virus (PDV) and CDV revealed that PDV emerged from the CDV several thousands of years ago by contact with terrestrial carnivores (Barret, 1999).

14.3 TRANSMISSION

CDV is a highly contagious airborne disease and the major transmission is through aerosolization of respiratory exudates containing a virus, although other body excretions and secretions (e.g., Urine) are also highly contagious to susceptible host if aerosolized (Deem et al., 2000). CDV is mainly transmitted through droplet nuclei and large size particle aerosol transmission. After aerosol infection of CDV, it is primarily replicated in lymphatic tissues of the respiratory tract and subsequently reaches the various portion of the body including the cells of lower respiratory and gastrointestinal parts, lymphoid system, urinary bladder, and the central nervous system (CNS). In general, nervous signs may be observed in the chronic form of distemper along with or without other clinical manifestations (Appel, 1987; Elia et al., 2006) CDV is also transmitted by transplacental mode in domestic dogs (Krakowka et al., 1977; Sykes, 2013) and vector-mediated transmission (Fleas) in mink (Trebbien et al., 2014). High population densities, predation, and crowding around water bodies is also a significant factor in CDV incidence among wildlife population species (Noon et al., 2003;Nagao et al., 2012)

14.4 PREVALENCE

Globally many researchers have documented the presence of CDV. In Texas (Guo et al., 1986), Santa Maria, Sal Cape Verde (Headley and Graca, 2000), Soul area, Korea (Kim et al., 2001), Warshaw, Poland (Jozwik and Frymus, 2002), Argentina (Calderon et al., 2007), Sao Paulo, Brazil (Alcalde et al., 2013), Chile (Acosta-Jamett et al., 2011; Garde et al., 2013; Acosta-Jamett et al., 2015), and India (Latha et al., 2007; Bora and Pathak, 2009; Pawar et al., 2011; Dongre et al., 2013; Ashmi et al., 2017; Buragohain et al., 2018).

14.5 DIAGNOSIS OF CDV

14.5.1 PCR

In the diagnosis of CD, reverse-transcriptase polymerase chain reaction (PCR) (RT-PCR) is routinely used and recommended by various researchers throughout the world (Frisk et al., 1999; Von Messling et al., 1999; Rzezutka and Mizak, 2002; Saito et al., 2006). Owing to the difficulties in virus isolation by cell culture, lymphocytes, and macrophages from the affected animal would help in increasing the sensitivity of CDV isolation through cell culture. In CD-infected animals for the purpose of antemortem and differentiation with other similar clinical sign diseases, PCR will be an appropriate test for diagnosis (Deem et al., 2000). The varieties of suitable biological sample usage in CDV molecular diagnosis and they stated that due to the high viral load present in the following samples, that is, urine, tonsil, conjunctival swab, and whole blood these are recommended to be used in real-time RT-PCR (Elia et al., 2006). In a descriptive study conducted by Frisk et al. (1999) and Saito et al. (2006) from the typical CD dogs CSF samples using PCR and they found negative for CD. They postulated that CDV is not necessary to be always present in CSF which may be the possibility of false negative results while using CSF samples.

In RT-PCR, the diagnosis can be made using the following samples ocular and nasal discharges, saliva, and feces (Kim et al., 2001). There are variety of diagnostic aid in CD disease that detection of antiCDV IgM antibody, Fluorescent Antibody test, and virus isolation. Due to time-consuming, lack of sensitivity, and cross-reaction with vaccinated animals sample makes it a low preference diagnostic aid among the above said diagnostic aids in the routine diagnosis of CD. RT-PCR is best available antemortem diagnostic tool in CD (Saito et al., 2006). In this diagnostic method, the only disadvantage is that the recent vaccination history also would give false positive which suggest that it's always necessary to ensure the recent vaccination history of animals while doing RT-PCR. Some authors have used urine as a sample of choice in diagnosing CD by RT-PCR (Gebara et al., 2004; Saito et al., 2006). Urine is the best sample of choice in the diagnosis of CD and it gives the diagnostic sensitivity near to CSF samples (Saito et al., 2006).

In a better way, to increase the diagnostic sensitivity the samples must be drawn from different anatomical locations of dogs consisting of blood, buffy coat, skin biopsies, urine, conjunctival swabs, and CSF (Sykes, 2013). However, the low level of RNA from clinical samples, due to RNA labile nature, degradation of viral nucleic acids during sample collection, and transport to the laboratory, the low incidence of disease may be reported by RT-PCR.

Throughout the world, the wild-virus strains are highly documented and, however, cross-neutralizing antibodies analysis revealed that it's not enough to state the change of existing vaccines in CDV (Sykes, 2013).

14.6 VACCINE-INDUCED CDV INFECTION IN WILD SPECIES

There are many reports with the incidence of vaccine-induced CDV infection in wild species, that is, Lesser Panda (Bush et al., 1976); European Mink (*Mustela lutreola*) (Sutherland-Smith et al., 1997; Ek-Kommonen et al., 2003); Manned Wolf (Thomas-Basker, 1985); Captive red pandas (Bush and Roberts, 1977); Black-footed ferrets (Carpenter et al., 1976); Gray foxes (Halbrooks et al., 1981); Kinkajous (*Potos flavus*) (Kazacos et al., 1981); South American bush dog (*Speothos venaticus*) (McInnes et al., 1992); African hunting dogs (*Lycaon Pictus*) (Durchfeld et al., 1990).

14.7 EPIDEMIOLOGICAL ATTRIBUTES

14.7.1 AGE

Vulnerability of young ones to CD was documented by several authors and they also added that due to sudden declining of MDA facilitate the infection when it's contacted with infected animals (Swango, 1989; Shell, 1990; Headley and Graca, 2000; Tarafder and Samad, 2010; Dongre et al., 2013). Some researchers found the increased susceptibility of CD between 4 and 24 months than other age groups (Patroneck et al., 1995; Ek-Kommonen et al., 1997; Headley and Graca, 2000; Martella et al., 2008; Acosta-Jamett et al., 2011; Garde et al., 2013). Active immunization may occur only when the MDA level is too low especially 3–6 months of age. Due to lack of vaccination or earlier vaccination of CDV may facilitate the chance of disease occurrence in 4–24 months age (Jozwik and Frymus, 2002; Latha et al., 2007). In contrary to this, some of the workers found that no significant association between age and the occurrence of CD (Twark and Dodds, 2000; Cattet et al., 2004). However, in endemic countries it's not easy to determine that period and age susceptibility might be due to the population community and exposure to the amount of pathogen and it is varied between time and location (Garde et al., 2013). On the other hand, the higher prevalence of CDV in adults than young ones might be due to following possible reasons, that is, increased constant endemic pressure, variation in the exposure level to different population, increase in disease exposure with age, and recent outbreak nature (Acosta-Jamett et al., 2015).

14.8 BREED

Dolichocephalic breeds are much prone to encephalitis than brachycephalic breeds (Gorham, 1995; Headley and Graca, 2000). There are many documentation of higher positivity of CD in mixed than purebred throughout the world (Alex and Dhanapalan, 1994; Patronek et al., 1995). Due to behavioral adaptation, the dogs roaming in streets may predispose the chance of exposure from infected dogs than purebreds (Patronek et al., 1995). In contrary to this, Twark and Dodds (2000) reported that there is no significant association between breed and CD occurrence.

14.9 SEX

Many researchers have documented that there is no influence of gender in the occurrence of CD (Gorham, 1966; Guo et al., 1986; Patronek et al., 1995; Headley and Graca, 2000; Twark and Dodds, 2000; Cattet et al., 2004; Eghafona et al., 2007; Latha et al., 2007). In other hand, Alex and Dhanapalan (1994) and Buragohain et al. (2018) reported the highest prevalence of CD in male animals which may be due to high male population in the city and behavioral character of male dog lead to getting more disease.

14.10 IMMUNE STATUS (VACCINATED/UNVACCINATED)

Annual vaccination in dogs to maintain protective antibody in suspected population is the recommended strategy in different parts of the world (McCaw et al., 1998; Latha et al., 2007; Jerret and Ramsey, 2001; Garde et al., 2013). The published data on maternal antibody interference with vaccination suggested that final CDV booster should be practiced around 18–20 weeks and 10 weeks of age, respectively, for raccoons and ferrets (Pare et al., 1999; Gorham, 1999). Domestic dog should be vaccinated every 3–4 week between 6th and 16th week of age whereas in colostrums deprived puppy, first dose should be given around 2 week of age followed by two booster with the interval of 3–4 weeks due to lack of maternally derived antibodies supply from dam and around 4–6 weeks of age (Appel, 1977; Deem et al., 2000). The variation in the CDV outbreaks recommended the importance of tailoring vaccination schedules required to particular species and modification of vaccination schedule needed during the epidemics of CD outbreaks.

Several reports were published in the occurrence of the CD even in vaccinated animals (Gouveia et al., 1987; Ek-Kommonen et al., 1997; Headley and Graca, 2000). The increased morbidity rate of canine distemper even in vaccinated animals alarm us to move with CDV regional-specific strain vaccines rather than prototypes (Ikeda et al., 2001).

14.11 SEASON/MONTH

In developing countries like India, there is a high chance of free-ranging dogs in rural and urban cities which make the dog population to be a sentinel to many infectious diseases to domestic as well as wildlife species (Jackman and Rowan, 2007; Ashmi et al., 2017). Though the vaccination measures were practiced in developing countries due to the higher population density of free-ranging dogs, it doesn't ensure adequate herd immunity (Jackman and Rowan, 2007). In this situation, the sudden stochastic disaster could be easier to transmit the disease to other suspected population (Latha et al., 2007; Bilxenkrone-Moller et al., 1993). Hence, an adequate understanding of infectious disease occurrence nature should be always studied continuously in all epidemiological circumstances (Garde et al., 2013).

Regarding the seasonal influence of CD epidemics, winter plays a major role in disease outbreaks (Ho and Babiuk, 1979; Alex and Dhanapalan, 1994; Gorham, 1966; Headley and Graca, 2000; Latha et al., 2007; Bora and Pathak, 2009; Tarafder and Samad, 2010). Due to virus, survival in the colder season may be favorable and also colder environment induces immunosuppression in neonates (Reeder and Kramer, 2005) and recently weaned animals which also contributed to the epidemics. Also due to the dispersal of dust and dirt more in a dry environment (winter) which makes easy transmission of the virus particle to distant places (Buragohain et al., 2018). In opposite to this higher positivity of CD during summer is also noted (Patronek et al., 1995).

14.12 THERAPY

Though it's not a curable disease in dogs, still many supportive therapies is tried to enhance the life span as empirical approach. The broad spectrum antibiotics and fluid therapy were regularly practiced. Some of the homeopathy medicines were tried in CD dogs, that is, *Conium Maculatum* also useful in reducing the post-hind paralysis and other neurological manifestations of CD (Karen, 2011). Even after using rapid diagnosis technique and high standard care given to treatment, it's not enough to say about the recovery of animals despite its lead to high mortality (Martella et al., 2008). In vitro efficacy of ribavirin and 5-ethynyl-1-β-D- ribofuranosylimidazole-4-carboxamide (EICAR) as antiviral therapy in CD also reported (Sykes, 2013). Zhao et al. (2017) conducted an analytical study on gut microbial fauna assessment in CDV-infected and noninfected Giant Panda from China and reported that CDV-infected animals gut fauna revealed the changes in gut microbial diversity with a reduced level of dominating bacteria, that is, *Escherichia* and *Clostridium*. This fauna microbial diversity identification may be useful in planning better nutrition to overcome this disease fatal nature with broad supportive nature.

14.13 CLINICAL SIGNS

Acute generalized CD infections were highly related to the respiratory and gastrointestinal systems and include conjunctivitis, pneumonia, diarrhoea (often hemorrhagic), anorexia, and severe dehydration and also they narrated the neurological manifestation of CD which may occur 1–3 weeks after recovery from acute generalized infection (Vandervelde and Cachin, 1993; Appel, 1987; Deem et al., 2000). In chronic progressive neurological dysfunction may occur in over 6 year age group of dogs mostly and also the least clinical signs of CD diseases as digital hyperkeratosis, optic neuritis, chorioretinitis, uveitis and in young dogs juvenile cellulitis, metaphyseal bone lesions, and enamel hypoplasia may be noted (Dubielzig et al., 1981; Malik et al., 1995).

In comparison with all other signs, the following signs are considered to be characteristic suggestive signs of CD, that is, myoclonus (Greene and Apple, 1998; Saito et al., 2006; Sykes, 2013) and "Chewing gum fits" (Breazile et al., 1966; Deem et al., 2000). In other hand, Garde et al. (2013) reported that "None of the clinical signs are considered to be pathognomic signs" due to the wide variation of CD nature pertaining to individual dog's immunity, type of virus infected, secondary bacterial infection involvement, and vaccination history. The enlisted influencing factor toward the clinical progression pattern of CDV in individual animals were virus strain virulence, environmental conditions, animals age, and immune status and infected species identity (Deem et al., 2000).

In domestic dogs, acute generalized CD infection would be routinely diagnosed by typical clinical signs whereas in nondomestic species suspected cases would be differentiated from rabies, feline panleukopenia, toxoplasmosis, canine parvovirus, lead poisoning, and bacterial enteritis (Deem et al., 2000). CD-infected ferrets and mink would have the suspected clinical signs like digital, nasal, and eyelid hyperkeratosis (Pearson and Gorham, 1987). The broad clinical manifestations of the CD with varying clinical signs lead to difficulty in differentiating with other respiratory diseases (Jones et al., 1997). The diagnosis of CD in acute or subacute stages can be done with typical clinical signs such as conjunctivitis, bronchitis, catarrhal pneumonia, gastroenteritis, and neurological disturbances. Some other diseases which mimic the similar signs may reduce the clear diagnosis arrival, that is, early stage of kennel cough, chronic distemper encephalitis, parainfluenza virus 2, canine coronavirus, infectious canine hepatitis virus, and canine parvovirus (Kim et al., 2001). CDV is the major immunosuppressant disease in dogs which makes host with increased vulnerability to secondary pathogen infection that lead to increased CDV associated death (Pawar et al., 2011).

Based on the virus strain, the proliferation of CDV in various brain cells was also one of the clinical sign variations reported instance Synder Hill strain mainly produce neurological signs (Sykes, 2013). In the ocular form of CDV characterized by uveitis, chorioretinitis, keratoconjunctivitis sicca (KCS), keratitis, and optic neuritis lead to blindness. The common copathogen in CDV in dogs was *Bordetella bronchiseptica* and other opportunistic pathogens were toxoplasmosis, salmonellosis, nocadiosis, and generalized demodecosis. Similarly, in mink, *Pneumocystis carinii* and in raccoon, neopsorosis were reported as a copathogenic infection with CDV (Sykes, 2013).

14.14 FELIDAE

Serological studies on CDV in cat populations of the United States revealed that 10% of cat populations had Viral Neutralizing Antibodies against CDV (Appel et al., 1974). The epidemics of CDV in 1992 from North America's large captive felids and they histopathologically confirmed CDV infection in African Lion (Panthera leo), tiger (Panthera tigris), leopard (Panthera pardus), and Jaguar (Panther onca) (Appel et al., 1994). In these epidemics, the severity pattern of CDV pathogenesis with major involvement of gastrointestinal, respiratory, and CNS systemic signs.

The CD infection with multispecies morbidity in the Serengeti ecosystem with 30% of a population of 3000 African lions and they hypothesized that "from unvaccinated dogs, CDV infection might be transmitted to spotted hyenas (*Crocuta crocuta*) which in turn to lion's CDV transmission" (Roelke-Parker et al., 1996). As per rare serological studies of CD in cat documented by Appel et al. (1974) and Ikeda et al. (2001), respectively, in 1974 at the United States and in 1982 at Japan suggested that CDV infection was associated with cats for a considerably longer time worldwide. Ikeda et al. (2001) conducted a serological study on CDV in Asian felids and revealed that cat with antibodies had likely been exposed to field strains rather than typical CDV vaccine strains. They also enlisted the risk factor associated with the prevalence of CDV antibodies in Asian felids and the factors were region, individual, and exposure to infected dogs. They additionally added the strong circumstantial evidence of typical CDV transmission from infected dogs to unaffected cats. In this analytical study, they found that there are no relationships between CDV seropositive cat populations with their physical condition except seropositive cases were likely to have an anemic issue than other cats. They also hypothesized the low pathogenic tropism of CDV in cats and the possibility of respiratory and hepatic involvement in seropositive CDV cats.

A phylogenetic and molecular characterization of CDV H gene isolates revealed that "CDV infections in felids are not only incidental or spill over of infections and also it is a part of the regular host spectrum of infectivity" (Terio and craft, 2013).

14.15 HOST RANGE

CD is a most important disease in all families of terrestrial carnivores due to its wide spectrum of host ranges, that is, Canidae, Felidae, Hyaenidae, Mustelidae, Procyonidae, Ursidae, and Viverridae (Deem et al., 2000). Wildlife spillover of CDV infection from domestic dogs prevalence reported by various researcher with epidemiological investigations strongly recommended that the "Domestic dogs with CDV infection may serve as a reservoir for free-ranging wildlife" (Alexander et al., 1995; Roelke-Parker et al., 1996; Deem et al., 2000; Beineke et al., 2015).

A phylogenetic and molecular evolutionary analysis of CDV was conducted by various researchers throughout the world (McCarthy et al., 2007; Bieringer et al., 2013; Nikolin et al., 2012; Origgi et al., 2012; Sattler et al., 2014). They hypothesized that mutation affecting the binding site of the virus (H protein) entry receptors, that is, signaling lymphocytic activation molecule (SLAM), CD 150 and nectin are the associated factors in the emergence of CDV epidemics in new host species.

14.16 NATURAL INFECTION OF CDV IN NON-DOMESTIC CANID SPECIES

There are varieties of documentation of CDV in non-domestic canines, that is, African hunting dogs (*Lycaon Pictus*) (Van Heerden et al., 1990; Alexander et al., 1993; Alexander et al., 1994; Alexander et al., 1996;); Australian dingos (*Canis dingo*) (Armstrong and Anthony, 1942); Kit foxes (*Vulpes Macrotis Macrotis*) (Armstrong and Anthony, 1942); Bat-Eared foxes (*Otocyon megalotis*) (Hofmeyer, 1956; Moehlman, 1983); Raccoon dogs (*Nyctereutes procyonoides*) (Armstrong and Anthony, 1942; Machida et al., 1993); Coyotes (*Canis latrans*) (Gese et al., 1991; Gese et al., 1997; Cypher et al., 1998); Red foxes (*Vulpes vulpes*) (Armstrong and Anthony, 1942; Little et al., 1998) and Grey foxes (*Urocyon cinereargenteus*) (Armstrong and Anthony, 1942; Hoff and Bigler, 1974).

14.17 NONCARNIVORE

CD diseases were also reported in noncarnivore's, that is, Collared Peccaries (Javelina; *Pecari tajaui*) (Appel et al., 1991); Japanese Macaques (*Macaca fuscata*) (Yoshikawa et al., 1989), and Rhesus Monkey (Sun et al., 2010).

14.18 FELIDAE

There are many reports on wild felines CDV occurrence were reported, that is, Bengal tiger (Blythe et al., 1983); Snow leopard (*Panthera uncia*) (Fix et al 1989); Siberian tiger (Gould and Fenner, 1983); Lion (Piat, 1950; Appel et al., 1994; Truyen et al., 1998); Iberian lynx (*Lynx pardinus*) (Meli et al., 2010); Eurasian lynx (*Lynx lynx*) (Origgi et al., 2012); Canadian lynx (Lynx Canadensis) and Bobcats (Lynx rufus) (Daoust et al., 2009).

14.19 MUSTELIDAE

Among the wide spectrum of CDV infection, Mustelidae are the most susceptible species than others (Deem et al., 2000). Domestic ferrets and black-footed ferrets (*Mustela nigripes*) are the highly prone species of natural CDV with increased morbidity and case fatality rate than other *Mustelidae* genus species (Bernard et al., 1984; Davidson, 1986). In *Mustelidae* family many CD reports are documented throughout the world, that is, American badgers (Taxidea taxus) (Armstrong and Anthony, 1942; Goodrich et al., 1994); Striped skunk (*Mephitis mephitis*) (Diters and Nielsen, 1978); European mink (*Mustela lutrola*); American mink (*Mustela vison*) (Montali et al., 1994; Pearson and Gorham, 1987; Sutherland-Smith et al., 1997); Eurasian badgers (*Meles meles*) (Armstrong and Anthony, 1942); European otters (*Lutra lutra*) (Scott, 1979; Giesel, 1979). In a phylogenetic analysis of various carnivores CDV infection in central Europe revealed a distinct CDV lineage in ferrets, polecats, and martens which alarmed the importance of mustelid-adapted strains (Liermann et al., 1998).

14.20 URSIDAE

Many reports of CDV infection in Ursidae were reported throughout the world, that is, American black bears (*Ursus americanus*) (Cottrell et al., 2013; Stephenson et al., 2015); Asian black bears (*Ursus tibethanus*) (Nagao et al., 2012); Polar bears (*Ursus maritimus*) (Kirk et al., 2010); Grizzly bears (*Ursus arctos horribilis*) (Philippa et al., 2004), and Marsican brown bears (*Ursus arctos marsicanus*) (Di Francesco et al., 2015). After the first incidence of CDV infection in wild black bears (Ursus americanus) and further molecular sequencing found that CDV vaccine strain (Rockborn strain) is the etiological agent for that epidemics which suggested that importance of potential virus exchange between vaccinated domestic animals and wildlife (Cottrell et al., 2013).

14.21 VIVERRIDS

Some of the incidence of CDV in Viverrids also documented in various parts of world, that is, Binturong (*Arcticis binturong*) (Chandra et al., 2000); Masked palm civet (*Paradoxurus hermaphrodites*) and Asian palm civet (*Paradoxurus hermaphrodites*) (Techangamsuwan et al., 2014) and Genet (*Genetta Genetta*) (Lopez-Pena et al., 2001).

14.22 HYANEIDAE

Alexander et al. (1995) conducted a study on free ranging spotted hyenas in the Masai Mara, Kenya and showed a high titre of CDV with clinical signs or mortality during a trend in which domestic dog CD epidemics where noted. They hypothetically documented the association of CD epidemics in hyena and domestic dog.

14.23 MARINE MAMMALS

In marine mammals, PDV was documented with high similarity lineage between CDV (Duignan et al., 2014). In Siberia, Baikal Seals (*Phoca Sibirica*) reported with CDV also documented (Butina et al., 2010)

14.24 ZOONOTIC POTENTIAL

In an experimental study, reports of asymptomatic human CDV infection signifies the CDV viral host mutation and attachment which make future threads to human (Deem et al., 2000).

14.25 CONCLUSION

There are varieties of incidence of CDV were documented in both carnivores and noncarnivores. It's thought to be restricted in dogs; recent epidemics suggested increased chance of mutilation and wide host range. These hypothesized data possessing the future infection with human species too. Considering the fact continuous monitoring on CDV with molecular characterization possesses a huge background for combating future epidemics. In a nutshell, canine distemper is a deadly disease with multi-host range of clinical manifestation which needs a systemic and thorough understanding to overcome the epidemic curve peak of CDV.

KEYWORDS

- **morbillivirus**
- **CD**
- **multiple species**
- **combat-deadly disease**
- **contagious**
- **zoonotic**
- **multisystemic involvement**

REFERENCE

1. Acosta-Jamett, G.; Chalmers, W. S. K.; Cunningham, A. A.; Cleaveland, S.; Handel, I. G. Urban Domestic Dog Populations as a Source of Canine Distemper Virus for Wild Carnivores in the Coquimbo Region of Chile. *Vet. Microbiol.* **2011,** *152* (3–4), 247–257.
2. Acosta-Jamett, G.; Surot, D.; Cortes, M.; Marambio, V.; Valenzuela, C.; Vallverdu, A.; Ward, M. P. Epidemiology of Canine Distemper and Canine Parvovirus in Domestic Dogs in Urban and Rural Areas of the Araucanía Region in Chile. *Vet. Microbiol.* **2015,** *178* (3–4), 260–264.
3. Alcalde, R.; Kogika, M. M.; Fortunato, V. A. B.; Coelho, B. M. P.; Lopes, L. R.; Paiva, P. B.; Durigon, E. L. Canine Distemper Virus: Detection of Viral RNA by Nested RT-PCR in Dogs with Clinical Diagnosis. *Braz. J. Vet. Res. Anim. Sci.* **2013,** *50* (1), 74–76.
4. Alex, P. C.; Dhanapalan, P. Distemper Encephalitis in Dogs: Incidence, Symptomathology and Electroencephalographic Findings. *J. Vet. Anim. Sci.* **1994,** *25*, 127–131.
5. Alexander, K. A.; Appel, M. J. G. African Wild Dogs (Lycaon pictus) Endangered by a Canine Distemper Epizootic among Domestic Dogs Near the Masai Mara National Reserve, Kenya. *J. Wildl. Dis.* **1994,** *30*, 481–485.
6. Alexander, K. A.; Conrad, P. A.; Gardner, L. A.; Parish, C.; Appel, M.; Levy, M. G.; Lerche, N.; Kat, P. Serologic Survey for Selected Microbial Pathogens in African Wild Dogs (Lycaon pictus) and Sympatric Domestic Dogs (Canis familiaris) in Masai Mara, Kenya. *J. Zoo Wildl. Med.* **1993,** *24*, 140–144.
7. Alexander, K. A.; Kat, P. W.; Frank, L. G.; Holekamp, K. E.; Smale, L.; House, C.; Appel, M. J. G. Evidence of Canine Distemper Virus Infection among Free Ranging Spotted Hyena (Crocuta crocuta) in the Masai Mara, Kenya. *J. Zoo Wildl. Med.* **1995,** *26*, 201–206.
8. Alexander, K. A.; Kat, P. W.; Munson, L. A.; Kalake, A.; Appel, M. J. G. Canine Distemper–Related Mortality among Wild Dogs (Lycaon pictus) in Chobe National Park, Botswana. *J. Zoo Wildl. Med.* **1996,** *27*, 426–427.
9. Appel, M. J. G. Canine Distemper Virus. In *Virus Infections of Carnivores*; Appel, M. J. G., Ed.; Elsevier Science Publishers B. V., New York: New York, 1987; pp 133–159.
10. Appel, M. J. G. Canine Distemper. In *Current Veterinary Therapy;* Kirk, R. W., Ed.; Small Animal Practice, W.B. Saunders: Philadelphia, Pennsylvania, 1977; Vol. 6, pp 1308–1313.
11. Appel, M. J. G.; Sheffy, B. E.; Percy, D. H.; Gaskin, J. M. Canine Distemper Virus in Domesticated Cats and Pigs. *Am. J. Vet. Res.* **1974,** *35*, 803–806.
12. Appel, M. J.; Reggiardo, C.; Summers, B. A.; Pearce-Kelling, S.; Mare, C. J.; Noon, T. H.; Reed, R. E.; Shively, J. N.; Orvell, C. Canine Distemper Virus Infection and Encephalitis in Javelinas (*Collared peccaries*). *Arch. Virol.* **1991,** *119*, 147–152.
13. Appel, M. J.; Yates, R. A.; Foley, G. L.; Bernstein, J. J.; Santinelli, S.; Spelman, L. H.; Miller, L. D.; Arp, L. H.; Anderson, M.; Barr, M.; Pearce-Kelling, S. Canine Distemper Epizootic in Lions, Tigers, and Leopards in North America. *J. Vet. Diagn. Investi.* **1994,** *6* (3), pp 277–288.
14. Armstrong, W. H.; Anthony, C. H. An Epizootic of Canine Distemper in a Zoological Park. *Cornell Vet.* **1942,** *32*, 286–288.
15. Ashmi, J. M.; Thangavelu, A.; Senthilkumar, T. M. A.; Manimaran, K. Molecular Characterization of Canine Distemper Virus from Tamil Nadu, India. *Indian J. Anim. Sci.* **2017,** *87* (9), 1062–1067.

16. Barrett, T. Morbillivirus Infections, with Special Emphasis on Morbilliviruses of Carnivores. *Vet. Microbiol.* **1999,** *69*, 3–13.
17. Beineke, A.; Baumgartner, W.; Wohlsein, P. Cross-Species Transmission of Canine Distemper Virus—an Update. *One Health* **2015,** *1*, 49–59.
18. Bernard, S. L.; Gorham, J. R.; Ryland, L. M. Biology and Diseases of Ferrets. In *Laboratory Animal Medicine;* Fox, J. G., Cohen, B. J., Loew, F. M., Eds.; 1984.
19. Bieringer, M.; Han, J. W.; Kendl, S.; Khosravi, M.; Plattet, P.; Schneider-Schaulies, J. Experimental Adaptation of Wild-Type Canine Distemper Virus (CDV) to the Human Entry Receptor CD150. *PloS one* **2013,** *8* (3), 57488.
20. Blixenkrone-Moeller, M.; Svansson, V.; Have, P.; Orvell, C.; Appel, M.; Pedersen, I. R.; Dietz, H. H.; Henriksen, P. Studies on Manifestations of Canine Distemper Virus Infection in an Urban Dog Population. *Vet. Microbiol.* **1993,** *37* (1–2), 163–173.
21. Blythe, L. L.; Schmitz, J. A.; Roelke, M.; Skinner, S. Chronic Encephalomyelitis Caused by Canine Distemper Virus in a Bengal Tiger. *J. Am. Vet. Med. Assoc.* **1983,** *183* (11), 1159–1162.
22. Bora, H. K.; Pathak, D. C. Incidence of Diseases Involving the Central Nervous System in Animals. *Indian Vet. J.* **2009,** *86* (5), 523–524.
23. Breazile, J. E.; Blaugh, B. S.; Nail, N. Experimental Study of Canine Distemper Myoclonus. *Am. J. Vet. Res.* **1966,** *27* (120),1375.
24. Buragohain, M.; Goswami, S.; Boro, S. K.; Boruah, K.; Saikia, K. Occurrence of Canine Distemper Virus Infection in Guwahati (Assam). *J. Entomol. Zool. Stud.* **2018,** *6* (1), 1008–1011
25. Bush, M.; Montali, R. J.; Brownstein, O.; James, A. E.; Appel, M. J. G. Vaccine-Induced Canine Distemper in a Lesser Panda. *J. Am. Vet. Med. Assoc.* **1976,** *169*, 959–960
26. Bush, M.; Roberts, M. Distemper in Captive Red Pandas. *Int. Zoo Yearb.* **1977,** *17*, 194–196
27. Butina, T. V.; Denikina, N. N.; Belikov, S. I. Canine Distemper Virus Diversity in Lake Baikal Seal (Phoca sibirica) Population. *Vet. Microbiol.* **2010,** *144* (1–2), 192–197.
28. Calderon, M. G.; Remorini, P.; Periolo, O.; Iglesias, M.; Mattion, N.; La Torre, J. Detection by RT-PCR and Genetic Characterization of Canine Distemper Virus from Vaccinated and Non-Vaccinated Dogs in Argentina. *Vet. Microbiol.* **2007,** *125* (3–4), 341–349.
29. Carpenter, J. W.; Appel, M. T. J.; Erickson, R. C.; Novilla, M. N. Fatal Vaccine-Induced Canine Distemper Virus Infection in Black-Footed Ferrets. *J. Am. Vet. Med. Assoc.* **1976,** *169*, 961–964
30. Cattet, M. R.; Duignan, P. J.; House, C. A.; St. Aubin, D.J. Antibodies to Canine Distemper and Phocine Distemper Viruses in Polar Bears from the Canadian arctic. *J. Wildl. Dis.* **2004,** *40* (2), 338–342.
31. Chandra, A. S.; Ginn, P. E.; Terrell, S. P.; Ferguson, B.; Adjiri-Awere, A.; Dennis, P.; Homer, B. L. Canine Distemper Virus Infection in Binturongs (Arctictis binturong). *J. Vet. Diagn. Invest.* **2000,** *12* (1), 88–91.
32. Cottrell, W. O.; Keel, M. K.; Brooks, J. W.; Mead, D. G.; Phillips, J. E. First Report of Clinical Disease Associated with Canine Distemper Virus Infection in a Wild Black Bear (Ursus Americana). *J. Wildl. Dis.* **2013,** *49* (4), 1024–1027.
33. Cypher, B. L.; Scrivner, J. H.; Hammer, K. L.; O'Farrell, T. P. Viral Antibodies in Coyotes from California. *J. Wildl. Dis.* **1998,** *34*, 259–264.
34. Daoust, P. Y.; McBurney, S. R.; Godson, D. L.; Van De Bildt, M. W.; Osterhaus, A. D. Canine Distemper Virus-Associated Encephalitis in Free-Living Lynx (*Lynx canadensis*) and Bobcats (*Lynx rufus*) of Eastern Canada. *J. Wildl. Dis.* **2009,** *45* (3), 611–624.

35. Davidson, M. Canine Distemper, Virus Infection in the Domestic Ferret. *Compend. Contin. Educ. Pract. Vet.* **1986,** *8,* 448–453.
36. Deem, S. L.; Spelman, L. H.; Yates, R. A.; Montali, R. J. Canine Distemper in Terrestrial Carnivores: a Review. *J. Zoo Wildl. Med.* **2000,** *31* (4), 441–452.
37. Di Francesco, C. E.; Gentile, L.; Di Pirro, V.; Ladiana, L.; Tagliabue, S.; Marsilio, F. Serologic Evidence for Selected Infectious Diseases in Marsican Brown Bears (*Ursus arctos marsicanus*) in Italy (2004–09). *J. Wildl. Dis.* **2015,** *51* (1), 209–213.
38. Diters, R. W.; Nielsen, S. W. Toxoplasmosis, Distemper, and Herpesvirus Infection in a Skunk (Mephitis mephitis). *J. Wildl. Dis.* **1978,** *14* (1), 132–136.
39. Dongre, J.; Mehta, H. K.; Maheshwari, P. Incidence of Canine Distemper Infection in and Around Mhow Region of Madhya Pradesh. *Int. J. Agri. Sci. Vet. Med.* **2013,** *1* (4), 69–71.
40. Dubielzig, R. R.; Higgins, R. J.; Krakowka, S. Lesions of the Enamel Organ of Developing Dog Teeth Following Experimental Inoculation of Gnotobiotic Puppies with Canine Distemper Virus. *Vet Pathol.* **1981,** *18* (5), 684–689.
41. Duignan, P.; Van Bressem, M. F.; Baker, J.; Barbieri, M.; Colegrove, K.; De Guise, S.; de Swart, R.; Di Guardo, G.; Dobson, A.; Duprex, W. P.; Early, G. Phocine Distemper Virus: Current Knowledge and Future Directions. *Viruses* **2014,** *6* (12), 5093–5134.
42. Durchfeld, B.; Baumgartner, W.; Herbst, W.; Brahm, R. Vaccine-Associated Canine Distemper Infection in a Litter of African Hunting Dogs (*Lycaon pictus*). *Zentralbl Veterinarmed B.* **1990,** *37* (3), 203–212.
43. Eghafona, N. O.; Jacob, J.; Yah, S. C. Evaluation of Post-Vaccination Immunity to Canine Distemper and Parvoviruses in Benin City, Nigeria. *Afr. J. Biotechnol.* **2007,** *6* (16).
44. Ek-Kommonen, C.; Sihvonen, L.; Pekkanen, K.; Rikula, U.; Nuotio, L. Outbreak of Canine Distemper in Vaccinated Dogs in Finland. *Vet Rec.* **1997,** *141* (15), 380–383.
45. Ek-Kommonena, C.; Rudbackb, E.; Anttilab, C. M.; Ahoc, M.; Huovilainena, A. Canine Distemper of Vaccine Origin in European Mink, Mustela lutreola-a Case Report. *Vet Microbiol.* **2003,** *92*, 289–293.
46. Elia, G.; Decaro, N.; Martella, V.; Cirone, F.; Lucente, M. S.; Lorusso, E.; Di Trani, L; Buonavoglia, C. Detection of Canine Distemper Virus in Dogs by Real-Time RT-PCR. *J. Virol. Methods* **2006,** *136* (1–2), 171–176.
47. Fix, A. S.; Riordan, D. P.; Hill, H. T.; Gill, M. A.; Evans, M. B. Feline Panleukopenia Virus and Subsequent Canine Distemper Virus Infection in Two Snow Leopards (*Panthera uncia*). *J. Zoo Wildl. Med.* **1989,** 273–281.
48. Frisk, A. L.; Konig, M.; Moritz, A.; Baumgartner, W. Detection of Canine Distemper Virus Nucleoprotein RNA by Reverse Transcription-PCR Using Serum, Whole Blood, and Cerebrospinal Fluid from Dogs with Distemper. *J. Clin. Microbiol.* **1999,** *37* (11), 3634–3643.
49. Garde, E.; Perez, G.; Acosta-Jamett, G.; Bronsvoort, B. M. Characteristics of a Canine Distemper Virus Outbreak in Dichato, Chile Following the February 2010 Earthquake. *Animals* **2013,** *3* (3), 843–854.
50. Gebara, C. M. S.; Wosiacki, S. R.; Negrao, F. J.; De Oliveira, D. B.; Beloni, S. N. E.; Alfieri, A. A.; Alfieri, A. F. Detection of Canine Distemper Virus Nucleoprotein Gene by RT-PCR in Urine of Dogs with Distemper Clinical Signs. *Arquivo Brasileiro de Medicina Veterinária e Zootecnia* **2004,** *56* (4), 480–487.
51. Gese, E. M.; Schultz, R. D.; Johnson, M. R.; Williams, E. S.; Crabtree, R. L.; Ruff, R. L. Serological Survey for Diseases in Free-Ranging Coyotes (*Canis latrans*) in Yellowstone National Park, Wyoming. *J. Wildl. Dis.* **1997,** *33*, 47–56.

52. Gese, E. M.; Schultz, R. D.; Rongstad, O. J.; Andersen, D. E. Prevalence of Antibodies Against Canine Parvovirus and Canine Distemper Virus in Wild Coyotes in Southeastern Colorado. *J. Wildl. Dis.* **1991,** *27,* 320–323.
53. Giesel, O. Distemper in Otters. *Berl. Muench. Tieraerztl. Wochenschr.* **1979,** *92*, 304. (In German.)
54. Goodrich, J. M.; Williams, E. S.; Buskirk, S. W. Effects of a Modified-Live Virus Canine Distemper Vaccine on Captive Badgers (*Taxidea taxus*). *J. Wildl. Dis.* **1994,** *30* (4), 492–496.
55. Gorham, J. R. The Epizootiology of Distemper. *J. Am. Vet. Med. Assoc.* **1966,** *149* (5), 610–622.
56. Gou, W.; Evermann, J. F.; Foreyt, W. J.; Knowlton, F. F.; Windberg, L. A. Distemper Virus in Coyotes: a Serological Survey. *Am. J. Vet. Res.* **1986,** *189*, 1099–1100.
57. Gould, D. H.; Fenner, W. R. Paramyxovirus-Like Nucleocapsids Associated with Encephalitis in a Captive Siberian Tiger. *J. Am. Vet. Med. Assoc.* **1983,** *183* (11), 1319–1322.
58. Gouveia, A. M. G.; Magalhães, H. H.; Ribeiro, A. L. Cinomose canina: ocorrência em animais vacinados e distribuição por faixa etária. *Arquivo Brasileiro de Medicina Veterinária e Zootecnia* **1987,** *39* (4), 539–545.
59. Greene, C. E.; Apple, M. J. Canine Distemper. In *Infectious Disease of the Dog and Cat*, 2nd ed.; Greene, C. E., Ed.; WB Saunders: Philadelphia, **1998**; pp 9–22.
60. Greene, G. E.; Appel, M. J. Canine Distemper. In *Infectious Diseases of the Dog and Cat*; Greene, C. E., Ed.; W. B. Saunders: Philadelphia, Pennsylvania, **1990;** pp 226–241.
61. Halbrooks, R. D.; Swango, L. J.; Schnurrenberger, P. R.; Mitchell, F. E.; Hill, E. P. Response of Gray Foxes to Modified Live-Virus Canine Distemper Vaccines. *J. Am. Vet. Med. Assoc.* **1981**, *179*, 1170–1174.
62. Headley, S. A.; Graça, D. L. Canine Distemper: Epidemiological Findings of 250 Cases. *Braz. J. Vet. Res. Anim. Sci.* **2000,** *37* (2).
63. Ho, C. K.; Babiuk, L. A. A New Plaque System for Canine Distemper: Characteristics of the Green Strain of Canine Distemper Virus. *Can. J. Microbiol.* **1979,** *25* (6), 680–685.
64. Hoff, G. L.; Bigler, W. J. Epizootic of Canine Distemper Virus Infection among Urban Raccoons and Gray Foxes. *J. Wildl. Dis.* **1974,** *10*, 423–428.
65. Hofmeyer, G. F. B. Two Hundred and Eighty Four Autopsies at the National Zoological Gardens, Pretoria. *J. S. Afr. Vet. Med. Assoc.* **1956,** *27*, 263–282.
66. Ikeda, Y.; Nakamura, K.; Miyazawa, T.; Chen, M. C.; Kuo, T. F.; Lin, J. A.; Mikami, T.; Kai, C.; Takahashi, E. Seroprevalence of Canine Distemper Virus in Cats. *Clin. Diagn. Lab. Immunol.* **2001,** *8* (3), 641–644.
67. Jackman, J.; Rowan, A. Free-Roaming Dogs in Developing Countries: The Benefits of Capture, Neuter, and Return Programs. In *The State of the Animals*; Salem, D. J., Rowan, A. N., Eds.; Humane Society Press: Washington, DC, **2007,** 55–78.
68. Jerrett, O.; Ramsey, I. Vaccination in Manual of Canine and Feline Infectious Diseases; Ramsey, I., Tennant, B., Eds.; British Small Animal Veterinary Association: Gloucester, **2001**; pp 41–51.
69. Jones, L.; Tenorio, E.; Gorham, J.; Yilma, T. Protective Vaccination of Ferrets Against Canine Distemper with Recombinant Pox Virus Vaccines Expressing the H or F Genes of Rinderpest Virus. *Am. J. Vet. Res.* **1997,** *58* (6), 590–593.
70. Jozwik, A.; Frymus, T. Natural Distemper in Vaccinated and Unvaccinated Dogs in Warsaw. *J. Vet. Med. Series B.* **2002,** *49* (9), 413–414.
71. Karen. Conium Maculatum—Homeopup. **2011**. http://pets-homeopathy.com/additional_remedies/conium-maculatum/(accessed Feb 27, 2019).

72. Kazacos, K. R.; Thacker, H. L.; Shivaprasad, H. L.; Burger, P. P. Vaccination-Induced Distemper in Kinkajous. *J. Am. Vet. Med. Assoc.* **1981,** *179*, 1166–1169.
73. Kim, Y. H.; Cho, K. W.; Youn, H. Y.; Yoo, H. S.; Han, H. R. Detection of Canine Distemper Virus (CDV) Through One Step RT-PCR Combined with Nested PCR. *J. Vet. Sci.* **2001,** *2* (1), 59–63.
74. Kirk, C. M.; Amstrup, S.; Swor, R.; Holcomb, D.; O'Hara, T. M. Morbillivirus and Toxoplasma Exposure and Association with Hematological Parameters for Southern Beaufort Sea Polar Bears: Potential Response to Infectious Agents in a Sentinel Species. *Ecohealth* **2010,** *7* (3), 321–331.
75. Krakowka, S.; Hoover, E. A.; Koestner, A.; Ketring, K. Experimental and Naturally Occurring Transplacental Transmission of Canine Distemper Virus. *Am. J. Vet. Res.* **1997,** *38*, 919–922.
76. Latha, D.; Geetha, M.; Ramadass, P.; Narayanan, R. B. Evaluation of ELISA Based on the Conserved and Functional Middle Region of Nucleocapsid Protein to Detect Distemper Infection in dogs. *Vet. Microbiol.* **2007,** *120* (3–4), 251–260.
77. Liermann, H.; Harder, T. C.; Lo, M.; Von Messling, V.; Baumga, W.; Moennig, V.; Haas, L. Genetic Analysis of the Central Untranslated Genome Region and the Proximal Coding Part of the F Gene of Wild-Type and Vaccine Canine Distemper Morbilliviruses. *Virus Genes* **1998,** *17* (3), 259–270.
78. Little, S. E.; Davidson, W. R.; Howerth, E. W.; Rakich, P. M.; Nettles, V. M. Diseases Diagnosed in Red Foxes from the Southwestern United States. *J. Wildl. Dis.* **1998,** *34*, 620–624.
79. Lopez-Pena, M.; Vázquez, S.; Aleman, N.; Lopez-Beceiro, A.; Munoz, F.; Pereira, J. L.; Nieto, J. M. Canine Distemper in a Genet (Gennetta Gennetta), Associated with Endogenous Lipid Pneumonia. *J. Comp. Pathol.* **2001,** *124* (2–3), 207–211.
80. Malik, R.; Dowden, M.; Davis, P. E.; Allan, G. S.; Barrs, V. R.; Canfield, P. J.; Love, D. N. Concurrent Juvenile Cellulitis and Metaphyseal Osteopathy-an Atypical Canine-Distemper Virus Syndrome. *Aust. Vet. Pract.* **1995,** *25* (2), 62–67.
81. Martella, V.; Elia, G.; Buonavoglia, C. Canine Distemper Virus. *Vet. Clin. North. Am. Small. Anim. Pract.* **2008,** *38* (4), 787–797.
82. McCarthy, A. J.; Shaw, M. A.; Goodman, S. J. Pathogen Evolution and Disease Emergence in Carnivores. *Proc. R. Soc. Biol. Sci.* **2007,** *274* (1629), 3165–3174.
83. McCaw, D. L.; Thompson, M.; Tate, D.; Bonderer, A.; Chen, Y. J. Serum Distemper Virus and Parvovirus Antibody Titers among Dogs Brought to a Veterinary Hospital for Revaccination. *J. Am. Vet. Med. Assoc.* **1998,** *213* (1), 72–75.
84. Mclnnes, E. F.; Burroughs, R. E.; Duncan, N. M. Possible Vaccine-Induced Canine Distemper in a South American Bush Dog (*Speothos venaticus*). *J. Wildl. Dis.* **1992,** *28*, 614–617.
85. Meli, M. L.; Simmler, P.; Cattori, V.; Martinez, F.; Vargas, A.; Palomares, F.; Lopez-Bao, J. V.; Simón, M. A.; López, G.; León-Vizcaino, L.; Hofmann-Lehmann, R. Importance of Canine Distemper Virus (CDV) Infection in Free-Ranging Iberian Lynxes (Lynx pardinus). *Vet. Microbiol.* **2010,** *146* (1–2), 132–137.
86. Moehlman, P. D. Socioecology of Silverbacked and Golden Jackals (*Canis misomelas* and *Canis aureus*). In *Advances in the Study of Mammalian Behavior*; Eisenberg, J. F., Kleiman, D. G., Eds.; Special Publication 7. American Society of Mammalogists: Lawrence, Kansas, **1983**; pp 423–453.

87. Montali, R. J.; Cambre, R. C.; Kenny, D.; Sutherland-Smith, M.; Appel, M. J. G. Vaccination against Canine Distemper in Exotic Carnivores: Successes and Failures. *Proc. Am. Assoc. Zoo Vet.* **1994,** 340–344.
88. Nagao, Y.; Nishio, Y.; Shiomoda, H.; Tamaru, S.; Shimojima, M.; Goto, M.; Une, Y.; Sato, A.; Ikebe, Y.; Maeda, K. An Outbreak of Canine Distemper Virus in Tigers (Panthera tigris): Possible Transmission from Wild Animals to zoo Animals. *J. Vet. Med. Sci.* **2012,** *74* (6), 699–705.
89. Nikolin, V. M.; Wibbelt, G.; Michler, F. U. F.; Wolf, P.; East, M. L. Susceptibility of Carnivore Hosts to Strains of Canine Distemper Virus from Distinct Genetic Lineages. *Vet. Microbiol.* **2012,** *156* (1–2), 45–53.
90. Noon, T. H.; Heffelfinger, J. R.; Olding, R. J.; Wesche, S. L.; Reggiardo, C. Serologic Survey for Antibodies to Canine Distemper Virus in Collared Peccary (*Tayassu tajacu*) Populations in Arizona. *J. Wildl. Dis.* **2003,** *39*, 221–223.
91. Origgi, F. C.; Plattet, P.; Sattler, U.; Robert, N.; Casaubon, J.; Mavrot, F.; Pewsner, M.; Wu, N.; Giovannini, S.; Oevermann, A.; Stoffel, M. H. Emergence of Canine Distemper Virus Strains with Modified Molecular Signature and Enhanced Neuronal Tropism Leading to High Mortality in Wild Carnivores. *Vet. Pathol.* **2012,** *49* (6), 913–929.
92. Paré, J. A.; Barker, I. K.; Crawshaw, G. J.; McEwen, S. A.; Carman, P. S.; Johnson, R. P. Humoral Response and Protection from Experimental Challenge Following Vaccination of Raccoon Pups with a Modified-Live Canine Distemper Virus Vaccine. *J. Wildl. Dis.* **1999,** *35* (3), 430–439.
93. Patronek, G. J.; Glickman, L. T.; Johnson, R.; Emerick, T. J. Canine Distemper Infection in Pet Dogs: II. A Case-Control Study of Risk Factors During a Suspected Outbreak in Indiana. *J. Am. Anim. Hosp. Assoc.* **1995,** *31* (3), 230–235.
94. Pawar, R. M.; Raj, G. D.; Gopinath, V. P.; Ashok, A.; Raja, A. Isolation and Molecular Characterization of Canine Distemper Virus from India. *Trop. Anim. Health Prod.* **2011,** *43* (8), 1617–1622.
95. Pearson, R. C.; Gorham, J. R. Canine Distemper Virus. *Virus Infections of Carnivores.* Elsevier Science Publishers BV: New York. **1987,** 371–378.
96. Philippa, J. D. W.; Martina, B. E. E.; Kuiken, T.; Van de Bildt, M. W. G.; Osterhaus, A. D. M. E.; Leighton, F. A.; Daoust, P. Y.; Nielsen, O.; Pagliarulo, M.; Schwantje, H.; Shury, T. Antibodies to Selected Pathogens in Free-Ranging Terrestrial Carnivores and Marine Mammals in Canada. *Vet Rec.* **2004,** *155* (5), 135–140.
97. Piat, B. L. Susceptibility of Young Lions to Dog Distemper. *Bull. Serv. Elev. In Dustr. Anim. AOF* **1950,** *3*.
98. Reeder, D. M.; Kramer, K. M. Stress in Free-Ranging Mammals: Integrating Physiology, Ecology, and Natural History. *J. Mammal.* **2005,** *86* (2), 225–235.
99. Roelke-Parker, M. E.; Munson, L.; Packer, C.; Kock, R.; Gleaveland, S.; Carpenter, M.; O'Brien, S. J.; Pospischil, A.; Hofmann-Lehmann, R.; Lutz, H.; Mwamengele, G. L. M.; Mgasa, M. N.; Machange, G. A.; Summers, B.; Appel, M. J. G. A Canine Distemper Virus Epidemic in Serengeti Lions (*Panthera leo*). *Nature* **1996,** *379*, 441–445.
100. Rzeżutka, A.; Mizak, B. Application of N-PCR for Diagnosis of Distemper in Dogs and Fur Animals. *Vet. Microbiol.* **2002,** *88* (1), 95–103.
101. Saito, T. B.; Alfieri, A. A.; Wosiacki, S. R.; Negrao, F. J.; Morais, H. S. A.; Alfieri, A. F. Detection of Canine Distemper Virus by Reverse Transcriptase-Polymerase Chain Reaction in the Urine of Dogs with Clinical Signs of Distemper Encephalitis. *Res. Vet. Sci.* **2006,** *80* (1), 116–119.

102. Sattler, U.; Khosravi, M.; Avila, M.; Pilo, P.; Langedijk, J. P.; Ader-Ebert, N.; Alves, L. A.; Plattet, P.; Origgi, F. C. Identification of Amino Acid Substitutions with Compensational Effects in the Attachment Protein of Canine Distemper Virus. *J. Virol.* **2014,** *88* (14), 8057–8064.
103. Scott, W. A. Use of Vaccines in Exotic Species. *Vet. Rec.* **1979,** *104*, 199.
104. Shell, L. D. Canine Distemper. Compendium Continuing Education. *J. Small Anim. Pract.* **1990,** *12* (2), 173–179.
105. Shen, D. T.; Gorham, J. R.; Survival of Pathogenic Distemper Virus at 5°C and 25°C. *Vet. Med. Small Anim. Clin.* **1980,** *75* (1), 69–72.
106. Stephenson, N.; Higley, J. M.; Sajecki, J. L.; Chomel, B. B.; Brown, R. N.; Foley, J. E. Demographic Characteristics and Infectious Diseases of a Population of American Black Bears in Humboldt County, California. *Vector-Borne Zoonotic Dis*. **2015,** *15* (2), 116–123.
107. Sun, Z.; Li, A.; Ye, H.; Shi, Y.; Hu, Z.; Zeng, L. Natural Infection with Canine Distemper Virus in Hand-Feeding Rhesus Monkeys in China. *Vet. Microbiol.* **2010,** *141,* 374–378.
108. Sutherland-Smith, M. R.; Rideout, B. A.; Mikolon, A. B.; Appel, M. J. G.; Morris, P. J.; Shima, A. L.; Janssen, D. J. Vaccine-Induced Canine Distemper in European Mink, Mustela lutreola. *J. Zoo Wildl. Med.* **1997,** *28*, 312–318.
109. Swango, L. J. Canine Viral Diseases. In *Textbook of Veterinary Internal Medicine: Diseases of the Dog and Cat;* Ettinger, S. J., Feldman, E. C., Eds.; W. B. Saunders Co.: Philadelphia, Pennsylvania, **1995**; pp 398–409.
110. Swango, L. J. Canine Viral Diseases. In *Textbook of Veterinary Internal Medicine,* 3rd ed.; Ettinger, S. J., Eds.; W. B. Saunders: Philadelphia, **1989**; pp 301–303.
111. Sykes, J. E. Canine Distemper Virus Infection in Chapter 15. In *Canine and Feline Infectious Diseases-E-BOOK*. Elsevier Health Sciences, **2013**; pp 152–165.
112. Tarafder, M.; Samad, M. A. Prevalence of Clinical Diseases of Pet Dogs and Risk Perception of Zoonotic Infection by Dog Owners in Bangladesh. *Bangl. J. Vet. Med.* **2010,** *8* (2), 163–174.
113. Techangamsuwan, S.; Banlunara, W.; Radtanakatikanon, A.; Sommanustweechai, A.; Siriaroonrat, B.; Lombardini, E. D.; Rungsipipat, A. Pathologic and Molecular Virologic Characterization of a Canine Distemper Outbreak in Farmed Civets. *Vet. Pathol.* **2015,** *52* (4), 724–731.
114. Terio, K. A.; Craft, M. E. Canine Distemper Virus (CDV) in Another Big Cat: Should CDV be Renamed Carnivore Distemper Virus? *mBio*. **2013,** *4* (5), 00702–007013.
115. Thomas-Baker, B. Vaccination-Induced Distemper in Maned Wolves, Vaccination-Induced Corneal Opacity in a Maned Wolf. *Proc. Am. Assoc. Zoo Vet.* **1985,** *53.*
116. Trebbien, R.; Chriel, M.; Struve, T.; Hjulsager, C. K.; Larsen, G.; Larsen, L. E. Wildlife Reservoirs of Canine Distemper Virus Resulted in a Major Outbreak in Danish Farmed Mink (Neovison vison). *PLOS ONE* **2014,** *9* (1), 85598.
117. Truyen, U.; Muller, T.; Heidrich, R.; Tackmann, K.; Carmichael, L. E. Survey on Viral Pathogens in Wild Red Foxes (Vulpes vulpes) in Germany with Emphasis on Parvoviruses and Analysis of a DNA Sequence from a Red Fox Parvovirus. *Epidemiol. Infect.* **1988,** *121* (2), 433–440.
118. Twark, L.; Dodds, W. J. Clinical use of Serum Parvovirus and Distemper Virus Antibody Titers for Determining Revaccination Strategies in Healthy Dogs. *J. Am. Vet. Med. Assoc.* **2000,** *217* (7), 1021–1024.

119. Van Heerden, J.; Bainbridge, N.; Burroughs, R. E. J.; Kriek, N. P. J. Distemper-Like Disease and Encephalitozoonosis in Wild Dogs (*Lycaon pictus*). *J. Wildl. Dis.* **1990,** *25*, 70–75.
120. Vandevelde, M.; Cachin, M. The Neurologic Form of Canine Distemper. In *Kirk's Current Veterinary Therapy. XI. Small Animal Practice*; Bonagura, J. D., Kirk, R. W., Eds.; WB Saunders: Philadelphia, **1993**; pp 1003–1007.
121. Von Messling, V.; Harder, T. C.; Moennig, V.; Rautenberg, P.; Nolte, I.; Haas, L. Rapid and Sensitive Detection of Immunoglobulin M (IgM) and IgG Antibodies against Canine Distemper Virus by a New Recombinant Nucleocapsid Protein-Based Enzyme-Linked Immunosorbent Assay. *J. Clin. Microbiol.* **1999,** *37* (4), 1049–1056.
122. Yoshikawa, Y.; Ochikubo, F.; Matsubara, Y.; Tsuruoka, H.; Ishii, M.; Shirota, K.; Nomura, Y.; Sugiyama, M.; Yamanouchi, K. Natural Infection with Canine Distemper Virus in a Japanese Monkey (*Macaca fuscata*). *Vet. Microbiol.* **1989,** *20*, 193–205.
123. Zhao, N.; Li, M.; Luo, J.; Wang, S.; Liu, S.; Wang, S.; Lyu, W.; Chen, L.; Su, W.; Ding, H.; He, H. Impacts of Canine Distemper Virus Infection on the Giant Panda Population from the Perspective of Gut Microbiota. *Sci. Rep.* **2017,** *7*, 39954.

CHAPTER 15

Epigenomics, Epithelial Plasticity, Clinical Genetics, and Rare Diseases

FRANCISCO TORRENS[1*] and GLORIA CASTELLANO[2]

[1]*Institut Universitari de Ciència Molecular, Universitat de València, Edifici d'Instituts de Paterna, P. O. Box 22085, E-46071 València, Spain*

[2]*Departamento de Ciencias Experimentales y Matemáticas, Facultad de Veterinaria y Ciencias Experimentales, Universidad Católica de Valencia San Vicente Mártir, Guillem de Castro-94, E-46001 València, Spain*

**Corresponding author. E-mail: torrens@uv.es*

ABSTRACT

It is important to distinguish traditional from *epigenetic ages*. Applications of epigenomics in cancer are: makers diagnostic, of prognosis, of response, and therapeutic targets. Genetics and epigenetics: People cannot change genetics but epigenetics by healthy food and physical exercise. A rotation of 1 month is proposed for the internal resident physician in family medicine. The weird is significant. Research in the most rare diseases has the capacity of benefiting all people. Together, people are stronger—patients–physicians–researchers–science journalists. Systemic-lupus-erythematosus aetiology is multifactorial and includes contributions from the environment, stochastic factors, and genetic issues.

15.1 INTRODUCTION

Setting the scene: what epigenomics can do for people, epithelial plasticity in health and disease, clinical genetics (CG) and rare diseases (RDs) for primary care (PC), Societat Valenciana de Medicina Familiar i Comunitària (SOVAMFIC) Working Group, research in PC, RDs in hospital, cystic fibrosis (CF) as a paradigm of respiratory RDs (RRDs), past, present, and future.

It is important to distinguish traditional from *epigenetic ages*. Applications of epigenomics in cancer follow: diagnostic makers, makers of prognosis, makers of response and therapeutic targets. Genetics and epigenetics: People cannot change genetics but epigenetics *via* healthy food and physical exercise. A rotation of 1 month is proposed for the internal resident physician in family medicine. The weird is significant. Research in the most RDs has the capacity of benefiting all people. Together, people are stronger—patients–physicians–researchers–science journalists. The aetiology of systemic lupus erythematosus (SLE) is multifactorial and includes contributions from the environment, stochastic factors, and genetic issues.

In earlier publications it was informed the modeling of complex multicellular systems and tumor–immune cells competition,[1] information theoretic entropy for molecular classification of oxadiazolamines as potential therapeutic agents,[2] molecular classification of 5-amino-2-aroylquinolines and 4-aroyl-6,7,8-trimethoxyquinolines as highly potent tubulin polymerization inhibitors,[3] polyphenolic phytochemicals in cancer prevention, therapy, bioavailability vs. bioefficacy,[4] molecular classification of antitubulin agents with indole ring binding at colchicine-binding site,[5] molecular classification of 2-phenylindole-3-carbaldehydes as potential antimitotic agents in human breast cancer cells,[6] cancer, its hypotheses,[7] precision personalized medicine (PPM) from theory to practice and cancer.[8] It was reported how human immunodeficiency virus/acquired immunodeficiency syndrome (HIV/AIDS) destroy immune defenses, hypothesis,[9] 2014 emergence, spread, uncontrolled Ebola outbreak,[10,11] Ebola virus disease, questions, ideas, hypotheses, models,[12] clinical translational research, cancer, diabetes, and cardiovascular disease (CVD).[13] The present report reviews what epigenomics can do for people, epithelial plasticity in health and disease, CG and RDs for PC, SOVAMFIC Working Group, research in PC, RDs in hospital, CF as a paradigm of RRDs, past, present, and future. The aim of this work is to initiate a debate by suggesting a number of questions (Q), which can arise when addressing subjects of epigenomics, epithelial plasticity, CG, RDs, and PC, and providing, when possible, answers (A) and hypotheses (H).

15.2 WHAT CAN EPIGENOMICS DO FOR PEOPLE?

Sandoval del Amor proposed questions/answers/H on what epigenomics can do for people.[14]

Q1. What can epigenomics do for people?
Q2. Do genes determine people's future?
A2. Only that of kings.
H1. *Epigenetics and reinventing Lamarck.*[15]
Q3. In which animal does its social organization depend on genetics?
A3. Bees.
H2. Traditional and *epigenetic ages.*
Q4. *Deoxyribonucleic acid* (DNA) *methylation profiling in cancer discordant identical twins.*[16]
Q5. Which is the meaning of life?
Q6. What has to do epigenetics and basilisk?
A6. In basilisk, sex depends on temperature.
H3. Applications in cancer: makers diagnostic, of prognosis, of response; therapeutic targets.
Q7. If people catch this murderer, 6,000,000 deaths per year will be avoided.
A7. Tobacco. Ingredients: ^{210}Pu, etc.
Q8. Is there a marker to distinguish two groups of patients?
Q9. Better theory: either nothing (Cx) + chemotherapy (CT) or CT?
Q10. In the next future in cancer?
A10. Liquid biopsy: Circulating (c) DNA and tumor cells.

He provided the following conclusion (C).

C1. Genetics and epigenetics: People cannot change genetics but epigenetics by healthy food/physical exercise.
Q11. What does it activate people's differences?
Q12. Should people be careful of the names?
Q13. Cosmetics?
Q14. Pharmacy?
Q15. What does it cure a disease?
A15. The National Health System.

15.3 EPITHELIAL PLASTICITY IN HEALTH AND DISEASE

Epithelial homeostasis is crucial to maintain tissue architecture and it needs to be tightly regulated in the adult. By contrast, embryonic cells show a high degree of epithelial plasticity required for proper morphogenesis and, in particular, for the implementation of massive cell movements that occur during gastrulation, neural crest delamination, etc. Nieto group was interested in the analysis of cell movements, plasticity, and epithelial to mesenchymal transitions (EMTs), and found that the reactivation of developmental EMT-like programs in adult cells leads to several pathologies (e.g., tumor progression, organ degeneration).[17] While the epithelial and mesenchymal cells are considered as extreme phenotypes, intermediate EMT states exist. Under the circumstances, cells depict a hybrid phenotype expressing epithelial and mesenchymal markers, and from which they reverse to the original state or move to a more mesenchymal phenotype. Hybrid transitory states favor coordinated cell migration or wound healing, but they enable the formation of clusters of migratory cancer (circulating tumor) cells (CTCs) with increased metastatic potential. However, in contrast to cancer, the intermediate phenotype holds promise for antifibrotic therapeutic approaches, as inhibiting EMT attenuates established fibrosis. They discussed different scenarios in which the intermediate phenotype is observed in development and disease, and referred to a developmental EMT that they found to be crucial for heart laterality and morphogenesis in vertebrates.

15.4 DAY CLINICAL GENETICS AND RARE DISEASES FOR PRIMARY CARE

Navarro et al. organized Day on *Clinical Genetics and Rare Diseases for Primary Care.* [18]

Soria Olivas proposed H on artificial intelligence (AI) opportunities in research in PC.[19]

- H1. (Agrawal et al., 2018). *Prediction Machines: The Simple Economics of* AI.[20]
- H2. Problems: classification, prediction, clustering, selection of characteristics, detection of outliers.
- H3. Algorithm for reinforcement learning.
- H4. Generative Adversarial Networks (GAN), 2014.
- H5. Google Duplex: An AI system for accomplishing real-world tasks over the phone.
- H6. (Topol, 2019). *High-performance medicine: The convergence of human and* AI.[21]
- H7. (Xia, 2019). *Evaluation and accurate diagnoses of pediatric diseases using* AI.[22]

15.5 ROUND TABLE 1: SOVAMFIC WORKING GROUP

Cuenca Valero proposed questions and H on what a family doctor should know of genetics.[23]

Q1. What should a family doctor know of genetics?
H1. Change of paradigm. Before: first to get pregnant; now: first genetic analysis.
Q2. How far to go?
H2. Limitation: To know how far to go.

She provided the following conclusion (C).

C1. We propose a rotation of 1 month for the internal resident physician in family medicine.

Ejarque Doménech proposed hypotheses/Q on *Guide of Preconception Advice from* PC.[24]

H3. (Ejarque Doménech, 2018). *Guide of Preconception Advice from* PC.[25]
H4. (Ejarque et al., 2007). *Family physicians and responsibilities in* CG.[26]
Q3. How is the process of genetic advice carried out?

Doménech Casasús raised a question on web pages of CG for the family physician.[27]

Q4. Need help?

15.6 ROUND TABLE 2: RESEARCH IN PC

Fernández-Nohales proposed questions and answer on the management of health research.[28]

Q1. What are institutes for health research?
Q2. How can one research?
A2. Promotion: outside (clinical assays, health services, collaborations); inside (research projects).

Dasí proposed questions/A/H on PC research and α-1 antitrypsin (AAT) deficiency.[29]

Q3. Can one research in PC of health?
A3. Yes, one can research in PC of health, for example, AAT deficiency.
Q4. Why to research in PC of health?
H1. (REDIAPP, 2012). *Report 2007–2012*.[30]

H2. RARE × RARE = RARE.[2]

Q5. Who do I ask?

H3. (Lara, 2017). ATT: *Patients* PC-*diagnosed with chronic obstructive pulmonary disease.*[31]

15.7 ROUND TABLE 3: RDs IN HOSPITAL

Dasí proposed the following questions, answer, and hypothesis on what RDs are.[32]

Q1. However, what are RDs?

H1. I have been diagnosed with an unprofitable disease, I have no salvation.

Q2. RDs: More common than you think?

A2. US: fewer than 200,000; EU: fewer than 1:2000, for example, Einstein (Asperger syndrome).

H2. Problems: infradiagnosis and phenotypic variability.

Q3. Why to research in RDs *is so difficult*?

A3. Patients are *rare*.

H3. However, RDs have contributed to *metabolic discoveries*, for example, drug statin.

Q4. In addition, what to say about *phenotypic variability*?

A4. For example, CF transmembrane conductance regulator (CFTR) gene.

He provided the following conclusion (C).

C1. The weird is important.

C2. Research in the most RDs has the capacity of benefiting all people.

C3. Together, people are stronger.

C4. Patients–physicians–researchers–science journalists.

Sancho proposed the following hypotheses on amyotrophic lateral sclerosis (ALS).[33]

H4. Case: Stephen Hawking.

H5. (Ministry of Health and Consumption, 2007). *Guide for* ALS *in Spain.*[34]

H6. (Sancho, 2010). *Noninvasive respiratory muscle aids* ALS *percutaneous endoscopic gastrostomy.*[35]

H7. (Sancho, 2019). *Guía Manejo Problemas Respiratorios de Esclerosis Lateral Amiotrófica.*[36]

Trenor proposed the following hypotheses and questions on SLE.[37]

H8. (Kelly, 2009). *Getting susceptibility to* SLE: *Fine mapping and genome-wide association*.[38]
Q5. (Boumpas, 2013). Diagnostic criteria for SLE: Has the time come?
H9. (Bertsias et al., 2013). *Diagnostic criteria for* SLE.[39]
H10. (Mankad and Gabriel, 2014). *Rheumatoid arthritis* (RA): *Treating* CVD *risk in* RA.[40]
Q6. (Deane and El-Gabalawy, 2014). At what site does autoimmunity occur?
H11. (Deane and El-Gabalawy, 2014). *Pathogenesis and prevention of rheumatic disease*.[41]
H12. (Muskardin and Niewold, 2018). *Type-I interferon in rheumatic diseases*.[42]
H13. (Eyre et al., 2017). *The genetics revolution in rheumatology*.[43]
Q7. (Pisetsky, 2017). Antinuclear antibody testing: Misunderstood or misbegotten?
H14. (Pisetsky, 2017). *Antinuclear antibody testing*.[44]
H15. (Schneider and Liang, 2015). *Connective tissue diseases:* SLE *classification*.[45]
H16. (Jorge et al., 2018). *Hydroxychloroquine retinopathy*.[46]

She provided the following conclusion (C).

C5. SLE aetiology is multifactorial and includes: environment, stochastic factors, genetic issues.

15.8 CF AS A PARADIGM OF RRDS: PAST, PRESENT, AND FUTURE

Escribano proposed Q/A/H on CF as a paradigm of RRDs, past, present, and future.[47]

Q1. Which has the experience or training been, in all this time, in CF?
H1. (Warwick, 2013). CF*: A Geriatric Problem*.[48]
H2. (17th century). Haunted boy: *Poor boy that on kissing his forehead it tastes like salt, he will soon die*.
Q2. What has all this transformation meant?
Q3. Which factors have influenced on survival?
Q4. Which have the fundamental pillars been in this disease?
A4. Pillars: eradication/control of *Pseudomonas aeruginosa*, good nutrition, and physical exercise.

15.9 FINAL REMARKS

From the present results and discussion, the following final remarks can be drawn.

1. It is important to distinguish traditional from *epigenetic ages*.
2. Applications of epigenomics in cancer follow: diagnostic makers, makers of prognosis, makers of response, and therapeutic targets.
3. Genetics and epigenetics: People cannot change genetics but epigenetics via healthy food and physical exercise.
4. A rotation of 1 month is proposed for the internal resident physician in family medicine.
5. The weird is important.
6. Research in the most RDs has the capacity of benefiting all people.
7. Together, people are stronger.
8. Patients–physicians–researchers–science journalists.
9. The aetiology of SLE is multifactorial and includes contributions from the environment, stochastic factors, and genetic issues.

ACKNOWLEDGMENTS

The authors thank support from Generalitat Valenciana (Project No. PROMETEO/2016/094) and Universidad Católica de Valencia *San Vicente Mártir* (Project No. UCV.PRO.17-18.AIV.03).

KEYWORDS

- **people**
- **health**
- **disease**
- **medical society**
- **family medicine**
- **community medicine**
- **primary care research**

REFERENCES

1. Torrens, F; Castellano, G. Modelling of Complex Multicellular Systems: Tumour–Immune Cells Competition. *Chem. Central J.* **2009,** *3* (Suppl. I), 75-1-1.
2. Torrens, F.; Castellano, G. Information Theoretic Entropy for Molecular Classification: Oxadiazolamines as Potential Therapeutic Agents. *Curr. Comput. Aided Drug Des.* **2013,** *9*, 241–253.
3. Torrens, F.; Castellano, G. Molecular Classification of 5-amino-2-aroylquinolines and 4-aroyl-6,7,8-Trimethoxyquinolines as Highly Potent Tubulin Polymerization Inhibitors. *Int. J. Chemoinf. Chem. Eng.* **2013,** *3* (2), 1–26.
4. Estrela, J. M.; Mena, S.; Obrador, E.; Benlloch, M.; Castellano, G.; Salvador, R.; Dellinger, R. W. Polyphenolic Phytochemicals in Cancer Prevention and Therapy: Bioavailability Versus Bioefficacy. *J. Med. Chem.* **2017,** *60*, 9413–9436.
5. Torrens, F.; Castellano, G. Molecular Classification of Antitubulin Agents with Indole Ring Binding at Colchicine-Binding Site. In *Molecular Insight of Drug Design*; Parikesit, A. A., Ed.; IntechOpen: Vienna, 2018; pp 47–67.
6. Torrens, F.; Castellano, G. Molecular Classification of 2-Phenylindole-3-carbaldehydes as Potential Antimitotic Agents in Human Breast Cancer Cells. In *Theoretical Models and Experimental Approaches in Physical Chemistry: Research Methodology and Practical Methods*; Haghi, A. K., Thomas, S., Praveen, K. M., Pai, A. R., Eds.; Apple Academic/CRC: Waretown, NJ, in press.
7. Torrens, F.; Castellano, G. Cancer and Hypotheses on Cancer. In *Molecular Chemistry and Biomolecular Engineering: Integrating Theory and Research with Practice*; Pogliani, L., Torrens, F., Haghi, A. K., Eds.; Apple Academic/CRC: Waretown, NJ, in press.
8. Torrens, F.; Castellano, G. Precision Personalized Medicine from Theory to Practice: Cancer. In *Green Chemistry and Biodiversity: Principles, Techniques, and Correlations*; Aguilar, C. N., Ameta, S. C., Haghi, A. K., Eds.; Apple Academic/CRC: Waretown. NJ, in press.
9. Torrens, F.; Castellano, G. AIDS Destroys Immune Defences: Hypothesis. *New Front. Chem.* **2014,** *23*, 11–20.
10. Torrens-Zaragozá, F.; Castellano-Estornell, G. Emergence, Spread and Uncontrolled Ebola Outbreak. *Basic Clin. Pharmacol. Toxicol.* **2015,** *117* (Suppl. 2) 38–38.
11. Torrens, F.; Castellano, G. 2014 Spread/Uncontrolled Ebola Outbreak. *New Front. Chem.* 2015, *24*, 81–91.
12. Torrens, F.; Castellano, G. Ebola Virus Disease: Questions, Ideas, Hypotheses and Models. *Pharmaceuticals* **2016,** *9*, 14-6-6.
13. Torrens, F.; Castellano, G. Clinical Translational Research: Cancer, Diabetes, and Cardiovascular Disease. In *Physical Biochemistry, Biophysics, and Molecular Chemistry: Applied Research and Interactions*; Torrens, F., Mahapatra, D. K., Haghi, A. K., Eds.; Apple Academic/CRC: Waretown, NJ, in press.
14. Sandoval del Amor, J. Personal Communication.
15. Sandoval, J. Epigenetics and Reinventing Lamarck. *Gen. Eng. Biotechnol. News* **2011,** *31* (17), 46–49.
16. Heyn, H.; Carmona, F. J.; Gomez, A.; Ferreira, H.; Bell, J.; Sayols, S.; Kelly, K.; Stefansson, O.; Moran, S.; Sandoval, J.; Eyfjörd, J.; Spector, T.; Esteller, M. DNA Methylation Profiling in Breast Cancer Discordant Identical Twins Identifies DOK7 as Novel Epigenetic Biomarker. *Carcinogenesis* **2013,** *34*, 102–108.

17. Nieto, M. Á. Personal Communication.
18. Navarro, J.; Dasí, F.; Ejarque, I.; Castillo, S., Eds. Book of Abstracts, Jornada *Genética Clínica y Enfermedades Raras para Atención Primaria*, València, Spain, February 27, 2019, Hospital Clínico Universitario de Valencia: València, Spain, 2019.
19. Soria Olivas, E. Book of Abstracts, Jornada *Genética Clínica y Enfermedades Raras para Atención Primaria*, València, Spain, February 27, 2019, Hospital Clínico Universitario de Valencia: València, Spain, 2019; RT-2.
20. Agrawal, A.; Gans, J.; Goldfarb, A. *Prediction Machines: The Simple Economics of Artificial Intelligence*; Harvard Business Review: Brighton, MA, 2018.
21. Topol, E. J. High-Performance Medicine: The Convergence of Human and Artificial Intelligence. *Nat. Med.* **2019,** *25*, 44–56.
22. Liang, H.; Tsui, B. Y.; Ni, H.; Valentim, C. S. C.; Baxter, S. L.; Liu, G.; Cai, W.; Kermany, D. S.; Sun, X.; Chen, J.; He, L.; Zhu, J.; Tian, P.; Shao, H.; Zheng, L.; Hou, R.; Hewett, S.; Li, G.; Liang, P.; Zang, X.; Zhang, Z.; Pan, L.; Cai, H.; Ling, R.; Li, S.; Cui, Y.; Tang, S.; Ye, H.; Huang, X.; He, W.; Liang, W.; Zhang, Q.; Jiamg, J.; Yu, W.; Gao, J.; Ou, W.; Deng, Y.; Hou, Q.; Wang, B.; Yao, C.; Liang, Y.; Zhang, S.; Duan, Y.; Zhang, R.; Gibson, S.; Zhang, C. L.; Li, O.; Zhang, E. D.; Karin, G.; Nguyen, N.; Wu, X.; Wen, C.; Xu, J.; Xu, W.; Wang, B.; Wang, W.; Li, J.; Pizzato, B.; Bao, C.; Xiang, D.; He, W.; He, S.; Zhou, Y.; Haw, W.; Goldbaum, M.; Tremoulet, A.; Hsu, C. N.; Carter, H.; Zhu, L.; Zhang, K.; Xia, H. Evaluation and Accurate Diagnoses of Pediatric Diseases Using Artificial Intelligence. *Nat. Med.* **2019,** *2019*, 1–8.
23. Cuenca Valero, C. Book of Abstracts, Jornada *Genética Clínica y Enfermedades Raras para Atención Primaria*, València, Spain, February 27, 2019, Hospital Clínico Universitario de Valencia: València, Spain, 2019; RT-1.
24. Ejarque Doménech, I. Book of Abstracts, Jornada *Genética Clínica y Enfermedades Raras para Atención Primaria*, València, Spain, February 27, 2019, Hospital Clínico Universitario de Valencia: València, Spain, 2019; RT-1.
25. Ejarque Doménech, I.; Castelló López, M. I.; Sorlí Guerola, J. V. *Guía de Asesoramiento Preconcepcional desde Atención Primaria*; Societat Valenciana de Medicina Familiar i Comunitària: València, Spain, 2018.
26. Ejarque, I.; García-Ribes, M.; Martín, V. El médico de familia y sus competencies en Genética Clínica. *Atención Primaria* **2007,** *39*, 113–114.
27. Doménech Casasús, C. Book of Abstracts, Jornada *Genética Clínica y Enfermedades Raras para Atención Primaria*, València, Spain, February 27, 2019 <Should I retain it>, Hospital Clínico Universitario de Valencia: València, Spain, 2019; RT-1.
28. Fernández-Nohales, P. Book of Abstracts, Jornada *Genética Clínica y Enfermedades Raras para Atención Primaria*, València, Spain, February 27, 2019, Hospital Clínico Universitario de Valencia: València, Spain, 2019; RT-2.
29. Dasí, F. Book of Abstracts, Jornada *Genética Clínica y Enfermedades Raras para Atención Primaria*, València, Spain, February 27, 2019, Hospital Clínico Universitario de Valencia: València, Spain, 2019; RT-2.
30. REDIAPP. *Memoria 2007–2012*; Red de Investigación en Actividades Preventivas y Promociín de la Salud (REDIAPP): Barcelona, Spain, 2012.
31. García-Palenzuela, R.; Timiraos Carraco, R.; Gómez-Besteiro, M. I.; Lavia, G.; Lago Pose, M.; Lara B. Detección del deficit de alfa-1 antitripsina: Estudio en pacientes con enfermedad pulmonar obstructiva crónica diagnosticados en atención primaria. *SEMERGEN—Medicina de Familia* **2017,** *43*, 289–294.

32. Dasí, F. Book of Abstracts, Jornada *Genética Clínica y Enfermedades Raras para Atención Primaria*, València, Spain, February 27, 2019, Hospital Clínico Universitario de Valencia: València, Spain, 2019; RT-3.
33. Sancho, J. Book of Abstracts, Jornada *Genética Clínica y Enfermedades Raras para Atención Primaria*, València, Spain, February 27, 2019, Hospital Clínico Universitario de Valencia: València, Spain, 2019; RT-3.
34. Ministerio de Sanidad y Consumo. *Guía para la Atención de la Esclerosis Lateral Amiotrófica (ELA) en España*; Ministerio de Sanidad y Consumo: Madrid, Spain, 2007.
35. Sancho, J.; Servera, E.; Chiner, E.; Bañuls, P.; Gómez-Merino, E.; Sancho-Chust, J. N.; Marín, J. Noninvasive Respiratory Muscle Aids During PEG Placement in ALS Patients with Severe Ventilatory Impairment. *J. Neurol. Sci.* **2010,** *297*, 55–59.
36. Sancho, J., Ed. *Guía para el Manejo de los Problemas Respiratorios de la Esclerosis Lateral Amiotrófica*; Sociedad Valenciana de Neumología–Sociedad Valenciana de Medicina Intensiva, Crítica y Unidades Coronarias–Sociedad Valenciana de Medicina Familiar y Comunitaria–Sociedad Valenciana Hospital a Domicilio: València, Spain, 2019.
37. Trenor, P. Book of Abstracts, Jornada *Genética Clínica y Enfermedades Raras para Atención Primaria*, València, Spain, February 27, 2019, Hospital Clínico Universitario de Valencia: València, Spain, 2019; RT-3.
38. Harley, I. T. W.; Kaufman, K. M.; Langefeld, C. D.; Harley, J. B.; Kelly, J. A. Getting Susceptibility to SLE: New Insights from Fine Mapping and Genome-Wide Association Studies. *Nat. Rev. Genet.* **2009,** *10*, 285–290.
39. Bertsias, G. K.; Pamfil, C.; Fanouriakis, A.; Boumpas, D. T. Diagnostic Criteria for Systemic Lupus Erythematosus: Has the time come? *Nat. Rev. Rheumatol.* **2013,** *9*, 687–694.
40. Mankad, R.; Gabriel, S. E. Rheumatoid Arthritis: Treating Cardiovascular Risk in RA Requires Multidisciplinary Care. *Nat. Rev. Rheumatol.* **2014,** *10*, 202–204.
41. Deane, K. D.; El-Gabalawy, H. Pathogenesis and Prevention of Rheumatic Disease: Focus on Preclinical RA and SLE. *Nat. Rev. Rheumatol.* **2014,** *10*, 212–228.
42. Muskardin, T. L. W.; Niewold, T. B. Type I Interferon in Rheumatic Diseases. *Nat. Rev. Rheumatol.* **2018,** *14*, 214–218.
43. Eyre, S.; Orozco, G.; Worthington, J. The Genetics Revolution in Rheumatology: Large Scale Genomic Arrays and Genetic Mapping. *Nat. Rev. Rheumatol.* **2017,** *13*, 421–432.
44. Pisetsky, D. S. Antinuclear Antibody Testing—Misunderstood or Misbegotten? *Nat. Rev. Rheumatol.* **2017,** *13*, 495–502.
45. Schneider, M.; Liang, M. H. Connective Tissue Diseases: SLE Classification: Plus ça change, plus c'est la même chose. *Nat. Rev. Rheumatol.* **2015,** *11*, 262–264.
46. Jorge, A.; Ung, C.; Young, L. H.; Melles, R. B.; Choi, H. K. Hydroxychloroquine Retinopathy—Implications of Research Advances for Rheumatology Care. *Nat. Rev. Rheumatol.* **2018,** *14*, 693–703.
47. Escribano, A. Book of Abstracts, Jornada *Genética Clínica y Enfermedades Raras para Atención Primaria*, València, Spain, February 27, 2019, Hospital Clínico Universitario de Valencia: València, Spain, 2019; CC-1.
48. Warwick, W. J. *Cystic Fibrosis: A Geriatric Problem*; Minnesota Cystic Fibrosis Center: Minneapolis, MN, 2013.

CHAPTER 16

Scientific Creativity, Social Stability, and Attention to Diversity

FRANCISCO TORRENS[1*] and GLORIA CASTELLANO[2]

[1]*Institut Universitari de Ciència Molecular, Universitat de València, Edifici d'Instituts de Paterna, P. O. Box 22085, E-46071 València, Spain*

[2]*Departamento de Ciencias Experimentales y Matemáticas, Facultad de Veterinaria y Ciencias Experimentales, Universidad Católica de Valencia San Vicente Mártir, Guillem de Castro-94, E-46001 València, Spain*

[*]*Corresponding author. E-mail: torrens@uv.es*

ABSTRACT

If a social framework tending to stability exists, this acquires for the individual priority status and reduces psychological stress necessary to scientific creativity, which situation generalizes in Spain and one can foresee a decay in quality. The best way to avoid this situation is not to pass via stress situations or *obligations to science*, the circumstance of the one that, being little demanded, suffers for a contradiction, especially when it is compared to the sacrifices, and instabilities of those that work in counties rich in science and, why not, economic resources. People cannot simplify philosophy and science. People cannot obviate philosophy and science metaphors and paradoxes. People cannot ignore the social contexts of philosophy and science. People cannot ignore publics, considering them as clients. People cannot ignore the shift of time between available and rights. People cannot ignore philosophy and science ethical consequences. People should move in metaphors and paradoxes. No exit exists from paradoxes. The question of ends has no exit. The exit is social and political. A survey on university diversity attention measures: (1) the socio-demographic information, (2) beliefs, attitudes, and practices of attention to diversity in universities, (3) measuring of ideological attitudes, and (4) measuring of personality.

16.1 INTRODUCTION

Setting the scene: Scientific creativity and social stability, some metaphors that have made history in philosophy, attention to diversity and inclusive education in universities, its diagnosis and evaluation of institutionalization indicators. Present science can set itself up as a metaphilosophical discourse, one can *think from science*. It is not about claiming a *unique language* that limits the possibilities of others. By the contrary, it is raised that, although science passes *via* paths that take it toward an interpretation unequivocal, unique if one wants, of the indescribable, not because of it the possibility to interpretative elaborations of other nature closes. Together with other discourses, science adds capacity for seeing and transforming that external world that is appreciated from Plato's cavern, but leads to existential loneliness and melancholy.

If a social framework tending to stability exists, this acquires for the individual, unconsciously, priority status, and reduces psychological stress necessary to scientific creativity, which situation generalizes in Spain and one can foresee, so, a decay in quality although not, seemingly, quantity. The best way to avoid the situation above is not to pass via stress situations or *obligations to science*, the circumstance of the one that, being very little demanded, suffers for a contradiction, especially when it is compared to the sacrifices and instabilities of those that work in counties rich in science and, why not, economic resources. People can simplify neither philosophy nor science. People can obviate neither the metaphors nor paradoxes of philosophy and science. People cannot ignore the social contexts of philosophy and science. People cannot ignore publics and consider them as clients. People cannot ignore the shift of time between available and rights. People cannot ignore the ethical consequences of philosophy and science. People should move in metaphors and paradoxes. No exit exists from paradoxes. The question of ends has no exit. The exit is social and political. A survey on university diversity attention measures what is important to know: (1) the socio-demographic information, (2) beliefs, attitudes, and practices of attention to diversity in universities, (3) measuring of ideological attitudes, and (4) measuring of personality. The psychological features of lecturers and researchers are interesting for the sitting of the flows of academic knowledge, transference channels, and personality traits of researchers. Sign of prejudice should not exist in a literary work. It remains to be solved the knowledge of the factors hindering the translation of science education research to teaching practice. As educators of the professionals of tomorrow, professors have the ethical duty of providing them an integral training where teachers should include environment protection.

In earlier publications, it was reported the periodic table of the elements (PTE),[1–3] quantum simulators,[4–12] science, ethics of developing sustainability *via* nanosystems, devices,[13] *green nanotechnology* as an approach toward environment safety,[14] molecular devices, machines as hybrid organic–inorganic structures,[15] PTE, quantum biting its tail, sustainable chemistry,[16] quantum molecular *spintronics*, nanoscience, and graphenes.[17] It was informed cancer, its hypotheses,[18] precision personalized medicine from theory to practice, cancer,[19] how human immunodeficiency virus/acquired immunodeficiency syndrome (HIV/AIDS) destroy immune defenses, hypothesis,[20] 2014 emergence, spread, uncontrolled Ebola outbreak,[21,22] Ebola virus disease, questions, ideas, hypotheses, models,[23] metaphors that made history, and reflections on philosophy, science, deoxyribonucleic acid (DNA),[24] scientific integrity, ethics, science communication, and psychology.[25] In the present report, it is reviewed some reflections on scientific creativity, social stability, some metaphors that have made history in philosophy, attention to diversity, and inclusive education in universities, its diagnosis and evaluation of institutionalization indicators. The aim of this work is to initiate a debate by suggesting a number of questions (Q), which can arise when addressing subjects of scientific creativity, social stability, metaphors, attention to diversity, and inclusive education in universities. The general objectives are to explore the beliefs, attitudes, and social behaviors of university [teaching–researcher (TRS), administration and services (ASS)] staff round diversity attention and education inclusion degree present in university practices. The research purpose is to diagnose diversity attention in universities for elaborating an institutionalization proposal of inclusive approach in Spanish context.

16.2 SCIENTIFIC CREATIVITY AND SOCIAL STABILITY

If one is interested in some other thing, he should chase after it.[26] One could be wrong, but the impression results that social pressure, in order to be financially secure, presents powerful influences over scientific-creativity psychology. If a social framework tending to stability exists, this acquires for the individual, unconsciously, priority status, and reduces psychological strain necessary to scientific creativity, which situation generalizes in Spain and one can foresee, so, a decay in quality although not, seemingly, quantity. The best way to avoid such a situation is not to pass via strain situations or *obligations to science*, the circumstance of the one that, being very little demanded, suffers for a contradiction, especially when it is compared to the sacrifices and instabilities of those that work in counties rich in science and, why not, economic resources.

16.3 SOME METAPHORS THAT HAVE MADE HISTORY IN PHILOSOPHY

Strauss published an introduction to philosophy.[27] Table 16.1 gives some metaphors that have made history in philosophy.[28]

TABLE 16.1 Some Metaphors That Have Made History in Philosophy.

Metaphor
Humans are born to die.
We all die in the end.
(Cervantes). As much true, Sancho, as someday we will be dead.
(Keynes). In the long run we are all dead.
(Louis L'Amour). The more one learns the more he understands his ignorance.
(Albert Einstein). The more I learn, the more I realize how much I don't know.
(Javier Sábada). The delimitation that supposes the beach between the ocean of the unattainable and the rationality of the land.

16.4 ATTENTION TO DIVERSITY IN UNIVERSITIES: DIAGNOSIS AND EVALUATION

Córdoba University organized an online survey that presents as general objectives: to explore the beliefs, attitudes, and social behaviors of university (TRS and ASS) staff round diversity attention and education inclusion degree present in university practices.[29] The information collection is situated in the framework of Project *Attention to Diversity and Inclusive Education in Universities: Diagnosis and Institutionalization Indicators Evaluation*. The research purpose is to diagnose diversity attention in universities for elaborating an institutionalization proposal of inclusive approach in Spanish context. The survey consists of four parts: (1) socio-demographic information, (2) beliefs, attitudes, and practices of attention to diversity in universities, (3) measuring of ideological attitudes, and (4) measuring of personality.

Socio-demographic information included the following questions:

Q1. Did you feel personally discriminated, in the last 12 months, by the following?

- Belonging to an ethnic minority.
- Belonging to a minority of migrated persons.
- Having a minority sexual orientation.
- My sex.
- Belonging to a poblational group of advanced age (greater than 55 years).
- Belonging to a poblational group of young age (lesser than 30 years).
- Practicing Catholic religion.
- Practicing other Christian religion.
- Practicing Moslem religion.
- Practicing Jewish religion.
- Practicing a minority religion not previously mentioned.
- Being atheist.
- Being agnostic.
- Being indifferent vs. religion.
- Belonging to a group of affiliated or sympathizers of a right-wing political party.
- Belonging to a group of affiliated or sympathizers of a left-wing political party.
- Belonging to a minority with disability.
- Belonging to a minority with a chronic or infectious disease.
- Belonging to a socio-economic group of high income.
- Belonging to a socio-economic group of low income.
- Belonging to a linguistic minority.
- Other collective to which I belong, by which I felt discriminated and that was not shown before.

Q2. With which of these collectives do you recognize yourself or do you feel identified? With:

- An ethnic minority.
- A minority of migrated persons.
- A minority sexual orientation.
- My sex.
- The poblational group of advanced age (greater than 55 years).
- The poblational group of young age (lesser than 30 years).
- The Catholic religion collective.

- A collective of other Christian religion.
- The Moslem religion collective.
- The Jewish religion collective.
- A collective of other minority religion not previously mentioned.
- The atheistic collective.
- The agnostic collective.
- A collective indifferent vs. religion.
- The group of affiliated or sympathizers of a right-wing political party.
- The group of affiliated or sympathizers of a left-wing political party.
- A minority with disability.
- A minority with a chronic or infectious disease.
- The socio-economic group of high income.
- The socio-economic group of low income.
- A linguistic minority.
- Other:

Beliefs, attitudes, and practices of attention to diversity in universities included questions.

Q1. The diversity concept means different ethnic group, race, nationality, or culture.
Q2. The diversity concept means persons with different thoughts and ideas.
Q3. The diversity concept means different education level.
Q4. The education system should emphasize learning on other cultural, religious, or gender groups.
Q5. Diversity/inclusion/fairness are an essential part of education that should be attended in university.
Q6. Diversity/inclusion/fairness are an institutional question but also individual, of every member.
Q7. One of the university purposes is to train students to manage in a more diverse society.
Q8. One of the university purposes is to train students to manage in a more diverse labor market.
Q9. In university curricula one should add specific subjects focused on women and minorities role.
Q10. It is important for universities to train persons to be successful in diverse world/prepare them in technical/academic competences.
Q11. Universities should develop specific actions to attend for diversity in the students collective.

Q12. Directive team of this university promotes actions of attention to diversity/inclusion/fairness.
Q13. In my university established trend exists favoring diversity/inclusion/fairness in curricular offers.
Q14. I develop research that in form/contents reflect commitment with diversity/inclusion/fairness.
Q15. I incorporate into research designs elements that favor cultures diversity/inclusion/fairness.
Q16. In my university, training courses related to diversity, inclusion, and fairness are offered.
Q17. I participate in education innovation projects explicitly reflecting diversity/etc. compromise.
Q18. I design teaching innovation projects incorporating matters of attention to gender diversity.
Q19. I invest time/effort on education practices promoting competences taking success of all student body.
Q20. I design objectives of my teaching focused on diversity, inclusion and fairness.
Q21. I offer support to help student body to develop individualized plans allowing getting learning.
Q22. I incorporate in lectures different teaching–learning methods to attend student body diversity.
Q23. I offer resources to meet student body necessities and attend inclusive-teaching development.
Q24. I tackle digital learning/cooperative activities to favor learning of student body with diverse needs.

Measuring of ideological attitudes included the following questions:

Q1. Our society needs strong leaders that could eradicate extremism/immorality that prevail in it.
Q2. Our society needs freethinkers with resoluteness to confront conventionalisms if it be annoying.
Q3. Traditions and old values already continue to indicate us the best way of living.
Q4. Our society would be better if we show tolerance and understanding for different ideas/values.
Q5. Laws punishing abortion/pornography and contributing protecting marriage must be observed.
Q6. Society needs to show opening to persons thinking by themselves and different to authorities.

Q7. Many persons dare State/criticize religion/ignore life forms without stopping being good.
Q8. One would admire/respect more our ancestors by their contribution to construction of society.
Q9. Many radical or immoral persons exist that try to ruin things; society should slow them down.
Q10. In defense of freedom of speech, we should allow publication of literature considered bad.
Q11. Society would improve if agitators were treated with humanity/ trying to make them see sense.
Q12. Every good citizen should help to remove the evil that poisons our country from inside.
Q13. An ideal society requires that some groups be at the peak and others be at the bottom.
Q14. Some social groups are simply inferior than other groups.
Q15. No group should dominate in the society.
Q16. Groups that are at the bottom are deserving of the same options and things than at the peak.
Q17. To work for group equality should not be our main objective.
Q18. It is unfair to try to do that the groups be equal.
Q19. We should do what we can to make equal the conditions that are offered to different groups.
Q20. We should work to give all the groups the same opportunity for being successful.

Measuring of personality included the following questions:

Q1. When I read a poem, or gaze at a work of art, I feel deep emotion or excitement.
Q2. Poetry has low or no effect on me.
Q3. I have a great variety of intellectual interests.
Q4. The forms that I found in art and nature arouse my curiosity.
Q5. I find boring philosophical discussions.
Q6. I have a lot of fantasy.
Q7. I like to concentrate on a dream/fantasy and, letting it grow/ develop, explore all its possibilities.
Q8. I have not much interest in being thinking about the nature of the universe or human condition.
Q9. I lose interest when people talk about abstract and theoretical questions.
Q10. I experience a great variety of emotions and feelings.
Q11. Frequently, I taste food new or from other countries.

Q12. I rarely experience strong emotions.
Q13. I tend to think the best of people.
Q14. I intimidate or flatter people in order that they do what I want.
Q15. I get with cunning that people do what I want.
Q16. If somebody starts to fight with me, I am also willing to fight.
Q17. When I have been offended, what I try is to forgive and forget.
Q18. My first reaction is to trust people.
Q19. Some persons think about me that I am cool and calculating.
Q20. I have a lot of faith in human nature.
Q21. I try to be humble.
Q22. I think that most of the people with which I treat are honest and reliable.
Q23. If necessary I can be sarcastic and sharp.
Q24. The persons that beg do not inspire sympathy in me.

16.5 FINAL REMARKS

From the present results and discussion, the following final remarks can be drawn.

1. If a social framework tending to stability exists, this acquires for the individual, unconsciously, priority status, and reduces psychological strain necessary to scientific creativity, which situation generalizes in Spain and one can foresee, so, a decay in quality although not, seemingly, quantity.
2. The best way to avoid the situation above is not to pass via strain situations or *obligations to science*, the circumstance of the one that, being very little demanded, suffers for a contradiction, especially when it is compared to the sacrifices and instabilities of those that work in counties rich in science and, why not, economic resources.
3. People can simplify neither philosophy nor science. People can obviate neither the metaphors nor paradoxes of philosophy and science. People cannot ignore the social contexts of philosophy and science. People cannot ignore publics and consider them as clients. People cannot ignore the shift of time between available and rights. People cannot ignore the ethical consequences of philosophy and science. People should move in metaphors and paradoxes. No exit exists from paradoxes. The question of ends has no exit. The exit is social and political.

4. It is important to know: (1) the socio-demographic information, (2) beliefs, attitudes, and practices of attention to diversity in universities, (3) measuring of ideological attitudes, and (4) measuring of personality. The psychological features of lecturers and researchers are interesting for the sitting of the flows of academic knowledge, transference channels, and personality traits of researchers. Sign of prejudice should not exist in a literary work.
5. It remains to be solved the knowledge of the factors hindering the translation of science education research to teaching practice. As educators of the professionals of tomorrow, professors have the ethical duty of providing them an integral training where teachers should include environment protection.

ACKNOWLEDGMENTS

The authors thank support from Generalitat Valenciana (Project No. PROMETEO/2016/094) and Universidad Católica de Valencia *San Vicente Mártir* (Project No. 2019-217-001).

KEYWORDS

- **metaphor**
- **education inclusion degree**
- **university practice**
- **diagnosis**
- **evaluation**
- **institutionalization indicator**
- **Spain**

REFERENCES

1. Torrens, F.; Castellano, G. Reflections on the Nature of the Periodic Table of the Elements: Implications in Chemical Education. In *Synthetic Organic Chemistry*; Seijas, J. A., Vázquez Tato, M. P., Lin, S. K., Eds.; MDPI: Basel, Switzerland, 2015; Vol. 18; pp 1–15.

2. Torrens, F.; Castellano, G. Nanoscience: From a Two-Dimensional to a Three-Dimensional Periodic Table of the Elements. In *Methodologies and Applications for Analytical and Physical Chemistry*; Haghi, A. K., Thomas, S., Palit, S., Main, P., Eds.; Apple Academic/CRC: Waretown, NJ, 2018; pp 3–26.
3. Torrens, F.; Castellano, G. Periodic Table. In *New Frontiers in Nanochemistry: Concepts, Theories, and Trends*; Putz, M. V., Ed.; Apple Academic/CRC: Waretown, NJ, in press.
4. Torrens, F.; Castellano, G. Ideas in the History of Nano/Miniaturization and (Quantum) Simulators: Feynman, Education and Research Reorientation in Translational Science. In *Synthetic Organic Chemistry*; Seijas, J. A., Vázquez Tato, M. P., Lin, S. K., Eds.; MDPI: Basel, Switzerland, 2015; Vol. 19; pp 1–16.
5. Torrens, F.; Castellano, G. Reflections on the Cultural History of Nanominiaturization and Quantum Simulators (Computers). In *Sensors and Molecular Recognition*; Laguarda Miró, N., Masot Peris, R., Brun Sánchez, E., Eds.; Universidad Politécnica de Valencia: València, Spain, 2015; Vol. 9; pp 1–7.
6. Torrens, F.; Castellano, G. Nanominiaturization and Quantum Computing. In *Sensors and Molecular Recognition*; Costero Nieto, A. M., Parra Álvarez, M., Gaviña Costero, P., Gil Grau, S., Eds.; Universitat de València: València, Spain, 2016; Vol. 10; pp 31-1-5.
7. Torrens, F.; Castellano, G. Nanominiaturization, Classical/Quantum Computers/Simulators, Superconductivity, and Universe. In *Methodologies and Applications for Analytical and Physical Chemistry*; Haghi, A. K., Thomas, S., Palit, S., Main, P., Eds.; Apple Academic/CRC: Waretown, NJ, 2018; pp 27–44.
8. Torrens, F.; Castellano, G. Superconductors, Superconductivity, BCS Theory and Entangled Photons for Quantum Computing. In *Physical Chemistry for Engineering and Applied Sciences: Theoretical and Methodological Implication*; Haghi, A. K., Aguilar, C. N., Thomas, S., Praveen, K. M., Eds.; Apple Academic/CRC: Waretown, NJ, 2018; pp 379–387.
9. Torrens, F.; Castellano, G. EPR Paradox, Quantum Decoherence, Qubits, Goals and Opportunities in Quantum Simulation. In *Theoretical Models and Experimental Approaches in Physical Chemistry: Research Methodology and Practical Methods*; Haghi, A. K., Ed.; Apple Academic/CRC: Waretown, NJ, 2018; Vol. 5, pp 317–334.
10. Torrens, F.; Castellano, G. Nanomaterials, Molecular Ion Magnets, Ultrastrong and Spin–Orbit Couplings in Quantum Materials. In *Physical Chemistry for Chemists and Chemical Engineers: Multidisciplinary Research Perspectives*; Vakhrushev, A. V., Haghi, R., de Julián-Ortiz, J. V., Allahyari, E., Eds.; Apple Academic/CRC: Waretown, NJ, in press.
11. Torrens, F.; Castellano, G. Nanodevices and Organization of Single Ion Magnets and Spin Qubits. In *Chemical Science and Engineering Technology: Perspectives on Interdisciplinary Research*; Balköse, D., Ribeiro, A. C. F., Haghi, A. K., Ameta, S. C., Chakraborty, T., Eds.; Apple Academic/CRC: Waretown, NJ, in press.
12. Torrens, F.; Castellano, G. Superconductivity and Quantum Computing via Magnetic Molecules. In *New Insights in Chemical Engineering and Computational Chemistry*; Haghi, A. K., Ed.; Apple Academic/CRC: Waretown, NJ, in press.
13. Torrens, F.; Castellano, G. Developing Sustainability via Nanosystems and Devices: Science–Ethics. In *Chemical Science and Engineering Technology: Perspectives on Interdisciplinary Research*; Balköse, D., Ribeiro, A. C. F., Haghi, A. K., Ameta, S. C., Chakraborty, T., Eds.; Apple Academic/CRC: Waretown, NJ, in press.

14. Torrens, F.; Castellano, G. Green Nanotechnology: An Approach towards Environment Safety. In *Advances in Nanotechnology and the Environmental Sciences: Applications, Innovations, and Visions for the Future*; Vakhrushev, A. V.; Ameta, S. C.; Susanto, H., Haghi, A. K., Eds.; Apple Academic/CRC: Waretown, NJ, in press.
15. Torrens, F.; Castellano, G. Molecular Devices/Machines: Hybrid Organic–Inorganic Structures. In *Research Methods and Applications in Chemical and Biological Engineering*; Pourhashemi, A., Deka, S. C., Haghi, A. K., Eds.; Apple Academic/CRC: Waretown, NJ, in press.
16. Torrens, F.; Castellano, G. The Periodic Table, Quantum Biting its Tail, and Sustainable Chemistry. In *Chemical Nanoscience and Nanotechnology: New Materials and Modern Techniques*; Torrens, F., Haghi, A. K., Chakraborty, T., Eds.; Apple Academic/CRC: Waretown, NJ, in press.
17. Torrens, F.; Castellano, G. Quantum Molecular Spintronics, Nanoscience and Graphenes. In *Molecular Physical Chemistry*; Haghi, A. K., Ed.; Apple Academic/CRC: Waretown, NJ, in press.
18. Torrens, F.; Castellano, G. Cancer and Hypotheses on Cancer. In *Molecular Chemistry and Biomolecular Engineering: Integrating Theory and Research with Practice*; Pogliani, L., Torrens, F., Haghi, A. K., Eds.; Apple Academic/CRC: Waretown, NJ, in press.
19. Torrens, F.; Castellano, G. Precision Personalized Medicine from Theory to Practice: Cancer. In *Molecular Physical Chemistry*; Haghi, A. K., Ed.; Apple Academic/CRC: Waretown, NJ, in press.
20. Torrens, F.; Castellano, G. AIDS Destroys Immune Defences: Hypothesis. *New Front. Chem.* **2014,** *23*, 11–20.
21. Torrens-Zaragozá, F.; Castellano-Estornell, G. Emergence, Spread and Uncontrolled Ebola Outbreak. *Basic Clin. Pharmacol. Toxicol.* **2015,** *117* (Suppl. 2), 38–38.
22. Torrens, F.; Castellano, G. 2014 Spread/Uncontrolled Ebola Outbreak. *New Front. Chem.* **2015,** *24*, 81–91.
23. Torrens, F.; Castellano, G. Ebola Virus Disease: Questions, Ideas, Hypotheses and Models. *Pharmaceuticals* **2016,** *9*, 14–6-6.
24. Torrens, F.; Castellano, G. Metaphors That Made History: Reflections on Philosophy/Science/DNA. In *Molecular Physical Chemistry*; Haghi, A. K., Ed.; Apple Academic/CRC: Waretown, NJ, in press.
25. Torrens, F.; Castellano, G. Scientific Integrity and Ethics: Science Communication and Psychology. In *Biochemistry, Biophysics, and Molecular Chemistry: Applied Research and Interactions*; Torrens, F., Mahapatra, D. K., Haghi, A. K., Eds.; Apple Academic/CRC: Waretown, NJ, in press.
26. Moya, A. *Pensar desde la Ciencia*; Trotta: Madrid, Spain, 2010.
27. Strauss, L. *El Gusto de Jenofonte: Una Introducción a la Filosofía*; Biblioteca Nueva: Madrid, Spain, 2018.
28. Pascual, S., Ed. *Per què Filosofia?* Neopàtria: Alzira, València, Spain, 2017.
29. Cordoba University. Personal Communication.

CHAPTER 17

Methods of Reducing the Content of Radionuclides in Agricultural Plants After the Accident at the Chernobyl Nuclear Power Plant

MICHAEL O. SMIRNOV*, GENRIETTA E. MERZLAYA* and RAFAIL A. AFANAS'EV*

Pryanishnikov All-Russian Scientific Research Institute of Agrochemistry, d. 31A, Pryanishnikova St., Moscow, 127550, Russia

**Corresponding author. E-mail: User53530@yandex.ru; lab.organic@mail.ru; rafail-afanasev@mail.ru*

ABSTRACT

In the field experience in connection with the Chernobyl accident in the area of radioactive contamination, effective countermeasures to reduce the content of ^{137}Cs specifically for each of the main crop rotation were identified. It is shown that important methods of reducing the content of cesium radionuclide in plant products are: soil liming in the production of barley grain, increased doses of potash fertilizers—for oat grains and winter rye, as well as manure—for potato tubers and green mass of lupine.

17.1 INTRODUCTION

With the beginning of the development of nuclear energy and mass testing of nuclear weapons in the atmosphere (50–60 years of the 20th century), the problem of radioactive contamination of natural components, including soil, became urgent. On April 26, 1986 there was an accident at the Chernobyl nuclear power plant (NPP). This is the largest nuclear accident in the world, and we can say that the Chernobyl disaster is the largest man-made disaster of the century. Chernobyl NPP is located in the Ukrainian city of Pripyat, near the center of Chernobyl, almost at the junction of Ukraine, Belarus, and Russia.

The Chernobyl accident is a disaster for the whole of Europe. The fission products of nuclear fuel released from the destroyed reactor into the atmosphere were dispersed by air flows to large areas, causing their radioactive contamination not only near NPPs within the borders of Ukraine, Russia, and Belarus, but also for hundreds and even thousands of kilometers from the accident site. The territories of many countries have been exposed to radioactive contamination. As a result of the accident, the territories of 17 European countries with a total area of 207.5 thousand square kilometers were exposed to radioactive contamination of cesium-137 with levels above 1 Ci/km^2 (37 kBq/m^2). Significantly, contaminated with cesium-137 was the territory of Ukraine (37.63 thousand square kilometers), Belarus (43.5 thousand square kilometers), the European part of Russia (59.3 thousand square kilometers).

In Russia, 19 subjects were exposed to radiation contamination with cesium-137. The most polluted areas are the following regions of Russia: Bryansk (11.8 thousand square kilometers of contaminated areas), Kaluga (4.9 thousand square kilometers), Tula (11.6 thousand square kilometers), and Orel (8.9 thousand square kilometers).

About 60 thousand square kilometers of areas contaminated with cesium-137 with levels above 1 Ci/km^2 are outside the former USSR. The territories of Austria, Finland, Germany, Italy, Great Britain, Sweden, Norway, and a number of other Western European countries were contaminated. To demonstrate this, it is enough to cite the following table, published by media after the Chernobyl accident (Table 17.1).

TABLE 17.1 Radiation Measured in Various Cities Around the World After the Chernobyl Accident.

City	Country	The power of radiation in Mr/HR	Date
Pripyat	Ukraine (then USSR)	1,170,000	April 28, 1986
Kiev	Ukraine (then USSR)	6200	April 30, 1986
Novozybkov	Russia (then USSR)	2200	April 29, 1986
Gomel	Belarus (then USSR)	80	April 27, 1986
Minsk	Belarus (then USSR)	60	April 28, 1986
Salzburg	Austria	1400	May 2, 1986
Tavastehus	Finland	1400	April 29, 1986
Munich	Germany	2500	April 30, 1986

It is noteworthy that in the first days after the accident in the German city of Munich, the irradiation capacity exceeded that in the Russian city of Novozybkov, and the Austrian city of Salzburg and the Finnish city of Tavastehus received more radiation than the Belarusian cities of Gomel and Minsk.

If we take the entire area affected by the Chernobyl disaster for 100%, the distribution of radioactivity was approximately as follows: Russia: 30%, Belarus: 23%, Ukraine: 19%, Finland: 5%, Sweden: 4.5%, Norway: 3.1%, Austria: 2.5%. A significant part of the territory of Russia, Ukraine, and Belarus was contaminated at a level exceeding 5 Ci/km^2 (185 kBq/m^2). Approximately 50 thousand square kilometers of agricultural land were affected by cesium-137 and strontium-90, with half-lives of 30 and 28 years, respectively.

Land pollution is an increase in the concentration of radioactive substances in the soil profile. Pollution of agricultural land and the resulting production and consumption of products with high radionuclide content is one of the main sources of internal exposure of the population.[1–3] In the remote period after the accident, the probability of agricultural product production with a high level of pollution remains. This is largely due to the soil-geochemical characteristics of the contaminated areas, primarily, the presence of sod-podzolic soils of light granulometric composition in the soil cover; these soils are characterized by high rates of radionuclide migration.

So, the negative consequences of radioactive contamination of the soil are:

- Direct effects of ionizing radiation on soil and vegetation components.
- Limiting the use of contaminated soils in agriculture, as the products derived from such soils tend to have levels of concentration exceeding the permissible levels. Absorbing complex of soils absorbs radionuclides and preserves them for a long time.

The behavior of radionuclides in soils is determined by a set of factors:

- Chemical properties of radioactive elements.
- Physical and chemical forms of radionuclide compounds in deposition.
- Composition and properties of soils.
- Landscape features.
- Climate indicators.

Strontium is absorbed metabolically, other radionuclides by the formation of soluble humates, phosphates, carbonates, sulfates, often replacing calcium in these compounds.

The radionuclides transferred to the soil surface in the composition of the aerosols, particles of dispersed fuel, minerals, etc. The soluble fraction of radionuclides is mainly represented by strontium and cesium.

The time factor has a great influence on the behavior of radionuclides. The period of maximum accumulation of radionuclides increases dynamically as the solubility of radioactive fallout decreases. Soluble organic substances of the medium increase the migration of radionuclides.

It should be noted that before the Chernobyl disaster in the world there was no experience of returning radioactive contaminated land in circulation. At present, it can be argued that an effective system of protective measures against radioactive contamination has largely developed in Russia, that is, it was Russian experts who worked after the Chernobyl disaster, just played a leading role in the study of this issue. It should be said that in many respects this happened not only due to the above-described features of the geochemical behavior of radiocesium, but also due to the extensive countermeasures taken in the territories of the affected agricultural lands. These countermeasures and their effectiveness in the cultivation of specific crops are discussed below on the basis of experimental work.

The mobility of radionuclides in the soil medium depends on the granulometric, mineralogical composition of the soil, the properties of organic matter and the reaction of the environment, the presence of geochemical barriers in the soil profile. The Chernobyl accident contaminated a large part of the territory of Russia, Ukraine, and Belarus with long-lived Cs and Sr radionuclides. The main radioactive element that poses a danger to humans is cesium in the form of its two radioactive isotopes ^{134}Cs and ^{137}Cs. Taking into account the relatively short decay period of ^{134}Cs (90% in about 6 years), all protective measures in the pollution zones are based on the soil content of the radioactive isotope ^{137}Cs, the half-life of which is about 30 years. At the same time, the main task of agriculture in radionuclide-contaminated areas is to obtain products containing a minimum amount of radionuclides corresponding to radiological standards.

Currently, Russia, the Republic of Belarus and Ukraine, whose territories have been contaminated with radionuclides, have accumulated some positive experience in studying the translocation of ^{137}Cs in the soil and its transformation in food chains, as well as the use of protective measures to reduce the accumulation of radioactive substances in crop production.

The question is that it is necessary to study the impact of these components of the environment and on this basis to develop a set of protective measures for specific climatic conditions of the radioactive contamination zone.

Depending on the soil characteristics, the degree of contamination of the soil with radionuclides, biological characteristics of cultivated crops, methods of further use of crop production, various protective techniques can be used that reduce radioactive contamination of agricultural products many times.[4] There are also types of protective techniques: mechanical, agrotechnical, chemical, agrochemical, and biological.[5–8] All methods are divided into two groups: methods established in agricultural production, which lead to an increase in soil fertility, an increase in yield and at the same time to a decrease in the translocation of radionuclides into plants and special methods, the main purpose of which is only to reduce the receipt of radionuclides in crop production. This separation of techniques conventionally, as the conventional methods under certain conditions can become special. When using certain techniques it is quite difficult to distinguish between mechanical and agrotechnical techniques,[9–10] chemical and agrochemical, agrotechnical and biological,[11–17] etc. At the same time, the organization and implementation of measures to reduce the intake of radionuclides in crop production is carried out as a result of not one reception, but, as a rule, a set of techniques, technologically related to each other. Therefore, it is more correct to combine them into a certain number of basic systems for the integrated reduction of the radionuclide transition from soil to plants, including both conventional and special (soil treatment; the use of mineral and organic fertilizers; the introduction of new varieties and types of crops in the crop rotation; the management of irrigation regimes, as well as the introduction of special substances and compounds). We are more interested in agrochemical methods to reduce the translocation of ^{137}Cs, which include the introduction of organic and mineral fertilizers,[11–14] liming,[15–16] introduction of selective sorbent into the soil,[17] or the use of bacterial preparations.[18] The mechanism of action of certain countermeasures is clarified. Radionuclides mainly enter the environment in the form of insoluble and hardly soluble particles. Gradually, when interacting with oxygen and water, they form water-soluble forms. The acidic reaction of the soil solution contributes greatly to this process. It is noted that ^{90}Sr and ^{137}Cs pass into plants in large quantities on acidic soils than on slightly acidic and neutral ones. Therefore, in the practice of agriculture is widely used reception liming acidic soils, which not only creates favorable conditions for the growth and development of plants, but also reduces the flow of artificial radionuclides from the soil to plants.[19] The main component introduced into the soil during liming is calcium, which is in the lime material in the form of oxide, hydroxide, and calcium carbonate. An increase in soil calcium concentration leads to an increase in the so-called "strontium coefficient," which reduces the transition of ^{90}Sr from soil to plants. The reception of liming of radioactively contaminated acidic soils is one of the main methods that inhibit the process of ^{90}Sr transition from soil to plants. The effectiveness of potash fertilizers is due to the fact that the concentration and properties of isotopic and non-isotopic carriers (for ^{137}Cs-stable cesium and potassium[20]) influence the sorption of radionuclide by soils and its translocation into plants.

17.2 METHODS

With each studied culture, a series of experiments was carried out in three variants: 1—lime; 2—increased doses of potash fertilizers ($N_{120}P_{90}K_{180}$); 3—manure. The methods adopted in agrochemistry were used.[21]

Specific activity of ^{137}Cs in the studied plant and soil samples was determined at the universal spectrometric complex "Gamma Plus" (Russia), the measurement error did not exceed 30%.[22]

17.3 DISCUSSION

In Russia, the radioactive cloud covered four regions: Bryansk, Kaluga, Orel, and Tula. As a result, the accident at the Chernobyl NPP in the Bryansk region was the most contaminated with artificial radionuclides in Russia, both in terms of the area of contamination and the number of radionuclides in the soil.[23–25] In this regard, this chapter considers the problem on the example of the Bryansk region, where the Novozybkov experimental station of Pryanishnikov All-Russian Scientific Research Institute of Agrochemistry. After the fall of artificial radionuclides in 1986, the radiation situation on the soils of agricultural lands deteriorated sharply, 1756.7 thousand hectares were classified as contaminated, including 1267.4 thousand hectares of arable land and 489.3 thousand hectares of hayfields and pastures. According to the degree of soil contamination of agricultural land located in the following series: 1054.0 thousand ha up to 37 kBq/m^2; 401.5 thousand ha—37–185 kBq/m^2; 186.6 thousand ha—185–555 kBq/m^2; 97.6 thousand ha—555–1480 kBq/m^2; 17.0 thousand ha—more than 1480 kBq/m^2.[2, 26] At the same time, the largest area of agricultural land was with the smallest area of pollution.

It was assumed that three decades after the Chernobyl disaster, radioactive contamination of agricultural products has decreased significantly, since 30 years is the approximate half-life of the main pollutants. In fact, over the years, the content of cesium in crop products of the Bryansk region really decreased by 20–30 times. About the soil clearly this cannot be said. Thirty years is a period approximately equal to the half-life of cesium-137, the main component of Chernobyl pollution. It is this long-lived radionuclide of technogenic origin that massively entered the terrestrial and aquatic ecosystems of the European part of Russia after the Chernobyl accident. Due to the fact that cesium is well fixed in clay minerals, it weakly passes into plants. It is estimated that in plants it is 100–10,000 times less active than in soils. Otherwise things are on sandy soils and peat lands, which are common in the Bryansk region. There radionuclides are much more mobile and better converted into plant biomass.

Despite the considerable amount of information on overcoming the consequences of the Chernobyl disaster, after the Chernobyl accident, the issues of optimizing the doses of fertilizers, choosing the most effective counteraction for a specific culture and the optimal ratio of mineral fertilizers with their joint application remained poorly studied. The literature provides a variety of figures on the reduction of radionuclides in plants. It is reported that due to the use of potash fertilizers, the transition of ^{137}Cs from soils of different types to agricultural plants is allegedly reduced in the range from 2 to 20 times.[27–30] However, there is no information what activities are optimal for certain crops. In this regard, we set a goal to find an effective way to reduce the content of cesium radionuclide specifically for each of the main crop rotations in the area of radioactive contamination. As a result of the research carried out by the staff of our Institute, effective measures have been developed to reduce the content of cesium-137 in the production, depending on the biological characteristics of the crops grown. At the same time, it was found that the role of potash fertilizers and a certain ratio of them with nitrogen fertilizers are especially great in reducing the intake of ^{137}Cs in plants.

According to the research conducted on the basis of the Novozybkov experimental station of Pryanishnikov All-Russian Scientific Research Institute of Agrochemistry in 1991–2000 (Tables 17.2–17.6), it was found that in order to obtain products with contamination below the permissible level, it is necessary to make mineral fertilizers in doses of $N_{120}P_{90}K_{180}$, that is, at a ratio of nitrogen: potassium 1:1.5. It was found that this ratio of nitrogen and potash fertilizers is optimal for reducing radioactive contamination of crops that can accumulate high doses of radiocesium, or crops growing in fields with a constantly high level of pollution (25 Ci/km^2 or more). It is established that high doses of potash fertilizers with the specified ratio of nitrogen and potash fertilizers are the best measure to reduce the content of radionuclides in oat grains and winter rye.

TABLE 17.2 The Effectiveness of Protective Measures on Contaminated sod-Podzolic Sandy Soil When Growing Winter Rye (Grain).

Measure	The multiplicity of reducing the accumulation Cs-137 in production
Liming	1.3
Increased doses of potash fertilizers	2.0
Manure	1.1

TABLE 17.3 The Effectiveness of Protective Measures on Contaminated Sod-Podzolic Sandy Soil When Growing Barley (Grain).

Measure	The multiplicity of reducing the accumulation Cs-137 in production
Liming	2.4
Increased doses of potash fertilizers	1.2
Manure	1.2

TABLE 17.4 The Effectiveness of Protective Measures on Contaminated Sod-Podzolic Sandy Soil When Growing Oats (Grain).

Measure	The multiplicity of reducing the accumulation Cs-137 in production
Liming	1.3
Increased doses of potash fertilizers	3.0
Manure	1.5

TABLE 17.5 Effectiveness of Protective Measures on Contaminated Sod-Podzolic Sandy Soil When Growing Potatoes (tubers).

Measure	The multiplicity of reducing the accumulation Cs-137 in production
Liming	1.2
Increased doses of potash fertilizers	2.6
Manure	4.8

TABLE 17.6 The Effectiveness of Protective Measures on Contaminated Sod-Podzolic Sandy Soil When Growing Lupine (Green Mass).

Measure	The multiplicity of reducing the accumulation Cs-137 in production
Liming	1.6
Increased doses of potash fertilizers	2.3
Manure	4.2

It was established experimentally that from acidic low-fertile soils radiocesium is much more active in plants and accumulates in them in greater quantities than when grown on highly fertile soils.

Liming, eliminates excessive acidity of soils, improves their agrophysical and agrochemical properties, and increases the efficiency of mineral fertilizers, promoting the transition of radionuclides to a state inaccessible to plants. In particular, liming was the most effective counter-measure to reduce the content of cesium radionuclide in barley grain—2.4 times, while for other crops it was less effective.

Applying manure was the most favorable measure to reduce the content of radionuclides in potato tubers and green mass of lupine.

Undoubtedly, there is also a higher economic effect of the event chosen by us. For example, it was found that the return on fertilizer yield of potato tubers when using organic fertilizers is about 2 times higher than when using mineral (in 1997, respectively, 23.3 and 11.5 kg/kg nutrients, in 2000, 18.0 and 11.5 kg/kg nutrients).

This consideration of the biological characteristics of food and feed crops is a prerequisite for the use of a particular agrochemical activity. Thus, the choice of a specific counter-action is due to the cultivation of a particular agricultural culture and its reaction to the method used. According to the data presented in Tables 17.2–17.6, liming is the optimal counter-measure to reduce the content of cesium radionuclide in barley grain, increased content of potash fertilizers for oats and winter rye grains, and manure—for potato tubers and green mass of lupine.

Therefore, for each crop the most effective measures are revealed, which maximally reduce the content of radiocesium in products, which makes it expedient to include them as mandatory in combination with other agrochemical methods. Also new is the establishment of the optimal ratio of nitrogen and potash fertilizers. Thus, the application of agrochemical measures for specific crops, defined as optimal, allows achieving a significant environmental and economic effect.

17.4 CONCLUSIONS

1. It is revealed that for each crop there are the most effective measures that minimize the content of nuclides in agricultural products.
2. Liming is an optimal counter-measure to reduce the content of cesium radionuclide in barley grain, increased content of potash fertilizers—for oat grains and winter rye, and manure—for potato tubers and green mass of lupine.

3. The optimal ratio of nitrogen and potash fertilizers (1:2), which reduces the content of nuclides in agricultural products, was established.
4. The application of agrochemical measures for specific crops, defined by us as optimal, allows us to achieve a significant environmental and economic effect.

KEYWORDS

- **pollution**
- **radionuclides**
- **isotopes**
- **counter-measures**
- **mineral and organic fertilizers**
- **liming**
- **crop rotation**
- **crop production**

REFERENCES

1. Priester B. S.; Omelyanenko N. P.; Perepelyatnikova L. V. Migration of Radionuclides in Soil and Transfer to Plants in the Zone of the Accident at Chernobyl. Moscow: Pochvovedenie ("Pedology" in Russian), 1996; Vol. 10, pp 51–60 (in Russian).
2. Markina, Z. N.; Prudnikov, P. V.; Kovalev, L. A.; Novikov, A. A. Radiological Situation in the Soils of Agricultural Lands in Bryansk Region and Ways in Obtaining of Normatively Clean Products. Moscow: Agrochimicheskii Vestnik ("Agrochemical messenger" in Russian), 2006; Vol. 2, pp 10–11 (in Russian).
3. Sanzharova, N. I. Radiological Monitoring of Agroecosystems and Agriculture in the Area of Nuclear Power Plants: Autoref. DSc. All-Russian Research Institute of Radiology and Agroecology: Obninsk, 1997; p 52 (in Russian).
4. Alexakhin, R.; Geras'kin, S. 25 Years after the Accident at the Chernobyl Nuclear Power Plant: Radioecological Lessons. *Radioprotection* **2011,** *46*, 595–600.
5. Yatsalo, B. I.; Aleksakhin, R. M. Methods of Analysis of Protective Measures in Agriculture in Radioactive Contaminated Areas. Effectiveness Evaluation, Intervention Levels and Comparison of Different Countermeasures. *Raditsionnaya biologiya. Radioekologiya* ("Radiation biology. Radioecology" in Russian) **1997,** *37*, (5), 812–822 (in Russian).
6. Aleksakhin, R. M.; Lunev, M. I. Technogenic Pollution of Agricultural Lands (Research, Control and Rehabilitation of Territories). *Plodorodie* ("Fertility" in Russian) **2011,** *3*, 32–35 (in Russian).

7. Sychev, V. G.; Lunev, V. I.; Orlov, P. M.; Belous, N. M. Chernobyl: Radiation Monitoring of Agricultural Land and Agrochemical Aspects of Reducing the Effects of Radioactive Contamination of Soils (to the 30th Anniversary of the Chernobyl Accident); Pryanishnikov All-Russian Scientific Research Institute of Agrochemistry: Moscow, 2016; p 184 (in Russian).
8. Beresford, N. A.; Beaugelin-Seiller, K.; Burgos, J.; Cujic, M.; Fesenko, S.; Kryshev, A.; Pachal, N.; Real, A.; Su, B. S.; Tagami, K.; Vives i Batlle, J.; Vives-Lynch, S.; Wells, C.; Wood, M. D. Radionuclide Biological Half-Life Values for Terrestrial and Aquatic Wildlife. J. Environ. Radioact. **2015,** *150*, 270–276.
9. Firsakova, S. K.; Grebenshchikov, N.; Novikov, A. A. The Efficiency of Different Methods of Handling the Sod by Reducing the Contamination of Meadow Grass Mixtures. Proc. Reports of the Third All-Union Conference on Agricultural Radiology. *Obninsk* **1990,** *1*, 95–96 (in Russian).
10. Belous, I. M.; Moiseenko, F. V.; Shapovalov, V. F.; Pirgunov, A. N. The Effectiveness of Methods of Rehabilitation of Floodplain Meadows Contaminated with Radionuclides, to Obtain the Standard Quality of Green Fodder//Proc. the Second Regional Scientific-Practical Conference-Fair “New Ideas, Technologies, Projects and Investments”. 28 Nov.–1 Dec. 2000. Bryansk Institute of advanced training of education workers: Bryansk. Part 2, 2000; pp 42–43 (in Russian).
11. Markusen, A. R.; Fedorov, E. A. Accumulation of Strontium-90 Field Crops under Different Conditions of Cultivation. *Agrohimiya* (“Agrochemistry” in Russian). **1978,** *8*, 125–128 (in Russian).
12. Iudintseva, I. U. V.; Levin, E. M. On the Role of Potassium in the Availability of Cs to Plants. *Agrohimiya* (“Agrochemistry” in Russian). **1982,** *4*, 75–81(in Russian).
13. Tulin, S. A.; Stavrova, N. G.; Vorob'ev, G. T.; Korovyakovskya, A. S. Potassium as a Factor to Reduce the Negative Consequences of Contamination of Agricultural Lands by Cesium-137 after the Chernobyl Accident. Proceedings of the International Seminar "the Issues of Mitigation of Consequences of the Chernobyl Disaster". Bryansk, 1993; pp 383–384 (in Russian).
14. Aleksakhin, P. M.; Sanzharova, N. I.; Fesenko, V.; Kurganov, A. A.; Mosharov, V. N. The Main Results of Work on Liquidation of Consequences of the Chernobyl Accident in the Field of Agro-Industrial Production. Chernobyl: 15 Years Later; Ed. Gerasimova, N.; Contact-cultura (“Contact-culture” in Russian): Moscow, 2001; pp 105–141 (in Russian).
15. Ratnikov, N.; Zhigareva, T. J. I.; Popova, G. I.; Popova, G. E. Influence of Means of Chemicalization on Decrease of Receipt of 137Cs in Agricultural Products in Some Areas Contaminated as a Result of the Chernobyl Accident //Third All-Union Conference on Agricultural Radiology. *Obninsk* **1990,** *4*, 10–11 (in Russian).
16. Bogdevich, I. M.; Smeyan, N. I.; Lapa, V. V. Guidelines for the Conduct of Agro-Industrial Production in Conditions of Radioactive Contamination of Soils. *Minsk* **2000,** *76* (in Russian).
17. Sudarkov V. A., E. A., Torubarov A. A., Kalinin N. F., Gelis V. M., Milyutin V. V., Penzin, R. A. The method for reducing the transfer of radioactive cesium from soil to plants by soil application of selective sorbent. Patent RU No. 2013913. (in Russian)
18. Chernenok, V. I.; Bokhan, N. I.; Kashcheev, V. I.; Saguro, Yu. N. The Method of Cultivation of Agricultural Crops on Soils Contaminated with Radionuclides, Including Pre-Sowing Treatment of Seed and Soil with Application of Mineral Nitrogen Fertilizers. Patent RU 2077831 (in Russian).

19. Anisina, J. A. Efficiency of Cultivation of Perennial Grasses Herbs in a Single Crop in the Contaminated Floodplains (on the Example of Bryansk Region); Authoref. diss.... kand. PhD: Bryansk, 2012; p 20 (in Russian).
20. Gulyakin, I. V.; Yudintseva, E. V. Agricultural Radiology. Kolos ("Ear" in Rus.) Publishing House: Moscow, 1973; p 272 (in Russian).
21. Agrochemicals Methods of Investigations of Soils. "Nauka" ("Science" in Rus.) Publishing House, 1975; p 656 (in Russian).
22. Guidelines for the Determination of Natural Radionuclides in Soils and Plants. Bryansk, 1985; pp 79–85 (in Russian).
23. Svetov, V. A. Agricultural Production on the Contaminated Territories of the Russian Federation. *Himizatziya selskogo hozyastva* ("Chemicalization of agriculture" in Russian) **1991,** *11*, 9–13 (in Russian).
24. Vorobiev, G. T.; Bobrovsky, A. I.; Prudnikov, P. V. Agrochemical Properties of Soils of the Bryansk Region and the Use of Fertilizers. Agrohimradiologiya ("Agrochemradiology" in Russian): Bryansk, 1995; p 121 (in Russian).
25. Belous, N. M. The Case of the Chernobyl. *Vestnik Bryanskoi selskohozyaistvennoi akadenii* ("Bulletin of Bryansk State Agricultural Academy" in Russian) **2016,***2* (54), 3–8 (in Russian).
26. Belous, N. M. The Development of Radioactive Contaminated Territories in Bryansk Region in a Separate Period after the Accident at the Chernobyl Nuclear Power Plant. *Vestnik Bryanskoi selskohozyaistvennoi akadenii* "Bulletin of Bryansk state agricultural Academy" in Russian) **2018,** *1* (65), 3–11 (in Russian).
27. Podolyak, A. G.; Arastovich, T. V. Economic and Radiological Assessment of the Effectiveness of Agrochemical Methods for Improving Meadows Contaminated with 137Cs and 90Sr as a Result of the Chernobyl Accident. *Ekologicheskaya antropologiya* ("Environmental Anthropology" in Russian) **2004,** 77–81 (in Russian).
28. Podolyak, L. G.; Timofeev, S. F.; Grebenshchikova, N. V.; Arastovich, T. V., Zhdanovich, V. Prediction of Accumulation of Cs-137 and Sr-90 in the Grasslands of the Main Types of Meadows of the Belarusian Polesie on Agrochemical Properties of Soils. *Raditsionnaya biologiya. Radioekologiya* ("Radiation biology. Radioecology" in Russian) **2005,** *45*, 1, 100–111 (in Russian).
29. Sychev, V. G.; Belous, N. M.; Smolski, E. V. Influence of Potash Fertilizers on the Content of Cesium-137 in the Green Mass of Natural Forage Lands with Surface Improvement. *Plodorodie* ("Fertility" in Russian) **2012,** *1*, 2–4 (in Russian).
30. Korenev, V. B.; Vorobyova, L. A. Influence of Increasing Doses of Potash Fertilizers on the Yield and Accumulation of Cs-137 Crops. Agrochimicheskii Vestnik ("Agrochemical messenger" in Russian): Moscow, 2016; Vol. 2, pp 20–22. (in Russian)

CHAPTER 18

Beyond the Molecular Frontier: Challenges for Chemistry and Chemical Engineering

AIDÉ SÁENZ-GALINDO*, ADALI O. CASTAÑEDA-FACIO, JOSÉ J. CEDILLO-PORTILLO, and KARINA G. ESPINOZA-CAVAZOS

School of Chemistry, Universidad Autónoma de Coahuila. Boulevard Venustiano Carranza and José Cárdenas s/n. Republica Oriente, C.P. 25280, Saltillo, Coahuila, México

Corresponding author. E-mail: aidesaenz@uadec.edu.mx

ABSTRACT

Molecular chemistry is the scientific study about the molecular structure of a lot of materials and the difference between these molecules, although it is a confusing area of study for some investigators; due to that its a science that has short time of deep study, it is just constant development. Molecular chemistry has opened important applications for science due to it is development in different areas that include the use of nanoscale materials; these materials have many uses thanks to the versatility to be applied in different compounds.

This work shows the importance of and applications to molecular chemistry in multiple areas.

18.1 INTRODUCTION

Chemistry is a science of great importance for the development and progress of different needs the human. It is divided into diverse areas like organic, inorganic, analytical, materials, polymers, etc. The molecular chemistry is derivate of organic chemistry and inorganic chemistry, it is based on processes of molecular assembly and self-assembly through intermolecular links and interactions, these types of molecular processes are carried out through the different functional groups that present the molecules involved, the arrangement of the molecules, and each of the atoms or active center, that make them give as results three-dimensional a structure, with a well-defined stereo chemical and reactive, this type of arrangement which can be predesigned with computer software in most cases and can be on macro, micro, or nanoscale, which makes it very interesting to explore. The obtained products present important applications in different areas such as biology, materials science catalysis, polymers, metals, etc. (Fig. 18.1).

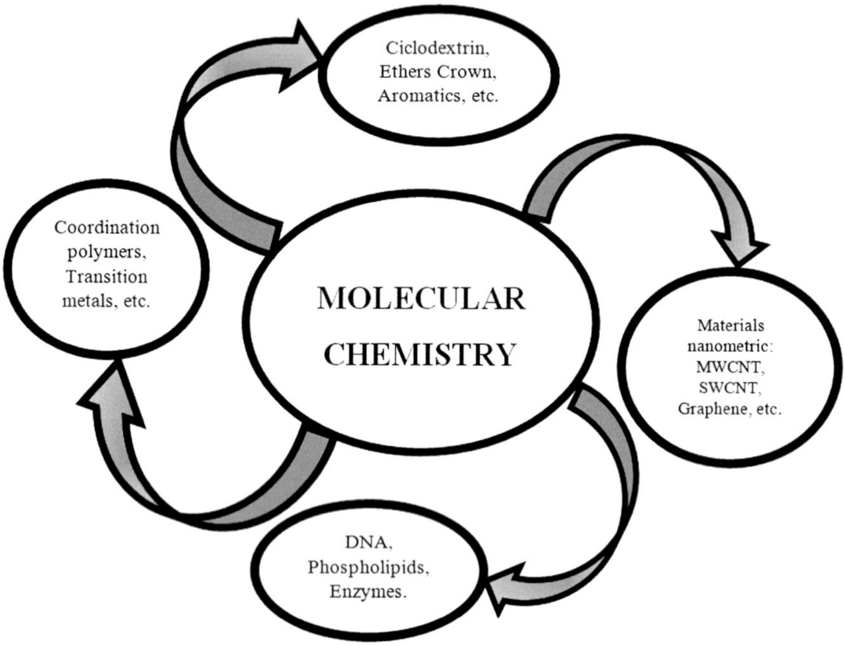

FIGURE 18.1 Examples of molecular chemistry.

In general, two types of molecules are used, one is known as a host and the second as a gust. The host usually be a cyclic molecule and have a cavity in its chemical structure, to be hosted by the host. In the present chapter, several molecular chemistry applications are presented in different areas, highlighting the advances that have been achieved, in catalysis, polymer, and biology. Currently, the molecular chemistry is of great interest because it studies organic and inorganic chemistry. The molecular chemistry is a subdivision according to the position of the elements in the periodic table.

18.2 APPLICATIONS

18.2.1 CATALYSIS

Catalysis is of great application in different areas, which has an impact on molecular chemistry. The catalysis is divided, according to the type of phase in homogeneous or heterogeneous catalysis, each type of catalysis has its veins and disadvantages, in general, the homogeneous catalysis is the most used because it is characterized by its low industrial costs. In molecular chemistry is present the system host-guest, where host are usually complex cyclical compounds. In Figure 18.2, examples of cyclic organic compounds, which can act as hosts, are presented in the host-host system, crown ether.

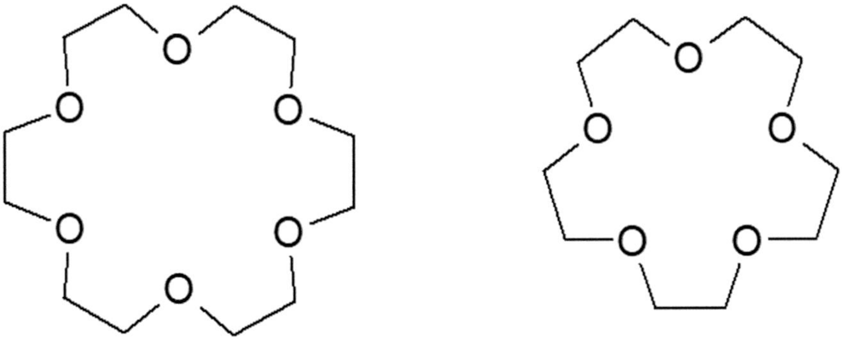

FIGURE 18.2 Examples of structure chemistry of cyclical compounds. Hosts.

Zeng et al. (2002) reported a study of the synthesis and design of the biomimetic catalytic used Schiff bases and transition-metal complexes, found that mono Schiff bases containing crow ether ring and transition metal complexes are active like biomimetic catalytic to oxidation for styrene to benzaldehyde whit 100% selectivity. The presence of the crow ether ring, dioxygen is important for the activity of the oxidation of the organic molecules. The biomimetic catalytic was formed whit crow ether ring whit transition-metal complexes like Co (II) and Mn (III), found that distance between the coordination center and crow ring is a parameter important to process oxidation.[1]

In 2002, Liu et al.,[2] published a study of the obtaining of the singles crystal of two calixarene derivatives with possible application in the synthesis of the catalysts this type of the compounds type of calixarene can act like a system host-guest complexes, reported the synthesis of 5,11,17,23-tetra-tert-butyl- 25,27-bis[2-[N-(3-methoxy-4-methoxybenzylidene)- amino] ethoxy]-26,28- dihydroxy calix[4]arene and 5,11,17,23-tetra-tert-butyl-25,27-bis[2-[3- pyridine carbonylamino] ethoxy]-26,28-dihydroxy calix[4]arene.

In 2017, reported a review of the study of the advances in click chemistry reactions, showing that catalysts based on Cu(I) ions and nanoparticles have the potential to mediate and activate the alkyne-azide cycloaddition reactions in mild conditions and with remarkable yields, in this report presents different applications in diverse area, where demonstrate the importance of molecular chemistry in different areas, highlighting the catalysis.[3]

Chelucci published a report very interesting about the synthesis and design of various classes of structure of pyrroles and indoles from alcohols as reactive. These organic compounds are used in metal-induced dehydrogenation converts primary or secondary alcohols into the related carbonyl compounds as aldehydes or ketones; the metals that are active at this type of complexes can be Ru, Pd, Os, and Ir.[4]

Currently was reported a study about of the different interactions noncovalents in metals complex catalysis, this type of interaction is emerging as a powerful catalytic method for compounds organic synthesis, one of the advantages of metal catalysis is that coordination compounds can promote divers organic reactions not only by coordination bound but also through noncovalent interactions, with different compounds or elements halogen (Cl, F, Br, etc.) hydrogen, cations-π, anion-π, etc. The catalytic transformations are a green alternative useful in diverse processes, in different areas, this type of catalytic transformation can make use of systems which have a donor and an acceptor, where the interactions are noncovalent, and this facilitates the action of the catalyst in the process, obtaining excellent results in time, costs, and viability.[5]

Recently was published a study about novel multidentate phosphine used ligant for Pd-catalyzed, highlighting that chemistry, structural and electronic properties of the phosphine play important role for Pd-catalized alkoxycarbonylation of alkynes, found that developed ionic tri-dentate phospine enabled Pd-catalyzed alkoxycarbonylation of composted type of alkynes, demonstrating that most efficiently while H_2O was used as an additive in comparison whit an acid.[6]

Lee et al. reported in 2018, the uses of dithiocarbamates and analogues as coordinated mono and di selenium binders, as an active center.[8] This type of coordination compounds present different applications, the dithiocarbamate ligands, whit mono and di-selenium analogues, this class of ligands and their metal complexes are ubiquitous in coordination chemistry, presented in different area of organic chemistry, materials chemistry, biology, etc. where the use of the type of catalyst accelerates various processes involving chemical transformations, obtaining a decreases in reaction times.[7]

In the 2019, Mahmudov et al. reported an interested review, where the importance of isocyanide metal complexes in catalysis was explained, highlighting the use of the isocyanides as complexes active in different reactions of organic transformations, where involve diverse metals like Mo, Rh, Fe, Ru, Ni, Co, etc. found that thanks to the catalysis it is possible to obtain different organic compounds, in process concerted.[8] The use of metals in the area of chemistry is of great interest; coordination chemistry or inorganic chemistry presents important applications in molecular chemistry.

18.2.2 MATERIAL SCIENCE POLYMER

The molecular chemistry presents different contributions in the area of materials, specifically in polymer which can be natural or synthetic. In natural polymer, different polymers have been studied, where the chitosan, Hernandez et al. in 2008, reported of study of Fe^{3+} coordination whit chitosan in acid aqueous solution, these authors studied the present species that are formed by changing the acid pH 2 to basic pH, finding that the most active species were detected at pH from 4 to 6. This active species presented complexation activity whit amino groups and hydroxyl groups present in chitosan.[9]

Cojocaru et al. reported in 2019, obtaining a new composite magnetic absorber bases on a biopolymer type polysaccharide, chitosan for removal of water-soluble anionic dye. The composite was prepared with chitosan doped spinel ferrite (15%), chitosan cross-linked with glutaraldehyde. This composite present applications as a remover of Acid Orange (AO7) dye from aqueous solution. Chitosan is considered a polycationic biopolymer containing free amino groups, responsible for different physical and chemical properties.[10]

In 2019, Rogina et al. studied the preparation of a chitosan-bases hydrogel modified by Cooper (II) and Zinc (II) ions through complexation interactions between the amine and hydroxyl groups of chitosan, obtaining that the indirect evaluation of the cytotoxicity of the hydrogel with Cooper and Zinc, reported that the interaction between amine and metals plays an important role in the stability of the hydrogel as well as its physical properties.[11]

Recently Tikhonov et al. reported in 2019, the formation of polymer complex (poly-1-vinyl-1,2,4-triazole), with Silver ions in the structure, observed that the presence of heterocycles with nitrogen in their structure allowed them the synthesis of stable polymers with silver, likewise, they reported that there is no change in the chemical structure of the polymer because the FT-IR spectroscopy does not show the appearance of new bands only the displacement in the bands of the triazole cycle showing that the metal compounds polymer maintain the properties of the polymer matrix and Silver because the structure is not modified, in addition, these complexes may be of interest for the design for new high-tech materials such as in the strategy of food storage or biomedical applications and are precursors of promising nanocomposites containing silver.[12]

Niu et al. reported in 2019, obtaining of two new complexes of crystalline silver coordination whit polymer using naphthridine ligands (BPAN-diimine, DPEN), finding that this type of polymer can present different values of the thermal conductivity.[13] Likewise in 2019, Cui et al., studied the synthetic route of new polymer of coordination whit Co(II) $\{[Co_3(L)_2(tib)2(H_2O)_2]*10\ H_2O*2HO*2O)\}$ n [L = 2, 6-di(4-carboxylphenyl) pyridine-4-carboxylic, tib = 1, 3, 5-tris(1-imidazolyl)benzene)] by methods solvothermal, this complexes present application in adsorbent in aqueous solutions to elimination methylene blue (MB) in water.[14]

Díaz et al. reported in 2019, the preparation of carbon mesoporous doped whit Co and Ni, from pyrolysis at 700°C and 900°C and acid leaching used a linear coordination polymer whit a complex ligand N-heteroatoms, obtaining that the material derived from Co 2,3-pyrazinedicarboxylate polymer. This type of the polymeric material exhibited a reduction in its capacitance of only 2% after 3000 charge–discharge cycle and found important results in the pore dispersion of the material, like wise reported that these results are relevant for the design of new capacitors.[15] The molecular chemistry presents diverse of application indirect on polymeric materials, especially in the area of conductor's polymer or doped polymer. A class of interesting polymer is highly branched polymers; these polymers present important properties, high solubility, low viscosity, and excellent thermal stability, in comparison with lineal polymers. These present different applications in the fields of drug delivery, catalyst carrier, liquid crystal, supramolecular chemistry, and nanophase materials.

In 2019, Zhang et al. reported a study of preparation of polysiloxane-based hyperbranched fluorescent materials used thiol-ene "click" chemistry. This type of novel polysiloxane-based hyperbranched fluorescent material has been prepared by rare earth ions coordination. The rare earth elements present unique fluorescence properties and applications in the fields of light, magnetism, and electricity. This type of material can be used in bioimaging.[16]

18.2.3 BIOLOGY

The area of biology, is an area of great importance for the needs of the human being and molecular chemistry, also has a presence. The "click chemistry" is an alternative viable whit different application as reported in 2017 by Yoon et al., which exposes an application of click chemistry in metabolic glycoengineering and biorthogonal click chemistry. The metabolic glycoengineering is an active technique that can introduce different and varied chemical groups to cellular glycan by treatment of unnatural monosaccharide. The importance of the "click chemistry" in biology is very important, with the use of the click chemistry can predict different process and design many modifications in numerous organic composed. This type of chemistry is very useful to a group of chemical reactions with high yield, fast reaction rate, nontoxic byproduct and can be done in aqueous condition or use green solvent. Yoon reported the investigation of the recent progress in molecular imaging based on the combination, product of the combination of the two technical "click chemistry," and metabolic glycoengineering. This combination has enabled many challenging trials for molecular imaging.[16]

Important area for humans is referent virology and click chemistry also has application; the whit use of click chemistry possibly introduces different bond C-C and C-halogen. By using click chemistry, the virus particle as well as viral protein and nucleic acids can be labeled. This type of investigation is present in application in bioscience fields, such as chemical biology, drug development, and bionanoparticles, as a promising tool to modify biomolecules, such as DNA, protein, and virions.[17]

In 2017, Finetti reported the study of novel gel for DNA electrophoresis using a classic "click chemistry" reaction whit copper (I)-catalyzed azide–alkyne cycloaddition (CuAAC), to cross-link functional polymer chains, found that this type of novel materials gel for DNA electrophoresis is useful, present, important and resulted in time and reagent consumption. The analysis gave results that the used classic reaction cop-per (I)-catalyzed azide–alkyne cycloaddition (CuAAC) reaction. This type of the reaction did not require UV initiator and it decreases the toxicity of monomers; found excellent resulted in separations of DNA. Conclude that is a good alternative green for separation of the DNA, used whit matrix of gel organic by "click chemistry."[18]

In 2019, was reported a study by Bartheldyová et al. about the applications of the click chemistry in synthesis and characterization of lipids, of mannan-coated nanoliposomes and in vitro stimulation of dendritic cells, found that a new route of synthesis and design of organic composed aminooxy lipid like a block for the formulation of functionalized nanoliposomes by microfluidic mixing, this organic composed aminooxy lipid present application like in vitro experiments on human and mouse dendritic cells demonstrate high selective.[19]

KEYWORDS

- **chemistry**
- **molecular**
- **applications**
- **materials**
- **nanoscale**

REFERENCES

1. Zeng, W.; Mao, Z.; Wei, X.; Li, J.; Hong, Z.; Qin, S. Synthesis, Dioxygen Affinities and Biomimetic Catalytic Oxidation Performance of Crown Ether-tethered Schiff Base Transition-Metal Complexes. *J. Supramol. Chem.* **2002,** *2*, 501–507.
2. Liu, Y.; Wang, H.; Zhang, H. Y.; Zhao, B. T.; Wang, Li. H. Novel Linear Molecular Aggregation Tethered by Hydrogen-Bonded Interaction Within the Crystalline Calix[4] arene Derivatives. *J. Supramol Chem.* **2002,** *2*, 515–519.
3. Tăbăcaru, A.; Furdui, B.; Ghinea, I. G.; Cârâc, G.; Dinic, R. M. Recent Advances in Click Chemistry Reactions Mediated by Transition Metal Based Systems. *Inorg. Chemica Acta.* **2017,** *455*, 329–349.
4. Chelucci, G. Metal-Catalyzed Dehydrogenative Synthesis of Pyrroles and Indoles from Alcohols. *Coord. Chem. Rev.* **2017,** *331*, 37–53.
5. Mahmudov, K. T.; Gurbanov, A. V.; Guseinov, F. I.; Guedes da Silva, M. F. Noncovalent Interactions in Metal Complex Catalysis. *Coord. Chem. Rev.* **2019,** *387*, 32–46.
6. Yang, D.; Liu, L.; Wang, D. L.; Lu, Y.; Zhao, X. L.; Liu, Y. Novel Multi-Dentate Phosphines for Pd-Catalyzed Alkoxycarbonylation of Alkynes Promoted by H_2O Additive. *J. Catal.* **2019,** 371, 236–244.

7. Lee, S. M.; Heard, P. J.; Tiekink, E. R. T. Molecular and Supramolecular Chemistry of Mono-and di- Selenium Analogues of Metal Dithiocarbamates. *Coord. Chem. Rev.* **2018,** *375*, 410–423.
8. Mahmudov, K. T.; Kukushkin, V. Y.; Gurbanov, A. V.; Kinzhalov, M. A.; Boyarskiy, V. P.; Guedes da Silva, M. F.; Pombeiro, A. J. Isocyanide Metal Complexes in Catalysis. *Coord. Chem. Rev.* **2019,** *384*, 65–89.
9. Hernández, R. B.; Franco, A. P.; Yola, O. R.; López-Delgado, A.; Felcman, J.; Recio, M. A.; Ramalho, A. L. Coordiantion Study of Chitosan and Fe^{3+}. *J. Mol. Struct.* **2008,** *877*, 89–99.
10. Cojocaru, C.; Samoila, P.; Pascariu, P. Chitosan-Based Magnetic Adsorbent for Removal of Water-Soluble Anionic Dye: Artificial Neural Network Modeling and Molecular Docking Insights. *Int. J. Bio. Macromol.* **2019,** *123*, 587–599.
11. Rogina, A.; Lončarević, A.; Antunović, M.; Marijanović, I.; Ivanković, M.; Ivanković. H. Tuning Physicochemical and Biological Properties of Chitosan through Complexation with Transition Metal Ion. *Int. J. Biol. Macromol.* **2019,** *129*, 645–652.
12. Tikhonov, N. I.; Khutsishvili, S. S.; Larina, L.; Pozdnyakov, A.; Emel′yanov, A.; Prozorva. A.; Vashchenko, A.; Vakul′skaya. Silver Polymer Complexes as Precursors of Nanocomposites Based on Polymers of 1-vinyl-1,2,4-Triazole. *J. Mol. Struct.* **2019,** *1180*, 272–279.
13. Niu, G. H.; Wentz, H. C.; Zheng, S. L.; Campbell, M. G. Silver (I) Coordination Polymers from Dinucleating Naphthyridine Ligands. *Inorganic Chem. Communication.* **2019,** *101*, 142–144.
14. Cui, L.; Li, Ch.; Qin, D.; Liu, F.; Long, J. Study on Adsorptive Removal for Methylene Blue of a Co (II) Coordination Polymer. *J. Mol. Struct.* **2019,** *1177*, 269–274.
15. Díaz-Duran, A. K., Montiel, G.; Viva, F.; Roncaroli, F. Co, N-Doped Mesoporous Carbons Cobalt Derived from Coordination Polymer as Supercapacitors. *Electro Chemica Acta.* **2019,** *299*, 987–988.
16. Zhang, Y.: Zuo, Y.; Yang, T.; Gou, Z.; Lin. W. Polysiloxane-Based Hyperbranched fluorescent Materials Prepared by Thiolene "Click" Chemistry as Potential Cellular Imaging Polymers. *Eur. Polym J.* **2019,** *112*, 515–523.
17. Yoon, H. Y.; Koo, H.; Kim, K.; Kwon, I. C. H. Molecular Imaging Based on Metabolic Glycoengineering and Bioorthogonal Click Chemistry. *Biomaterials* **2017,** *132*, 28–36.
18. Finettia, C. H.; Sola, L.; Elliott, L.; Chiaria, M. Synthesis of Hydrogel via Click Chemistry for DNA Electrophoresis. *J. Chromatogr. A.* **2017,** *1513*, 226–234.
19. Bartheldyová, E.; Knotigová, P. T.; Zachová, K.; Mašek, J.; Kulich, P.; Effenberg, R.; Zyka, D.; František Hubatka, F.; Kotouček, J.; Čelechovská, H.; Héžová, R.; Tomečková, A.; Mašková, E.; Fojtíková, M.; Macaulay, F.; Bystrický, P.; Paulovičová, L.; Paulovičová, E.; Ladislav Drož, L.; Ledvina, M.; Raška, M.; Turánek, J. N-Oxy Lipid-Based Click Chemistry for Orthogonal Coupling of Mannan Onto Nanoliposomes Prepared by Microfluidic Mixing: Synthesis of Lipids, Characterisation of Mannan-Coated Nanoliposomes and in vitro Stimulation of Dendritic Cells. *Carbohydr. Polym.* **2019,** *207,* 521–532.

CHAPTER 19

Neutrino: The Novel Messenger of the Universe

FRANCISCO TORRENS[1*] and GLORIA CASTELLANO[2]

[1]*Institut Universitari de Ciència Molecular, Universitat de València, Edifici d'Instituts de Paterna, P. O. Box 22085, E-46071 València, Spain*

[2]*Departamento de Ciencias Experimentales y Matemáticas, Facultad de Veterinaria y Ciencias Experimentales, Universidad Católica de Valencia San Vicente Mártir, Guillem de Castro-94, E-46001 València, Spain*

[*]*Corresponding author. E-mail: torrens@uv.es*

ABSTRACT

The 20th century witnessed tremendous progress thanks to the observation of electromagnetic radiation. Neutrino and gravitational waves are now the new messengers that allow opening a new window to the Cosmos that could revolutionize understanding of the universe. Neutrino and gravitational waves open unprecedented new windows in the cosmos. Neutrino and gravitational astronomies open two new windows to the observation of the universe, in addition to classical photon astronomy. No one knew of the neutrino when nuclear β-decay was first observed. When the total kinetic energy of the decay particles was measured, energy seemed to have disappeared because no one could detect the neutrino, let alone measure its energy. Quantum mechanics is so strange that some believed energy conservation might not strictly apply for the smallest distance scales. Pauli was convinced that energy conservation could never be violated, even for the submicroscopic atomic nuclei. He salvaged energy conservation postulating an invisible particle, which was later named the *neutrino* by Fermi, who also refined the theory. It is not surprising that some were sceptical of Pauli's particle and its unseen energy. Fortunately for physics, the neutrino is not quite invisible. In 1955, an experiment detected the elusive neutrino.

19.1 INTRODUCTION

Setting the scene: neutrino as the novel messenger of the universe, neutrino oscillation, neutrino astronomy, telescope, trembling particles, oscillations, searching for neutrino physics with liquid argon detectors, neutrinos, neutron-star merger, neutrinos in astrophysics, unveiling universe secrets with neutrinos, γ-rays and gravitational waves (GWs), the elusive neutrino, from quantum chromodynamics to medical imaging and story of a woman researcher.

Hahn, Schmidt and Meitner (1907) performed experiments on β-radiation and observed the fact, theoretically difficult to explain, that the electron is emitted from the nucleus not with a fixed energy but with a continuous energy distribution.[1] A peculiarity of β-decay is that the energy of the emitted electron has continuous values, not the discrete values that one would expect in view of the well-defined energy level of the nucleus. Pauli (1927) expressed the idea that a new, as yet unknown, particle emitted along with the electron was responsible for the differences in energy.

$$(A,Z) \rightarrow (A,Z+1) + \bar{e} + \nu_e$$

$$n \rightarrow p + \bar{e} + \bar{\nu}$$

Pauli thought that this particle was a constituent of the nucleus. Fermi (1932) formulated his β-decay theory, proposing the idea that the neutrino (ν) comes into being during the decay process, exactly as does the electron (*cf.* Fig. 19.1). From pictures taken in the Wilson cloud chamber (Csikai and Szalay, 1959), it is clear that with the neutrino hypothesis, momentum conservation can also be satisfied. Understandably, most physicists doubted the actual existence of such a particle because its existence was only supported by indirect evidence. They were hardly satisfied with the explanation that the neutrino has no observable positive effects because its interaction with matter is extremely weak. Direct experimental proof of the neutrino was by Reines and Cowan (1955, 1958). Neutrino inactivity does not mean that it is emitted with little energy: In a nuclear reactor with a power of 100 MW, 5 MW radiates away in the form of neutrinos. Because they do not interact much with matter, neutrinos tend to accumulate in the universe and add to its average mass density. The following questions (Q) were raised.

- Q1. Why does Universe consist of matter and not antimatter if at start equal amounts were created?
- Q2. How did universe form with the matter that people see?
- Q3. *Women* and *science*: A contradiction in terms?

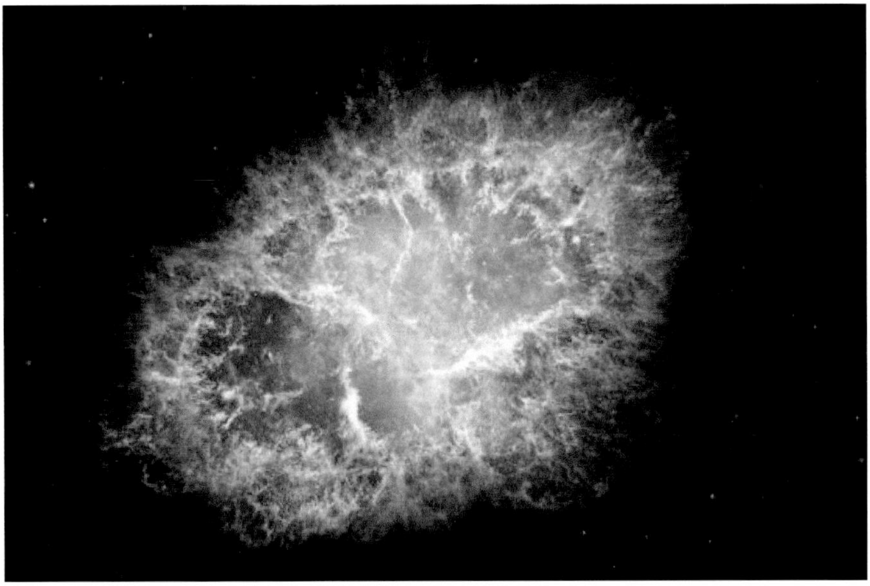

FIGURE. 19.1 Supernova 2018: a source of neutrinos.

In earlier publications, it was reported nuclear fusion, the American nuclear cover-up in Spain in Palomares disaster (1966),[2] Manhattan Project, *Atoms for Peace*, nuclear weapons, accidents[3] nuclear science and technology.[4] In the present report, it is reviewed some reflections on neutrino oscillation, neutrino astronomy, telescope, trembling particles, oscillations, searching for neutrino physics with liquid argon detectors, neutrinos, neutron-star merger, neutrinos in astrophysics, unveiling universe secrets with neutrinos, γ-rays and GWs, the elusive neutrino, from quantum chromodynamics to medical imaging and story of a woman researcher. The aim of this work is to initiate a debate by suggesting a number of questions, which can arise when addressing subjects of the elusive neutrino, from quantum chromodynamics to medical imaging and story of a woman researcher. It was provided, when possible, answers (A), facts (F), and hypotheses (H) on the elusive neutrino.

19.2 NEUTRINO OSCILLATION

In parallel to the development of neutrino-detection technology, neutrino-astronomy significance is growing: Neutrinos provide information from the interiors of stars. People's ideas about their structure (density, temperature distribution, etc.) can be verified or expanded. Measurements taken since

1970s with ever increasing precision yielded a stark discrepancy with the results calculated on the basis of the accepted solar model: The number of recorded neutrinos was at most half of the expected one. The current theory is that electron neutrinos created in the Sun change during their travels into muon or tauon neutrinos. This is called *neutrino oscillation*. For a zero-mass neutrino, however, such a transformation would be, in principle, forbidden. The particle physics Standard Model (SM) allows for, but does not explain, a nonzero mass for the neutrinos.

19.3 NEUTRINO ASTRONOMY, TELESCOPE, TREMBLING PARTICLES AND OSCILLATIONS

Kimball raised the following questions on neutrinos.[5]

Q1. Why do the neutrinos have mass?
Q2. Why are the masses so small?
Q3. Why are there three kinds of neutrinos, or are there more?
Q4. Do neutrinos possess time reversal symmetry?
Q5. Can neutrino properties explain the lack of antimatter?

Monrabal proposed questions/answers on neutrino telescope and neutrino astronomy.[6]

Q6. Is the proton stable?
Q7. Does neutrino become unpleasant?
Q8. What is a neutrino?
Q9. How to insert the neutrino in SM?
A9. The neutrino was inserted in SM with zero mass but it has mass and oscillates.
Q10. Where do neutrinos come from?
A10. They come from nuclear reactions (nuclear plants, stars).
Q11. If people do not know how a nuclear plant works, how does one know how a star works?
Q12. Why do people do astronomy with neutrinos?
A12. Photons stop by cosmic rubbish; charged particles (protons, electrons), by magnetic fields.
Q13. How do people detect neutrinos?
Q14. Why not to try to see the neutrinos from other stars?
Q15. What does the distance between detectors depend for visible communication?
A15. Distance is 2 m in tap water, 8 m in distilled water and 125 m in ice.

Q16. How do people search for neutrinos?
Q17. What does IceCUBE tell people?
A17. People can see neutrinos from out of the Solar System, possibly out of the Milky Way.
Q18. What objects can be source of neutrinos of so much energy?
Q19. Where do these superenergetic neutrinos come from?
Q20. What is the future?
A20. A bigger detector.
Q21. And this, what is it good for?
A21. Fermi paradox: *Where is everybody?*

Peña proposed the following questions, answers and hypotheses on neutrino astronomy.[7]

Q22. Where do neutrinos come from (*cf.* Fig. 19.2)?

FIGURE 19.2 Every second, 66,000 millions of neutrinos cross every square centimeter of one's skin.

Q23. What titanic processes did create those particles?
Q24. What is the origin of inertia?
A24. Today, people know that it is fundamentally in the force that binds nucleons in the nucleus.
Q25. What is the inertial mass?
A25. The inertia of the elemental particles is in the coupling of Higgs' field.

Q26. Is there something more?
Q27. How does the Sun shine?
Q28. Where are the neutrinos produced?
Q29. How are the neutrinos produced?
Q30. How many neutrinos come from the Sun?
Q31. What does produce galactic and extragalactic neutrinos?
Q32. Do γ-ray bursts (GRBs) produce galactic and extragalactic neutrinos?
Q33. Do black holes (BHs) produce galactic and extragalactic neutrinos?
H1. (Majorana, 1937). The neutrino is of a Majorana nature.
H2. Neutrinoless double-β (0νββ) decay

$$(A,Z) \rightarrow (A,Z+2) + 2\bar{e}$$

as well as a range of lepton-number violating (LNV) meson

$$K^+ \rightarrow \mu^+ \mu^+ \pi^-$$

and charged lepton (cL)

$$\tau^\pm \rightarrow (\mu_1\, \mu_2\, \mu_3)^\pm$$

decays are possible.
Q34. How are the experiments financed?
Q35. What resolution can be obtained in the detector of neutrinos.
A35. One degree (1°).
Q36. Do neutrinos of different origin travel at the same speed?
A36. Difference is small because they are relativistic particles: the more energy, the more speed.
Q37. What is the state of the detector in the Mediterranean Sea in France?
A37. ANTARES is small as old IceCUBE (1/30 IceCUBE); it has detected no extragalactic neutrino.
Q38. Did Majorana predict a neutrinoless double-β decay violating the lepton number?
A38. Yes. A number of experiments probing whether neutrino is a Majorana particle are at work.

Santamaria proposed questions/A on trembling *ghost* particles and neutrinos oscillations.[8]

Q39. What are neutrinos oscillations?
Q40. What is importance of such discovery to understand the Earth and evolution of the Universe?
Q41. Why have neutrinos mass?
Q42. Why do neutrinos transform into each other?
Q43. Why is no explanation of neutrinos in SM?

Q44. Periodic table of elements allows understanding atom; what does periodic table of particles?
Q45. How do neutrinos affect people's life?
Q46. Is the Canfranc Underground Astroparticle Laboratory used to detect neutrinos?
A46. Yes, detectors are underground; Canfranc is an abandoned tunnel, ideal for detecting neutrinos.
Q47. What about the fourth dimension?
A47. Problems are solved via additional dimensions but it is important to check that associated experiments exist to test.
Q48. Is antineutrino only a theoretical proposal?
A48. No, it is both theoretical/experimental, e.g., those from supernova SN1987A are antineutrinos.
Q49. How was the detection of neutrinos and light from SN1987A?
A49. Light allowed neutrinos time; in future, neutrinos will permit designing light observation.
Q50. Do neutrinos explain dark matter?
A50. No, SM needs some correction.
Q51. Is there change of mass in neutrinos conversion?
A51. No, every neutrino type has a constant mass; conversion reactions are at constant mass.
Q52. Why are there only three types of neutrinos?
A52. Experiment in Sudbury Neutrino Observatory (SNO) revealed 3 types with probability 100%.
Q53. Are there superluminic neutrinos?
A53. No, it was an experimental error.
Q54. Are there additional dimensions and curved space?
A54. If there are any additional dimension, people cannot see it; it must be rolled in itself.

Gómez Cadenas proposed questions, hypothesis and answers on searching neutrinos.[9]

Q55. Where does the matter come from?
Q56. Why does the universe exist?
H3. The neutrino is its own antiparticle.
Q57. If the universe is symmetric in matter and antimatter, why do not people see the antimatter?
A57. People not only do not see the antimatter but also they do not found a scrap of it left.
Q58. Where is the antimatter?

A58. If, in some place, there were antimatter, it had collided with matter generating enormous energy.
Q59. Why is not antimatter in the universe?
A59. All the antimatter generated during the Big Bang was annihilated on colliding with the matter.
Q60. However, then, where has all the matter, which forms the universe, come up from?
Q61. Why is there universe?
A61. At some moment, at first, something happened unbalancing matter–antimatter in favor of the former.
A62. Such something was a neutrino-like particle capable of acting as both matter and antimatter.
Q62. Why is the universe made from only matter?
Q63. In addition, explaining that the neutrino is its own antiparticle, how is it made?
Q64. What does it happen if the neutrinos are their own antiparticle?
A64. Alternative process exists in which one is particle, the other, antiparticle, and both annihilate: experiment NEXT (*cf.* Fig. 19.3).

FIGURE 19.3 Canfranc detector: Chamber inside, covered with ultraradiopure Cu to attenuate outer radioactivity.

19.4 SEARCHING FOR NEUTRINO PHYSICS WITH LIQUID ARGON DETECTORS

Guenette proposed questions/A on searching for ν physics with liquid argon (LAr) detectors.[10]

Q1. Why does matter dominate the Universe?
Q2. Why is the neutrino mass so small?
Q3. Are there any more types of neutrinos?
Q4. How can LAr detectors help people addressing some remaining puzzles of ν physics field?
Q5. How can people search for neutrino physics?
Q6. What does people know of neutrinos?
A6. Three neutrinos; two square mass differences; three mixing angles.

She proposed great questions on searching for neutrino physics with liquid argon detectors.

Q7. What are the absolute neutrino masses?
Q8. What is the nature of neutrinos (either Dirac or Majorana)?
Q9. What is the mass hierarchy (ordering)?
Q10. Is there charge–parity (CP) violation?
Q11. What is δ_{CP}?
Q12. Are there sterile neutrinos?

She proposed additional questions/answer on searching for ν physics with LAr detectors.

Q13. Why argon?
A13. Any inert gas is good but Ar is economic ≈1€/L but Xe ≈3000€/L.
Q14. Sterile neutrinos?
Q15. Protons $\rightarrow \nu_\mu \rightarrow \nu_e$?
Q16. Identity signal (either γ or electron)?
Q17. Either electron or photon?

19.5 NEUTRINOS AND NEUTRON-STAR MERGER: NEUTRINOS IN ASTROPHYSICS

Tamborra proposed the following questions and answer on neutrinos and neutron-star merger.[11]

Q1. Do neutrinos affect element production?

A1. Neutrino may play a major role especially for element production around the polar region.
Q2. What does it concern the element production around the polar region/early emission?

She raised the following questions on neutrinos in astrophysics.

Q3. Where are these neutrinos coming from?
Q4. New physics?
Q5. Do neutrinos affect element production?
Q6. First high-energy neutrino traced back to its birthplace?

She provided the following conclusions (Cs).

C1. Neutrinos offer an ideal window on the high-energy Universe.
C2. Sources poorly understood, multi-messenger methods powerful.
C3. Neutrino behavior in the sources remains to be unraveled.
C4. Neutrinos are unique in physics.

19.6 NEUTRINOS TELESCOPES

Institute for Corpuscular Physics proposed question/A/F on neutrinos telescopes (*cf.* Fig 19.4).[12]

Q1. How do people observe the Universe?
Q2. How are neutrinos produced?
A2. Cosmic neutrinos should be produced in cosmic rays–matter/electromagnetic radiation (EMR) interaction.
Q3. Where should one put the detector?
F1. Detector IceCube alerted a high-energy neutrino 170922A.[13]

19.7 UNVEILING UNIVERSE SECRETS: NEUTRINOS, Γ-RAYS, AND GRAVITATIONAL WAVES

Kajita unveiled Universe secrets with neutrinos (*cf.* Fig. 19.5), γ-rays and GWs raising questions[14]:

Q1. How were the masses of neutrinos discovered?
Q2. How can γ-rays and GWs shed some new light on our knowledge of the Universe?

He presented the neutrino program in Japan.

FIGURE 19.4 A blazar is accelerating protons that produce pions, which produce neutrinos and γ-rays. Neutrinos are always the result of a hadronic reaction, e.g., that displayed here. γ-Rays can be produced in hadronic and electromagnetic interactions. Source: IceCube/NASA.

FIGURE 19.5 Super-Kamiokande neutrinos detector.

19.8 THE ELUSIVE NEUTRINO

Fleming raised the following questions on the elusive neutrino.[15]

Q1. Are there more neutrinos out there than the three we have already observed?
Q2. Do neutrinos and anti-neutrinos behave in the same way?
Q3. Could this suggest that neutrinos are the reason we exist?

19.9 FROM QUANTUM CHROMODYNAMICS TO MEDICAL IMAGING: RESEARCHER

Ferrario proposed Q/A from quantum chromodynamics to medical imaging/woman researcher.[16]

Q1. Why do people want to photograph electrons?
A1. In order to detect the 2 electrons that leave double-β nuclear disintegration without neutrinos.
Q2. Where do neutrinos come from?
Q3. Why do not people notice the so many neutrinos that continuously go through them?
A3. Because the neutrino free halfway is much greater than a person's size.
Q4. Is the neutrino is own antiparticle?

19.10 DISCUSSION

Possible applications of neutrinos follow.

1. Neutrinos provide information from the interiors of stars (photons are coupled to plasma, bounce and do not go out in straight line).
2. They are proposed for unidirectional communication with submerged submarines.

19.11 FINAL REMARKS

From the present discussion, the following final remarks can be drawn.

1. The 20th century witnessed tremendous progress thanks to the observation of electromagnetic radiation. Neutrino and gravitational waves are now the new messengers that allow opening a new window to the cosmos that could revolutionize understanding of the Universe. Neutrino and gravitational waves open unprecedented new windows in the cosmos. Neutrino and gravitational astronomies open two new windows to the observation of the Universe, in addition to classical photon astronomy.
2. No one knew of the neutrino when nuclear β-decay was first observed. When the total kinetic energy of the decay particles was measured, energy seemed to have disappeared because no one could detect the neutrino, let alone measure its energy. Quantum mechanics is so strange that some (notably Bohr) believed energy conservation might not strictly apply for the smallest distance scales. Pauli was convinced that energy conservation could never be violated, even for the submicroscopic atomic nuclei. He salvaged energy conservation postulating an invisible particle, which was later named the *neutrino* by Fermi, who also refined the theory. It is not surprising that some were sceptical of Pauli's particle and its unseen energy. Fortunately for physics, the neutrino is not quite invisible. In 1955, an experiment detected the elusive neutrino. The experiment was far from easy. Today, neutrino measurements reveal new physics, but the experiments remain challenging because neutrinos usually pass via everything without leaving the faintest footprint.
3. Particle-physics problems hint at our unfinished search for the ultimate theory. The neutrino is a good illustration of mysteries yet to be solved. Its existence was postulated by Pauli long before this nearly invisible particle was *seen*. Pauli's neutrino was necessary for energy conservation in radioactive decay. Zero-mass neutrinos are an SM component. However, neutrino masses are not zero. Neutrinos mysteries seem more dramatic because they are the most abundant particles in the universe. They are as ubiquitous as photons and their total mass is compatible to everything else.
4. There is something wrong with quantum mechanics. It predicts probabilities for experiments, not certainties. However, the formal mathematics of quantum mechanics predicts the quantum state with certainty. The mathematical conclusion is a Schrödinger's cat that is both alive and dead. Those of people who observe Schrödinger's cat are also in a superposition of two states, but there is no way they

can be aware of their other half. It impresses many as too exotic that people themselves are in two (actually many) states and the probabilities of quantum mechanics are only artefacts of their existence in the form of many quantum states.

5. String theory opened the door to a unification of gravity and quantum mechanics but the door enters onto a mathematical jungle. Someone needs to open another door.
6. Supersymmetry might be the second door (or part of the first) above but so far experimental confirmation is lacking.
7. We do not share the faith of many scientists that the basic questions will eventually be answered. It may turn out that Wigner's idea is correct, and one is no more capable of solving the ultimate problems than a dog, of learning calculus. One may be barking up the wrong tree.
8. Neutrinos offer an ideal window on the high-energy Universe. Sources poorly understood, multi-messenger methods powerful. Neutrino behavior in the sources remains to be unraveled. Neutrinos are unique in physics.

ACKNOWLEDGMENTS

The authors thank support from Generalitat Valenciana (Project No. PROMETEO/2016/094) and Universidad Católica de Valencia *San Vicente Mártir* (Project No. 2019-217-001).

KEYWORDS

- **neutrino oscillation**
- **neutrino astronomy**
- **neutrino telescope**
- **trembling particle**
- **neutrino physics**
- **liquid argon detector**

REFERENCES

1. Furtado Valle, J.W.F.; Romão, J.C. *Neutrinos in High Energy and Astroparticle Physics*; Wiley–VCH: Winheim, Germany, 2015.
2. Torrens, F.; Castellano, G. Nuclear Fusion and the American Nuclear Cover-up in Spain: Palomares Disaster (1966). In *Engineering Technology and Industrial Chemistry with Applications*; Haghi, R., Torrens, F., Eds.; Apple Academic–CRC: Waretown, NJ, 2019, pp. 297-308.
3. Torrens, F.; Castellano, G. Manhattan Project, Atoms for Peace, Nuclear Weapons, and Accidents. In *Molecular Chemistry and Biomolecular Engineering: Integrating Theory and Research with Practice*; Pogliani, L., Torrens, F., Haghi, A. K., Eds.; Apple Academic–CRC: Waretown, NJ, in press.
4. Torrens, F.; Castellano, G. *Nuclear Science and Technology*. In *Chemistry and Industrial Techniques for Chemical Engineers*; Pogliani, L., Ameta, S. C., Haghi, A. K., Eds.; Apple Academic–CRC: Waretown, NJ, in press.
5. Kimball, J. *Physics: Curiosities, Oddities, and Novelties*; CRC: Boca Raton, FL, 2015.
6. Monrabal, F. Personal Communication.
7. Peña, C. Personal Communication.
8. Santamaria, A. Personal Communication.
9. Benavente, R.P. Juan José Gómez Cadenas: Físico de Partículas. *El Confidencial* 2015, 2015(Jan. 23) 1-1.
10. Guenette, R. Personal Communication.
11. Tamborra, I. Personal Communication.
12. Institute for Corpuscular Physics (IFIC), Personal communication.
13. IceCube Collaboration, Neutrino Emission from the Direction of the Blazar TXS 0506+056 prior to the IceCube-170922A alert. *Science* **2018,** *361*, 147–151.
14. Kajita, T. Personal Communication.
15. Fleming, B. Personal Communication.
16. Ferrario, P. Personal Communication.

CHAPTER 20

Infectious Bursal Disease

P. RAJA

Department of Animal Biotechnology, Madras Veterinary College, Tamil Nadu Veterinary and Animal Sciences University, Chennai 600007, Tamil Nadu, India.
E-mail: vetraja86@gmail.com

ABSTRACT

Infectious bursal disease (IBD) has been reported all over the world due to their high socio-economic significance. Infectious bursal disease is a highly contagious disease of young chickens in which the tissues of the immune system of birds, especially the bursa of Fabricius, resulting in immunosuppression and susceptibility to other infections, such as *E. coli, Salmonella, Mycoplasma*, coccidia, Marek's disease, and others. Moreover, the subclinical form of infection leads to immunosuppression which is often difficult to measure. The IBD virus genome is bi-segmented and is susceptible to genetic mutations through random recombination. The very virulent strains of IBDV (vvIBDV) have mutations that allow the virus to replicate faster and cause more severe disease. The diagnosis of different forms of IBD is by identifying a combination of characteristic clinical signs, the degree of flock mortality and post-mortem lesions. The laboratory confirmation of the disease is by identification of the virus in affected tissues from dead birds. The primary focus for prevention and control should be on biosecurity of poultry premises and proper implementation of vaccination schedule. Rigorous cleaning and disinfection between flocks are essential for minimizing potential disease spread.

20.1 INTRODUCTION

In India, poultry industry is growing at a very high pace and turning out to be a major contributor to the economy of rural and semi-urban India. India ranks seventeenth in the world poultry production index. Further, India is the third largest producer of eggs and fifth largest producer of poultry meat globally (Prabakaran, 2012). The major problems threatening the poultry industry such as low weight gain, low egg production, low meat production and heavy economic losses are basically due to nutrition, managemental practices, various infectious diseases and vaccination failures. Among them, infectious diseases which are immunosuppresive in nature are the major threat to susceptible chickens to subsequent infections. One of the most deadly immunosuppressive viral diseases of poultry is infectious bursal disease (IBD) which affects the flock even after administration of the respective vaccines and the reasons for occurrence of the disease in vaccinated birds could not exactly be pointed out or justified.

Infectious bursal disease virus (IBDV) is a member of the family *Birnaviridae*, genus *Avibirnavirus*. The earliest report showed that IBDV had affinity towards the kidney of infected chicken and caused severe kidney damage. Hence, the disease was first called as "Avian Nephrosis". Since the first outbreak was observed in the area of Gumboro, Delaware, USA the disease is also known as "Gumboro Disease" (Cosgrove, 1962). Isolation of the etiological agent of the disease in embryonated eggs was attempted by Winterfield and Hitchner in 1962 and they had succeeded. They referred it as "infectious bursal agent" as it caused more damage to bursa of Fabricius (BF). Finally, Hitchner in 1970, proposed the term "infectious bursal disease" as appropriate name for the disease due to their characteristic lesion in the cloacal bursa of the infected chickens. The disease remained in its classical form until early 1990 with mortality up to 30%. However, in the later part of 1990s, considerably high mortality due to IBD was reported and it remained one of the most important contributors to huge economic loss due to immunosuppressive nature to the poultry farmers across the world even after following proper vaccination schedule (Li et al., 2011). The disease may be fatal when highly virulent strains are involved. However, different antigenic variants of IBDV have been reported and they are responsible for subclinical form of disease that leads to various degrees of immunosuppression. As a consequence of immunosuppression, the birds may become extremely susceptible to other pathogenic agents, such as Newcastle disease virus (NDV), infectious bronchitis virus (IBV), infectious laryngotracheitis virus (ILT), *Salmonella* sp. or to opportunistic agents such as *Escherichia coli.*

Two IBDV serotypes namely serotype 1 and serotype 2 have been identified with subtypes based on the antigenic variation within each serotypes (Saif, 1984). Both serotypes naturally infect the chickens and turkeys, but only serotype I is pathogenic to chickens, causing clinical disease to the birds (Jackwood et al., 1984). Based on virulence and antigenicity, serotype 1 viruses have been classified as attenuated, intermediate virulent, classical virulent, variant, and very virulent (van den Berg, 2000). Currently, serotype I IBDV viruses are antigenically grouped as classic and variant strains based on virus neutralization assay (Eterradossi and Saif, 2008). The very virulent IBDV (vvIBDV) strains are considered to be the most antigenically similar to classic IBDV strains, despite their enhanced pathogenic properties and ability to penetrate maternal immunity induced by classical, mild IBDV vaccines (van den Berg and Meulemans, 1991). These findings suggest that vvIBDVs are continuously evolving in nature and further investigation is needed to determine the biological significance of these antigenic differences (Eterradossi et al., 2004).

Innate immune system is fully functional in the newly hatched chickens whereas the optimal adaptive immune responses develop following the first week after hatch. If chickens receive a vaccine at the time when the maternal antibody level is still high, the vaccine may be inactivated by maternal antibodies, which leads to a weak immune response. In contrast, if vaccination takes place long after the decline of maternal antibody levels, chickens would be susceptible to diseases prior to be immunized (Hair-Bejo et al., 2004). Damages to the BF due to late immunization of chickens at early susceptible days of age by IBDV or Marek's disease inhibit proper B cell maturation. This leads to immune susceptibility so that affected chickens will neither be able to raise immune response against attacking pathogens nor respond to immunization program, adequately. As a consequence, the right time for vaccination and boosting the immune system is critical for the poultry industry.

To establish the proper control procedures for IBD, it is important to characterize the antigenic and virulence properties of the IBDV strains prevalent in that geographic area; thus, it is necessary to develop rapid and accurate methods for identifying the virus and typing different strains of IBDV. The conventional precipitation assays such as agar gel immunodiffusion (AGID), single radial immunodiffusion (SRID) and counter immunoelectrophoresis (CIE) have been employed for many virus detection studies and are generally methods of choice in most of the laboratories with moderate facilities. With the advent of modern molecular biology, nucleic acid-based methods have evolved and proved useful just not only for detecting the viruses, but also for strain variation and other relevant molecular characterization studies.

20.2 ETIOLOGY

Infectious bursal disease virus (IBDV) is a non-enveloped virus, with a single capsid shell of icosahedral symmetry composed of 32 capsomers and a diameter of 55 to 60 nm. The virus genome consists of two segments, A and B. Segment A (3.2 kb) codes for structural proteins, VP2, VP3 and VP4, and non-structural protein, VP5. Segment B codes for non-structural protein VP1, the viral transcriptase. The VP2 gene is considered as the major host-protective virus antigen containing at least three neutralizing epitopes and the sites that determine the virus virulence. The hypervariable (HVR), is an ideal site for determining the antigenic variation by sequence analysis of different IBDV strains (Proffitt et al., 1999).

20.3 EPIDEMIOLOGY

After its first outbreak in poultry in Southern Delaware of US in 1962, a number of outbreaks were reported throughout the world with different percentage of morbidity and mortality. Factors such as the strain of the virus involved, age of birds, vaccination status, type of vaccine, season, concurrent infections, and immunosuppression status also influenced the occurance of the disease.

Winterfield and Hitchner (1962) described the differences between Infectious bronchitis virus (IBV) and the viral agent responsible for causing Gumboro disease. Incidences of clinical IBD were reported in birds between 21 and 35 days of age with high mortality when compared to other outbreaks with low mortality rates as reported by various workers. In India, the disease was first reported by Mohanty et al. (1971). However, the virus was isolated for the first time in the country by Jayaramaiah and Mallick (1974). Later on, the disease was reported in different forms from all the states. The existence of a second serotype of IBDV was reported by McFerran et al. (1980). According to Okoye (1984), the outbreaks of IBD were reported in almost all the poultry rearing parts of the world.

In 1986, a very virulent (vv) strain of IBDV (vvIBDV) was isolated in Netherlands. Thereafter, very severe clinical outbreaks with high mortality rates (90%–100%) caused by vvIBDV were reported in China, Japan, UK, USA, Taiwan, and many other countries of Europe and Asia. Giambrone (1987) stated that IBD in subclinical form was associated with a variant IBDV which varied in ability to cause mortality but invariably caused immunosuppression.

In 1984–1985, the antigenically variant IBDV started to appear in Delmarva peninsula of the USA where there was an increased mortality in broilers. This infection differed from classical serotype 1 strain in that it produced a very rapid bursal atrophy with less inflammatory response, whereas classical strains initially produced an increase in bursal size followed by atrophy (Cavanagh et al., 1992). Vaccination based on the classical strains did not protect against the variant IBDV strains. At the end of 1980s and during 1990s, hypervirulent or vvIBDV were reported in Europe and Asia. Unlike the variant strains, the vvIBDV were antigenically very similar to the classical strains but had a marked increase in virulence causing high mortality in infected flocks as reported by various workers.

The hypervirulent forms of the disease were first described in Europe in the late eighties followed by Japan in early nineties (Nunoya et al., 1992). Later, this form rapidly spread to all over the Asiatic and European continents including Indian subcontinent. Although very virulent strains were closely related to classical IBDV strains and no antigenic drift was detected, these strains caused increased mortality of up to 25% in broilers, 60% in layers and 100% in specific pathogen free (SPF) chickens (van den Berg et al., 1991). With the emergence of vvIBDV, the disease was found spreading to all parts of the country with alarming consequences. The acute IBD characterized by high mortality rate in broiler flocks in northern India followed by similar episodes in rest of the country throughout the year 1993–1994. The mortality was ranging from 10% to 75% in 3 to 16 weeks old White Leghorn and 10% to 40% in broiler flocks of 72 poultry farms in different states of the country.

Bekhit (1997) carried out investigations from 82 outbreaks of IBD in broiler chickens in Egypt, the mortality varied between 4.36% and 27.49%. Farooq et al. (2003) reported that average mortality due to IBD in 50 broiler farms in Mirpur and Koti districts of Kashmir (Pakistan) during the years 1997 and 1998 was 15.31%. Significantly, higher losses were recorded in winter (19.84%) than in the spring season (9.0%). Further, the losses were significantly higher (17.66%) in broilers of more than 32 days of age than at 19–23 days of age (12.42%). The losses due to IBD were more in the presence of coccidiosis than in those flocks without coccidiosis problem.

Zeleke et al. (2005) first time reported the occurrence of IBD in Ethiopia as the country was known to be free from infectious bursal disease. The disease affected 20–45 days old broiler and layer chickens. The mortality rate of the disease in broiler and layer chickens ranged from 45% to 50%. The overall mortality rate was 49.89%. Swai et al. (2011) studied the seroprevalence rates in healthy, non-vaccinated indigenous scavenging chickens in northern Tanzania and serum samples were screened for IBD virus antibodies using

ELISA. The study revealed high rates of IBDV antibodies, yielding an overall seropositive rate of 58.8%.

Tadesse and Jenbere (2014) reported seroprevalence of IBD in backyard chickens at Ethiopa. Higher seroprevalence of the infection 360 (94.5%) was recorded in 3–6 weeks of age groups. Zegeye et al. (2015) studied the seroprevelance of IBDV in chicken reared under backyard poultry production systems around Mekelle town, northern Ethiopia. Indirect ELISA was used to test the sera for IBD in chickens and the prevalence was found to be 45.05%. This study indicated the presence of IBD antibodies in unvaccinated backyard chickens due to field exposure to IBD viruses.

20.4 IBDV GENOME ORGANIZATION AND FUNCTIONS

IBDV is a nonenveloped virus with dsRNA genome (Dobos, 1979). The main open reading frame of segment A of IBDV encodes a polyprotein (NH_2-pVP2-VP4-VP3-COOH) that is cleaved by the virus encoded protease into pre-capsid virus protein (VP) VP2, VP4, and VP3 within infected cells. The precapsid VP2 undergoes defined sequential C-terminal cleavage by VP4 (viral protease), endopeptidase activity of VP2, and by host protease (puromycin-sensitive aminopeptidase) to release the mature VP2 protein.

The VP2 crystal structure indicated three domains: the base (B), shell (S), and projection (P) domains. The B and S domains were formed by the conserved N-and C-terminal stretches of VP2. The P domain was the middle part containing the host cell receptor binding motifs and the hypervariable region of VP2 (HVP2). The HVP2 harboured antigenic major hydrophilic peak A (aa 212-224) and B (aa 314-325) (Schnitzler et al., 1993) that formed the loops P_{BC} (aa 219-224) and P_{HI} (aa 316-324), respectively. The minor hydrophilic peak 1 (aa 248-254) and peak 2 (aa 279-290) in the HVP2 formed the loops P_{DE} (aa 249-254) and P_{FG} (aa 279-284). Another open reading frame on segment A encoded the VP5 protein that partially overlapped with the 5′ end of the polyprotein gene. It was a host membrane-associated and highly basic protein with a cytoplasmic N-terminus and an extracellular C-terminus domain (Lombardo et al., 2000).

Segment B encoded the RNA dependent RNA polymerase, which was present in the virion as a free protein or covalently linked to the 5′ ends of both genome segments (VPg) (Muller and Nitschke, 1987). The VP1 crystal structure revealed three domains namely the N-terminus (aa 1-167), central polymerase (aa 168-658), and C-terminal (aa 659-878) regions. The N-terminus of VP1 was involved in protein priming as it possessed the putative guanylation site. The central polymerase domain folded into a right-hand shape (fingers-palm-thumb) structure. The five RNA polymerase motifs (C, A, B, D, and E) were located in the palm region of the polymerase. Each of the RNA polymerase motifs had specialized function during virus replication. Motif A, B, and F involved in nucleotide recognition and binding, motifs A and C involved in phosphoryl group transfer, motif C involved in a metal ion like Mn^{2+} and Mg^{2+} binding, motif D involved in nucleotide guidance to active sites and motif E involves in primer gripping. Motif C formed the polymerase active site. The finger sub-domains had polymerase motifs F and G that were involved in virus replication (Pan et al., 2007).

20.5 HOST SUSCEPTIBILITY

McFerran et al. (1980) reported two serotypes of IBDV, namely serotype I and II in which serotype I virus affected the chicken. The natural hosts of IBDV were the chicken and the turkey. Turkeys, ducks, and ostriches could be naturally and experimentally infected with IBDV serotypes I and II, as evidenced by serological response and isolation; however, the infections were apathogenic. Several other avian species were also susceptible to infection including rooks, wild pheasants, crows, gulls, and falcons. But, only serotype I viruses replicated in lymphoid cells and were pathogenic to chickens. Among the chicken breeds, the most severe clinical signs and lesions and the highest mortality rates were observed in white leghorns (Eterradossi and Saif, 2008).

20.6 VIRAL REPLICATION

Even though, the main target organ for pathogenic serotype I IBDV was the bursa of Fabricius, source for B lymphocytes in avian species, the initial viral replication occured in the intestine. Following host entry via oral infection or inhalation, IBDV initiated replication in lymphocytes and macrophages of the gut-associated lymphoid tissues (GALT) such as the caecum, duodenum, and jejunum. This stage of viral replication marked the primary viremia. Within 5 h post-infection, viral antigen reached the liver, where it was trapped and phagocytosed by resident macrophages (Kupffer cells). Virus then entered the bloodstream where it was distributed to other tissues including the bursa of Fabricius. By 13 h post-infection, most bursal follicles were positive for virus and extensive necrosis of immature B-cells in the medullary and cortical regions of the bursal follicles ensued (Kim et al., 1998). By 16 hours post-infection, a second massive viremia occured with secondary replication in other B-lymphocyte-containing tissues including the spleen, thymus, Harderian gland, and cecal tonsils. Clinical signs and death might result from the acute phase (7–10 days) of IBD. Factors such as virulence of a strain, as well as, the chicken's age, breed, and immune status could influence the outcome and severity of infection.

20.7 IMMUNOSUPPRESSION

Immunosuppression caused by IBDV had a significant economic impact due to widespread nature of the disease in commercial chickens. If the infection occurred within the first two-three weeks of age, it would exhibit more severe form of immunosuppressive effects (Allan et al., 1972). Due to immunosuppressive nature, the infected chicken exhibited lower flock performance, more secondary infections, poor feed conversion, less protective response to vaccines and higher rate of carcass condemnation at the processing level (Sharma et al., 2000).

The immunosuppressive effects of IBDV were dependent on the strain of the virus. Chicken infected with IBDV at an earlier age succumbed to other infections like inclusion body hepatitis, reovirus, coccidiosis, Marek's disease, hemorrhagic-aplastic anemia and gangrenous dermatitis, infectious laryngotracheitis, infectious bronchitis, chicken anemia agent, salmonellosis, *Escherichia coli*, *Mycoplasma synoviae,* and *Eimeria tenella.*

20.8 REASSORTMENT IN IBDV

Genetic reassortment played a vital role in emergence of vvIBDV in the late 1980s in Europe. The time of appearance of the most recent common ancestor (tMRCA) of very virulent (vv) VP2 was approximated around 1960, whereas of vvVP1 around 1980. This newly appeared vvVP1 from an unidentified avian reservoir was suggested to recombine with an already existing vvVP2 to evolve to the vvIBDV genotype, which then caused massive mortality in Europe (Hon et al., 2006). This indicated the independent evolutionary history of the two segments of vvIBDV (Le Nouen et al., 2006). Recently, several natural reassortant IBDV isolates were reported all over the world. The most common reassortant IBDVs had segment A of vvIBDV and segment B from attenuated strains indicating the drawbacks of extensive application of live IBDV vaccines. Attempted experimental generation of reassortant viruses by co-infecting SPF chickens with vvIBDV and attenuated serotype 1 IBDV failed.

20.9 DIAGNOSTIC METHODS

20.9.1 ISOLATION OF IBDV

20.9.1.1 CHICKEN EMBRYO INOCULATION

The inoculation of bursal homogenates from IBDV infected chickens via the chorioallantoic membrane (CAM) route of 9–10 days old embryonated SPF chicken eggs was the most sensitive diagnostic method for virus isolation(Abdul, 2004). The embryos died mostly within three to five days in the case of classical and very virulent viruses and in case of variant strains embryos did not die but showed hepatic necrosis. Gross lesions in embryo might include cutaneous congestion, edematous distension of the abdomen, petechial hemorrhages along the feather tracts and occasional hemorrhages in the toe joints and cerebral area, necrosis of the liver, pallor of the heart, congestion and some necrosis of the kidneys, extreme congestion of the lungs, and pallor of the spleen with occasional small necrotic foci.

20.9.1.2 IN VITRO VIRUS PROPAGATION OF IBDV

Yamaguchi et al. (1996) demonstrated the method for adaptation of IBDV field isolates to a cell culture system by initially passaging several times in embryonated chicken eggs followed by subsequent passages in cell culture system to further attenuate the virus to the extent that these viruses do not induce bursal lesions. The cell culture adapted viruses replicated in primary avian cells such as CEFs and continuous cell lines of avian (QT35) and mammalian origins (Vero cells).

20.9.2 IMMUNOLOGICAL METHODS

20.9.2.1 AGAR GEL PRECIPITATION TEST (AGPT)

The test has been widely used over a long period of time throughout the world as it is easily adaptable to any laboratory condition. Kosters and Geissler (1971) performed AGPT using bursal homogenates from the infected chicken to detect IBDV. Faragher (1972) determined optimal conditions of immunodiffusion reactants associated with IBD and found them similar to those required by other avian systems. He found that precipitating antigen was organ specific and detected only in the bursa of the infected chickens.

20.9.2.2 DOT BLOT ASSAY

The dot blot assay is a good alternative to ELISA and IFAT in the serodiagnosis. Cruz-Coy et al. (1993) used monoclonal antibody (mAb) developed against a variant subtype of IBDV, to recognize all six serologic subtypes of IBDV and three untyped IBDV by dot blot method. Anil et al. (2002) reported that dot blot was as equally sensitive as RT-PCR.

20.9.2.3 COUNTER IMMUNO ELECTROPHORESIS (CIE)

In CIE, the antigen and antibody are driven by an electric current rather than passive diffusion. Durojaiye et al. (1985) confirmed eight outbreaks of IBD by CIE test in which precipitin lines were visible within 30 min when compared with AGPT which took 18–24 h. Somvanshi et al. (1985) compared AGPT and CIE test for detecting IBDV antigen and antibodies, and concluded that CIE was specific, faster, simple and more sensitive than AGPT.

20.9.2.4 ENZYME-LINKED IMMUNOSORBENT ASSAY (ELISA)

Howie and Thorsen (1981) showed that ELISA was a precise, sensitive and reproducible means of measuring IBDV antibodies in chicken and turkey sera. Viral antigen preparation was crucial to the precision of the ELISA test. Purified virus prepared from high titer seed virus was less non specific than that from low titer of seed virus for an ELISA. the VP2 gene of IBDV expressed in baculovirus was used as an antigen in an ELISA and could detect IBDV neutralizing antibodies from SPF chickens sera infected with IBDV strains. Wang et al. (2008) showed that recombinant VP3 expressed in *E. coli* was used as antigen in detecting the antibodies in field chicken sera.

20.9.2.5 VIRUS NEUTRALIZATION TEST (VNT)

The VNT was a more sensitive antibody detection tool and could be used for both antibody quantitation and differentiation of IBDV serotypes and subtypes. The test served to compare virus strains by evaluating the ability of antiserum raised against one IBDV strain to neutralize a heterologous strain. However, only IBDV strains that efficiently replicated in embryos or cell culture could be used in the VNT assay. Attempts to adapt the virus to a host might result in antigenic and pathological changes to the virus that might render VNT results as suspected.

20.9.2.6 LATEX AGGLUTINATION TEST (LAT)

The latex agglutination test is a simple, rapid and cost-effective test and thus quite applicable in developing countries. The test can be performed in 30 min and spot detection can be done with naked eye. Nakamura et al. (1994) performed a competitive agglutination test using the polystyrene latex bound monoclonal antibody to detect the serum antibody titer against IBDV.

20.9.3 MOLECULAR DIAGNOSTIC METHODS

20.9.3.1 POLYMERASE CHAIN REACTION

Reverse transcriptase–polymerase chain reaction (RT-PCR) was a molecular tool frequently applied for IBD diagnosis. Wu et al. (1997) developed the first PCR test for the detection of IBDV. A set of primers that specified a 150 bp fragment in segment A of IBDV genome was used to distinguish the IBDV from other infections in chicks. The PCR could detect 2 fg of IBDV RNA. A sensitive and specific multiplex polymerase chain reaction (mPCR) was developed and optimized for simultaneous detection and differentiation of avian reovirus (ARV), avian adenovirus group I (AAV-I), infectious bursal disease virus (IBDV) and chicken anemia virus (CAV) (Caterina et al., 2004). Zhang et al. (2002) detected IBDV during early stages of the disease by *in situ* RT-PCR using primers specific for VP4 region of segment A of IBDV. Kataria et al. (2004) performed RT-PCR on 17 bursal samples suspected for IBD collected from different parts of India. Multiple alignments with different strains suggested that simultaneous mutations in different regions were the most probable cause for strain variations. It was not possible to predict the most important amino acid residues without complete knowledge of the three-dimensional structure of the viral proteins.

20.10 GENOTYPIC CHARACTERIZATION OF IBDV ISOLATES

20.10.1 GENOTYPIC CHARACTERIZATION OF IBDV ISOLATES BY RT-PCR-RE/RFLP

Reverse transcription-polymerase chain reaction was a molecular tool frequently applied in IBDV diagnosis. Most of the RT-PCR protocols were based on VP2 nucleotide sequences. Reverse transcription-polymerase chain reaction followed by digestion with multiple restriction enzymes (RE) or restriction fragment length polymorphism (RFLP) and nucleotide sequencing of VP2 gene were used for differentiation of IBDV strains. The most widely used typing method was the amplification of VP2 gene of 743 bp followed by digestion with restriction enzymes. Different strains could be grouped into genotypes based on restriction fragment length polymorphism (RFLP) patterns.

20.10.2 GENOTYPIC CHARACTERIZATION OF IBDV BY VP2 PARTIAL SEQUENCING

Nucleotide sequencing of RT-PCR products was widely used for further characterization of IBDV strains. Amplification of IBDV-VP2 gene and linking genetic variation found in this region with antigenic variation was the major focus for strain identification in recent years. Among five proteins of IBDV, the VP2 was the major host-protective antigen that induced serotype-neutralizing antibodies (Fahey et al., 1989), the middle third of which had a highly variable region (HVR) that ranged from amino acid (aa) positions 206 to 350. It included two hydrophilic regions referred to as VP2 major hydrophilic peak A (aa 212–224) and peak B (aa 314–324). Mutations within these hydrophilic coding regions were thought to be responsible for the evolution of antigenic variants and virulent serotype 1 strains (Heine et al., 1991 and Schnitzier et al., 1993). Three additional minor hydrophilic peaks at aa positions 248–252, 279–290, and 299–305 were considered to influence

IBDV antigenicity (van den Berg et al., 1991). At aa position 326–332, there was a serine-rich heptapeptide (SWSASGS), which had correlation with the virulence of IBDVstrains (Heine et al., 1991) and concluded that this might be one of the virulence markers of IBDV strains. The aa residues at positions 253, 279, and 284 of VP2 were also involved in the virulence, and VP1 was considered as another important virulence marker of IBDV.

20.11 CONTROL AND PREVENTION

Prevention of IBD necessitates hygiene measures and medical prophylaxis. No vaccine can solve the problem if major sanitary precautions are not taken. These precautions include "all-in/all-out" farming methods, cleaning and disinfection of premises, and observance of a "down time". All farm equipment must be disassembled and stored in cleaning rooms located outside the farm buildings. The buildings, immediate surroundings and farm equipment must be dry-cleaned first, in order to eliminate all dust, and then washed using hot water (60 °C) with a detergent, at a pressure of 80–150 bar. A second disinfection of the full premises must be performed before the introduction of the chicks.

In addition to strict compliance with rules of hygiene and disinfection, the success of vaccination depends on the choice of the vaccine strain and on the vaccination schedule. These must take account of the existence of certain pathotypes and the presence of antigenic variants in certain regions.

20.12 CONCLUSION

Infectious bursal disease virus presents a certain number of characteristics that are of importance in the diagnosis and control of IBD. The disease is caused by a small, nonenveloped virus, highly resistant to the outside environment. Substantial economic losses result from both the clinical and subclinical (or immunosuppressive) forms of the disease. Infectious bursal disease virus has a high mutation rate and may thus give rise to viruses of modified antigenicity or increased virulence. Although satisfactory protection may be provided by the induction of high neutralising antibody titres, interference of parental antibodies in vaccination has become the most important obstacle in the establishment of control programs.

KEYWORDS

- **infectious bursal disease**
- **VP2 hypervariable region**
- **immunosuppresive disease**
- **reassorted IBDV strains**
- **molecular characterization**

REFERENCES

1. Abdul, A., Isolation and Pathological Characterization of IBD Isolate from an Outbreak of IBD in a Rural Poultry Unit in Bangladesh [M.S. thesis], The Royal Veterinary and Agricultural University, Copenhagen, Denmark, 2004.
2. Allan, W. H.; Faragher, J. T.; Cullen, G. A. Immunosuppression by the Infectious Bursal Agent in Chickens Immunized Against Newcastle disease. *Vet. Rec.* **1972,** *90*, 511–512.
3. Anil, T. M.; Shankar, K. M.; Mohan, C. V. Monoclonal Antibodies Developed for Sensitive Detection and Comparison of White Spot syndrome Virus Isolates in India. *Dis. Aquat. Org.* **2002,** *51*, 67–75.
4. Bekhit, A. B. A. Highly Virulent form of Infectious Bursal Disease in Egypt: Some Epidemiological Observations. *Ind. J. Anim. Sci.* **1997,** *67*, 363–366.
5. Caterina, K. M.; Salvatore, F. Jr.; Girshick, T.; Khan, M. Development of a Multiplex PCR for Detection of Avian Adenovirus, Avian Reovirus, Infectious Bursal Disease Virus, and Chicken Anemia Virus. *Mol. Cell. Probe.* **2004,** *18*(5), 293–298.
6. Cavanagh, D.; Davis, P. J.; Cook, J. K. A. Infectious Bronchitis Virus: Evidence for Recombination within the Massachusetts Serotype. *Avian Pathol.* **1992,** *21*, 401–408.
7. Cosgrove, A. S. An Apparently New Disease of Chickens-Avian Nephrosis. *Avian Dis.* **1962,** *6*, 385–389.
8. Cruz-Coy, J. S.; Giambrone, J. J.; Panangala, V. S. Production and Characterization of Monoclonal Antibodies Against Variant A Infectious Bursal Disease Virus. *Avian Dis.* **1993,** *37*, 406–411.
9. Dobos, P. Peptide Map Comparison of the Proteins of Infectious Bursal Disease Virus. *J. Virol.* **1979,** *32*, 1046–1050.
10. Durojaiye, O. A.; Adene, D. F.; Owoade, A. A. Counter Immuno-Osmo Electrophoresis for Diagnosis of IBD of Poultry. *Trop. Anim. Hlth. Prod.* **1985,** *17*, 255–263.
11. Eterradossi, N.; Gauthier, C.; Reda, I.; Comte, S.; Rivallan, G.; Toquin, D.; de Boisseson C.; Lamande, J.; Jestin, V.; Morin, Y.; Cazaban, C.; Borne, P. M. Extensive Antigenic Changes in a Typical Isolate of Very Virulent Infectious Bursal Disease Virus and Experimental Clinical Control of This Virus with an Antigenically Classical Live Vaccine. *Avian Pathol.* **2004,** *33*, 423–431.
12. Eterradossi, N; Saif, Y. M. Infectious Bursal Disease. In Y. M. Saif, A. M. Fadly, J. R. Glisson, L. R. McDougald, L. K. Nolan, D. E. Swayne (eds.). *Diseases of Poultry*, 12th ed., Blackwell Publishing Professional: Ames, IA, 2008, p. 185–208.

13. Fahey, K. J.; Erny, K.; Crooks, J. A Conformational Immunogen on VP2 of Infectious Bursal Disease Viruś that Induces Virus-Neutralizing Antibodies that Passively Protect Chickens. *J. Gen. Virol.* **1989,** *70*, 1473–1481.
14. Faragher, J. T. Infectious Bursal Disease of Chickens. *Vet. Bull.* **1972**, *42*, 361–369.
15. Farooq, M.; Durrani, F. R.; Imran, N.; Durrani, Z. Prevalence and Economic Losses due to Infectious Bursal Disease in Broilers in Mirpur and Koti Districts of Kashmir. *Int. J. Poult. Sci.* **2003,** *2*, 267–270.
16. Giambrone, J. J.; Ewert, D. L.; Eidson, C. S. Effect of Infectious Bursal Disease Virus on Immunological Responsiveness of Chicken. *Poult. Sci.* **1977,** *56*, 1591–1594.
17. Hair-Bejo, M.; Ng, M. K.; Ng, H. Y. Day Old Vaccination Against Infectious Bursal Disease in Broiler Chickens. *Int. J. Poult. Sci.* **2004,** *3*, 124–128.
18. Heine, H. G.; Haritou, M.; Failla, P.; Fahey, K.; Azad, A. Sequence Analysis and Expression of the Host-Protective Immunogen VP2 of a Variant Strain of Infectious Bursal Disease Virus Which Can Circumvent Vaccination with Standard Type I Strains. *J. Gen. Virol.* **1991,** *72*, 1835–1843.
19. Howie, R. I.; Thorsen, J. Identification of a Strain of Infectious Bursal Disease Virus Isolated from Mosquitos. *Can. J. Com. Med.* **1981s** *45*, 315–320.
20. Jackwood, D. J.; Saif, Y. M.; Moorhead, P. D.; Bishop, G. Failure of Two Serotype II Infectious Bursal Disease Viruses to Affect the Humoral Immune Response of Turkeys. *Avian Dis.* **1984,** *28*, 100–116.
21. Jayaramiah, B.; Mallick, B. B. Identification of Infectious Bursal Disease Virus in India. *Ind. J. Anim.Hlth.* **1974**, *13*, 99–2103.
22. Kataria, R. S.; Tiwari, A. K.; Butchaiah, G.; Das, S. K. Diagnosis of Infectious Bursal Disease by RT-PCR. *Ind. Vet. J.* **2004,** *81*, 12–15.
23. Kim, I. J.; Karaca, K.; Pertile, T. L.; Erickson, S. A.; Sharma, J. M. Enhanced Expression of Cytokine Genes in Spleen Macrophages during Acute Infection with Infectious Bursal Disease Virus in Chickens. *Vet. Immunol. Immunop.* **1998,** *61*, 331–341.
24. Kosters, J.; Geissler, H. H. Serologische Untersuchungen uber die verbreitung der infektiosen Bursitis der Junghennen (Gumboro disease), tierarztl. *Umsch.* **1971,** *26*, 573–575.
25. Le Nouen, C. L.; Toquin, D.; Muller, H.; Raue, R.; Kean, K. M.; Langlois, P.; Cherbonnel, M.; Eterradossi, N. Different Domains of the RNA Polymerase of Infectious Bursal Disease Virus Contribute to Virulence. *PLoS One.* **2012,** *7*, e28064.
26. Li, Y.; Wang, C.; Cheng, X.; Wu, T.; Zhang, C. Synonymous Codon Usage of the VP2 Gene of Very Virulent Infectious Bursal Disease Virus Isolate Serial Passaged in Chicken Embryos. *Biosystems* **2011,** *104*, 42–47.
27. Lombardo, E.; Maraver, A.; Espinosa, I.; Fernandez-Arias, A.; Rodriguez, J. F. VP5, the Nonstructural Polypeptide of Infectious Bursal Disease Virus, Accumulates within the Host Plasma Membrane and Induces Cell Lysis. *Virology* **2000,** *277*, 345–357.
28. McFerran, J. B.; McNutly, M.; McKilliop, E. R.; Conner, T. J.; McCracken, R. M.; Collins, D. S.; Allan, G. M. Isolation and Serological Studies with Infectious Bursal Disease Virus from Fowl, Turkeys and Ducks: Demonstration of a Second Serotype. *Avian Pathol.* **1980,** *9*, 395–404.
29. Mohanty, G. C.; Pandey, A. P.; Rajya, B. S. Infectious Bursal Disease in Chickens. *Curr. Sci.* **1971,** *40*, 181–184.
30. Muller, H; Nitschke, N. The Two Segments of Infectious Bursal Disease Virus Genome are Circularized by a 90,000 Da protein. *Virology* **1987,** *159*, 174–177.

31. Nakamura, T.; Otaki, Y.; Lin, Z.; Nunoya, T.; Hoshi, S.; Kato, A. Direct Correlation Between the Titer of Infectious Bursal Disease Virus VP2-specific Antibody and Protection. *Avian Dis.* **1994,** *38*, 251–255.
32. Nunoya, T.; Otaki, Y.; Tajima, M.; Hiraga, M.; Saito, T. Occurrence of Acute Infectious Bursal Disease with High Mortality in Japan and Pathogenicity of Field Isolates in SPF Chickens. *Avian Dis.* **1992,** *36*, 597–609.
33. Okoye, J. O. A. Infectious Bursal Disease of Chicken. *Vet. Bull.* **1984,** *54*, 425–436.
34. Pan, J.; Vakharia, V. N.; Tao, Y. J. The Structure of a Birnavirus Polymerase Reveals a Distinct Active Site Topology. *Proc. Natl. Acad. Sci.* **2007,** *104*, 7385–7390.
35. Prabakaran, R. *Overview of Poultry Production in India vis-a-vis Global Scenario.* Proceeding of XXIX IPSACON, 5-7 Dec., 2012, Hyderabad, 2012, p. 3–20.
36. Proffitt, J. M.; Bastin, D. A.; Lehrbach, P. R. Sequence Analysis of Australian Infectious Bursal Disease Viruses. *Aust. Vet. J.* **1999,** *77*, 186–188.
37. Saif, Y. M.. *Infectious Bursal Disease Virus Types.* In: Proceedings of the 19th National Meeting of Poultry Health Condemnations, Ocean City, MD, 1984, p. 105–107.
38. Schnitzler, D.; Bernstein, F.; Muller, H.; Becht, H. J. The Genetic Basis for the Antigenicity of the VP2 Protein of the Infectious Bursal Disease Virus. *J. Gen. Virol.* **1993,** *74*, 1563–1571.
39. Sharma, J. M.; Kim, I. J.; Rautenschlein, S.; Yeh, H. Y. Infectious Bursal Disease Virus of Chickens: Pathogenesis and Immunosuppression. *Dev. Comp. Immunol.* **2000,** *24*, 223–235.
40. Somvanshi, R.; Mohanty, G. C.; Kataria, J. M.; Verma, K. C. Detection of Infectious Bursal Disease Virus Antigen and Antibodies by Counter Immunoelectrophoresis. *Ind. J. Poult. Sci.* **1985,** *20*, 66–67.
41. Swai, E. S.; Kessy, M. J.; Sanka, P. N.; Mtui, P. N. A Serological Survey for Infectious Bursal Disease Virus Antibodies in Free-Range Village Chickens in Northern Tanzamia. *J. Vet.Med.* **2011,** *82*,32–35.
42. Tadesse, B.; Jenbere, S. Sero-Prevalence of Infectious Bursal Disease in Backyard Chickens at Selected Woredas of Eastern Ethiopia. *J. Biol. Agric. and Healthcare.* **2014,** *4*, 71–75.
43. van den Berg, T. P. Acute Infectious Bursal Disease in Poultry: a Review. *Avian Pathol.* **2000,** *29*, 175–194.
44. van den Berg, T. P.; Gonze, M.; Meulemans, G. Acute infectious Bursal Disease in Poultry and Characterization of a Highly Virulent Strain. *Avian Pathol.* **1991,** *20*, 133–143.
45. Wang, M. Y.; Hu, H. L.; Suen, S. Y.; Chiu, F. Y.; Shien, J. H.; Lai, S. Y. Development of an Enzyme-Linked Immunosorbent Assay for Detecting Infectious Bursal Disease Virus (IBDV) Infection Based on the VP3 Structural Protein. *Vet. Microbiol.* **2008,** *131*, 229–236.
46. Winterfield, R. W.; Hitchner, S. B. Etiology of an Infectious Nephritis-Nephrosis Syndrome of Chickens. *Am. J. Vet. Res.* **1962,** *23*, 1273–1279.
47. Wu, C. C.; Lin, T.; Aydemir, A. Quantitative Competitive Polymerase Chain Reaction for Detection and Quantification of Infectious Bursal Disease Virus cDNA and RNA. *J. Virol. Method.* **1997,** *66*(1), 29–38.
48. Yamaguchi, T.; Ogawa, M.; Inoshima, Y.; Miyoshi, M.; Fukushi, H.; Hirai, K. Identification of Sequence Changes Responsible for the Attenuation of Highly Virulent Infectious Bursal Disease Virus. *Virology.* **1996,** *223*, 219–223.

49. Zegeye, S., Tsegaye, Y.; Abreha, H.; Awol, N., Sero-Prevalence of Infectious Bursal Disease in Backyard Chickens around Mekelle, Northern Ethiopia. *Afr. J. Biotechnol.* **2015,** *14*(5), 434–437.
50. Zeleke, A.; Gelaye, E.; Sori, T.; Ayelet, G.; Sirak, A.; Zekarias, B. Investigation on Infectious Bursal disease Outbreak in Debre Zeit, Ethiopia. *Int. J. Poult. Sci.* **2005,** *4*, 504–506.
51. Zhang, M. F.; Huang, G. M.; Sulan, Q. Early Stages of Infectious Bursal Disease Virus Infection in Chickens Detected by in situ Reverse Transcriptase-Polymerase Chain Reaction. *Avian Pathol.* **2002,** *31*, 593–597.

CHAPTER 21

Biofilm Formation in the Food Industry

A. SANGEETHA

Department of Veterinary Public Health and Epidemiology, Veterinary College and Research Institute, Orathanadu, Thanjavur-614625, Tamil Nadu, India
E-mail: sangeetha.a@tanuvas.ac.in

ABSTRACT

Biofilms are a group of sessile microorganisms that are irreversibly attached to the surface and embedded in polymeric substances. Biofilm formation on the substances occurs in four stages and involves bacterial attachment, microcolony formation, bacterial biofilm maturation, and dispersion. The formation of biofilm can be influenced by many factors such as surface characteristics (properties of contact surface, surface chemistry, and topography of surface), bacterial characteristics (flagella, pili, and fimbriae), environmental characteristics (nutrient availability and contact time), gene regulation, and quorum sensing. Biofilm formed provides physical, chemical, and mechanical protection against desiccation and cleaning by sanitizing agents. Biofilms formed on food industry can result in equipment damage, product contamination, energy losses, and food-borne infections. Biofilm formation can be prevented and inhibited by use of chemical, physical, biological, or combination of more than one methods.

21.1 BIOFILM

Biofilm is defined as a group of sessile community consisting of viable and nonviable microorganism of more than one species that are irreversibly attached to the surface and embedded in extracellular polymeric substances (EPSs).[1,2] EPSs may contain polysaccharides, proteins, phospholipids, teichoic acid, extracellular nucleic acid, and other polymeric substances which adhere to each other and adhere to the surface of an object.[3,4] Biofilms are found extensively on moist surfaces such as food processing equipment, water pipelines, industrial piping, ventilation, medical devices, pathological human tissues and organs, and so on. The types of surfaces in food plants ranges from plastic, glass, rubber, metal, wood, and food products.[6]

21.2 STEPS IN BIOFILM FORMATION

Biofilm formation includes four stages: bacterial attachment, microcolony formation, bacterial biofilm maturation, and dispersion[5] (Fig. 21.1). During biofilm formation, initial attachment may occur on pipeline surfaces, surfaces of fruit products, and food processing utensils. The attachment is influenced mostly by the surface's properties, such as roughness and hydrophobic interactions. The biofilm maturation and are mainly regulated by signals communicated by different bacteria in quorum sensing. Further, biofilm formed is more stable due to accumulation-associated protein (Aap), extracellular DNA, polysaccharide intercellular adhesion factors. Finally, biofilm dispersion occurs with the disruptive factors like proteases and nucleases.[5]

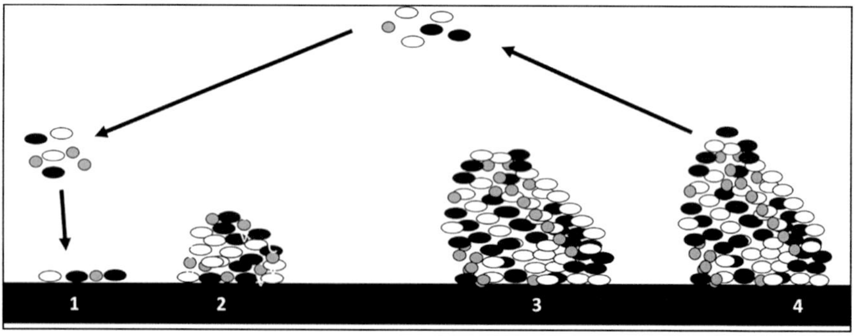

FIGURE 21.1 Four stages of biofilm formation involving (1) bacterial attachment, (2) microcolony formation, (3) biofilm maturation, and (4) biofilm dispersion.

1. Bacterial Attachment:

- Bacterial adherence to a contact surface can be active or passive depending on cell motility.[7] Active adhesion is driven by bacterial cell surface properties such as flagella, pili, adhesin protein, capsules, and surface charge whereas passive adhesion is driven by gravity, diffusion, and fluid dynamics.[7,8]

- Attachment to the surface often occurs within 5–30 s and occurs in two stages: reversible followed by irreversible adhesion.[9]
- Initial reversible attachment is an initial weak interaction of bacteria that involves van der Waals and electrostatic forces and hydrophobic interactions. Bacteria exhibit Brownian motion during this stage and can be easily removed by application of mild shear force.[7]
- The next crucial step is the irreversible attachment. Irreversible attachment occurs as a result of anchoring of appendages and/or the production of extracellular polymers. The change from reversible to irreversible attachment changes weak interaction of the bacteria with the surface to a permanent bonding with the presence of EPSs. Removal of irreversibly attached cells is difficult and requires application of strong shear force (scrubbing or scraping) or chemical breaking of the attachment forces through the application of enzymes, detergents, surfactants, sanitizers, and/or heat.[10,11]

2. *Microcolony Formation:*

Microcolony formation occurs after irreversible attachment during the presence of appropriate growth conditions. Microcolony formation results from concurrent aggregation and growth of microorganisms with the production of EPS. Quorum sensing occurs mainly in the microcolony formation stage.[5] Quorum sensing secretes the signal molecules for expression of corresponding genes needed for secretion of EPSs for formation of biofilm.[12]

3. *Biofilm Maturation*

In this step, it develops into an organized structure by utilizing the nutrients from the surrounding environment. In order to reach structural maturity, periods of 10 days or more are required. The composition of biofilms is highly heterogeneous due to the colonization of different microorganisms possessing different nutritional requirements. The mature biofilm may consist of a single layer of cells in porous extracellular polymer or multilayered loosely packed microcolonies held together with EPS and interspersed with water channels.[7] Cell lysis and released extracellular

DNA are critical for the initial biofilm attachment whereas the released extracellular DNA remains an important matrix component in biofilm maturation.[5]

4. *Dispersion*

The outermost part of biofilm is colonized by free unbound bacteria which are free to float to other areas and form a new biofilm. Biofilm bacteria can be detached by disruptive factors like catabolite repression, nutrient limitation, and secretary proteins.[5] The reasons for biofilm separation include external environmental effects such as increased shear stress and a lack of nutrient supply and internal biochemical changes in bacteria, such as endogenous enzyme degradation, EPS, or surface-binding protein releasing. The detachment from the biofilm plays a role in the spread of pathogens.[5,13]

21.3 FACTORS INFLUENCING FORMATION OF BIOFILM ON CONTACT SURFACES

- **Contact Surface Characteristics:**

  The character of surface like carrier interface, electrostatic, hydrophobic, interface roughness, and morphological characteristics influences the bacterial attachment and biofilm formation.[5]

  i. **Properties of contact surfaces:** Microbial attachment to food contact surfaces depends on availability of high surface energy and wettability of a surface. Surfaces with high free surface energy like stainless steel and glass are more hydrophilic and allow more bacterial attachment than hydrophobic surfaces like Teflon, nylon, tin, buna-N rubber, and fluorinated polymers. Contact surfaces also influences the attachment pattern of bacteria. Bacteria that attach to hydrophilic surfaces form a uniform monolayer whereas that attach to hydrophophobic tend to adhere in clumps.

  ii. **Surface chemistry and topography of contact surfaces**: Topography of contact surfaces noticed in food industry influences the biofilm formation. Surfaces with abrasions and defects are associated with increased biofilm formation. The surfaces like stainless steel, rubber, Teflon, and nylon are used for fabricating equipment and utensil, used for gaskets, various utensils, and equipment parts. These surfaces can get abrasions after repeated uses. Abraded surfaces entrap more bacteria and this protects the bacteria from cleaning and sanitation.[15]

iii. **Substratum preconditioning**: Clean surfaces submerged in solution are rapidly changed by the adsorption of organic molecules and charged ions. This process is called preconditioning. Adsorption of an organic layer onto a substratum can occur within seconds of exposure to an aqueous environment. Bacterial attachment occurs best on preconditioned surfaces in the presence of ions and organic molecules with high free energy. Initiation of bacterial attachment is dependent on the surface properties of the preconditioned substrate.[7]

- **Bacterial Characteristics:**

The initial attachment of bacteria involves casually sticking to the carrier surface by flagellum, pili, and fimbriae. Bacterial adherence to food contact surfaces is influenced by physicochemical properties of the cells surface which depends on microbial growth rate, growth medium, and culture conditions.[5]

i. In general, bacteria with high surface charge and hydrophobicity promotes increased attachment. Spores adhere better to food contact surfaces than vegetative cells due to the high hydrophobicity of their hairy surfaces.
ii. Growth media pH and nutrients influence the surface charge of bacteria. Glucose and lactic acid in the growth medium decreased the electro negativity of L. monocytogenes' cell wall through the neutralization of the surface charge and the production of acid stress proteins.
iii. High growth temperature is also associated with increased attachment ability due to the production of heat stress proteins associated with the cell surface.
iv. Attachment ability is controlled by surface proteins, flagella, structures that protrude from cell membranes as lipopolysaccharide (LPS), adhesins and other proteins, and lipoteichoic acids. Loss of these cell appendages changes surface properties leading to decreased attachment to food contact surfaces. Pili act like velcro to anchor bacteria to some surfaces and also act as chemoreceptors directing cells to move to specific attachment sites.

v. Adhesion of bacteria to substratum is influenced by ionic strength and alkali conditions. Adhesion is higher at higher ionic strength while it is low under alkali conditions. Electrostatic repulsion between cell and substratum can be overcome by suspending solution that provide conditioning layer at the attachment surface.

- **Environmental Characteristics:**

Environmental factors regulate bacterial biofilm formation through several pathways. Mature biofilms change with the environmental conditions in order to obtain the best nutrients to survive and reproduce. At either extreme, nutrient-rich, or very nutrient-poor conditions, biofilms will be inclined to be planktonic.[5,16]

i. **Nutrient availability**: Low nutrient availability elicits phenotypic changes of planktonic (free living) cells to the sessile (attached) form. In aquatic systems, nutrient availability is limited especially in drinking water distribution systems. Reduction of nutrients in raw water reduces the occurrence of biofilm formation in water distribution systems.
ii. **Contact time:** Contact time between the bacterial cell and the food contact surface is required for irreversible attachment.
iii. **Presence of organic contaminant on the surfaces**: Poor sanitation of food contact surfaces, equipment, and processing environments promote soil buildup and contribute to the development of bacterial biofilms.

- **Gene Regulation:**

Gene regulation plays a significant role in formation of biofilm on food industry surfaces. Most of the biofilm formation is mediated by production of polysaccharide intercellular adhesion. Some bacteria may form a protein biofilm regulated by fibronectin binding proteins (FnBPs) which forms PIA-dependent and PIA-independent biofilms.[5,17,18]

i. **PIA-independent biofilms**: The fibronectin binding protein mediated biofilm does not exist in fibrous, net-like structures and it is induced by ethanol. For formation of protein-mediated biofilm, upregulation of *Aap* encoding the accumulation-associated protein Aap and down regulation of Bap-homologous protein gene *bhp* is essential.[5]

ii. **PIA-dependent biofilms:** PIA/poly-N-acetylglucosamine (PNAG) with positive charge can easily with negative charge on the surface of the bacteria. PIA-mediated formation of biofilm is encoded by the ica operon and the icaADBC locus includes four genes (A, D, B, C). The induction of ica operon is affected by environmental conditions. The *IcaA* gene encodes the N-acetylglucosamyl transferase responsible for the synthesis of PIA and the co-expression of the *icaD* in vitro increases the activity. IcaB gene encodes deacetylase which is responsible for the maturation of the PIA, and the transmembrane protein of icaC appears to be involved in the polysaccharides involved in externalization and extension. Transcriptional level of ica which encodes the essential enzyme for PIA biosynthesis is reduced after treatment with thiols. Another gene, sarA can activate the development of biofilm by both enhancing the ica operon transcription and by suppressing the transcription of either the protein involved in the turnover of PIA/PNAG or a repressor of its synthesis. In Gram-negative bacteria, biofilm formation is regulated by Pel gene cluster encoded in Pel operon. The pel operon associated with quorum-sensing regulation regulates the chemical complementation of 3-oxo-dodecanoyl homoserine lactone to restore the biofilm formation ability. Transcription of the pel operon is greatly reduced in the absence of lasI and rhlI genes.[5,32,33]

- **Quorum Sensing**

  Quorum sensing plays an important role in biofilm formation. Quorum sensing is a cell-to-cell communication process that enables bacteria to secrete special signals in response to changes in the cell density of the surrounding microbial community. The quorum sensing can be turned "off" with a few bacteria in the environment. When the number of bacteria reaches a certain threshold, the activation of the corresponding genes are expressed in cells, and as more signal molecules release, it can lead to bacteria secreting toxins, biofilm formation, and bioluminescence with the quorum sensing "on."[5]

  ○ Quorum sensing regulates the whole stages of biofilm formation, which activates certain genes in the bacteria to secrete the extracellular matrix, such as EPS and protein, and gradually a full biofilm structure is formed, in which the bacteria have drug resistance. The normal operation of a quorum-sensing system requires the participation of signal molecules, and different types of bacteria secrete different signal molecules, such as AHL secreted by Gram-negative bacteria, AIP secreted by Gram-positive bacteria, and AI-2 secreted both by Gram-negative and Gram-positive bacteria.[5]

- The regulation of biofilm formation is complex and involves several regulatory mechanisms. AI-2 and LuxS act as a regulator of biofilm formation and LPS synthesis in sessile bacterial cells.[5]

21.4 ADVANTAGES OF BIOFILM FORMATION IN FOOD INDUSTRY

- Protection against physical resistance against desiccation by increased binding of water molecules that reduces the possibility of dehydration.
- Biofilm provides mechanical resistance against liquid streams in pipelines.
- It confers chemical protection against various chemicals, antibiotics, biocides, antimicrobials, and disinfectants used in the industry.
- It facilitates plasmid transfer for exchange of genetic material due to proximity of various bacteria in the biofilm. Plasmid transfer in biofilms may occur at a high rate due to the close proximity of cells. This is beneficial to the cells since the transferred DNA may carry useful capabilities that enhance survival.

21.5 SIGNIFICANCE OF BIOFILMS IN FOOD INDUSTRY

- Biofilms harbor and protect various pathogens and spoilage causing microorganisms thus compromising sanitation standards of food processing plants.
- Microbial biofilms on surfaces result in equipment damage, product contamination, energy losses, and medical infections.
- Food-borne diseases associated with bacterial biofilms arise via intoxications or infections.
- The presence of biofilm or attached cells on foods and food contact surfaces often adversely impacts food safety, especially in minimally processed foods and raw foods.

21.6 STRATEGIES FOR CONTROL AND PREVENTION OF BIOFILM FORMATION

1. **Prevention of biofilm formation**
   a. **Antibacterial compounds:** The best way to prevent biofilm formation is the inhibition of bacteria growth. Many different methods using bactericidal agents are available to sterilize food and processing equipment. Food additives have become one the most important type of antibacterial agents. Some food additives such as colorants, surfactants and flavorants like cinnamaldehyde possess strong antibacterial activity against foodborne pathogens. The addition and use of these food additives reduces the foodborne pathogens thereby decreasing the possibilities of biofilm formation.[5]

   b. **Inhibition of biofilm**
   - Measures to inhibit biofilm formation on contact surfaces:
     - Maintaining dry and clean environment.
     - Removing and preventing the accumulation of food residues and other compounds facilitating biofilm formation. Nutrient limitation is one way to control biofilm development without increasing the use of disinfectant.
     - Effective cleaning of equipment used in the industry. If cleaning in place (CIP) is followed, caustic and nitric acid can be added to enhance the removal of bacterial biofilm.[19]
     - Cleaning and disinfection are universal methods to disrupt the initial attachment of biofilms without sterilization. Sodium hypochlorite (NaClO), hydrogen peroxide (H_2O_2), and ozone (O_3) can be used as disinfectants.[5]
     - Essential oils can be used for inhibiting or slowing the growth of bacteria. Example: essential oil of *Satureja thymbra*, cassia, peru balsam, and red thyme.[5,20]

- Measures taken to reduce quorum sensing for inhibiting biofilm formation: Disturbing the production or transportation of quorum-sensing molecules that regulate biofilm using quorum-sensing inhibitors or quorum quenchers.[5,86]
- Measures aimed at environmental factors for inhibiting biofilm formation: Temperature, pH, and water hardness are important factors affecting

the effectiveness of disinfectants. Maintaining high temperature and low pH are helpful in removal of biofilm.[5]

- Measures aimed at food utensils: Biofilm formation on food utensils can be prevented by coating the surfaces with antimicrobial nanoparticles.[5]

2. Removal of biofilm

a. Physical treatment:

Physical treatments like heat shock treatment, shear stress, electric field, and ultrasound are widely used for removal of biofilm.[23–25] They can be effectively used for removal of initially attached bacterial cells. To remove the matured biofilm, it can be combined with anti-biofilm compounds.[5]

b. Chemical treatment

- Chemical reagents play an important role in the removal of biofilm and bacterial cells. Both physical treatment and chemical reagents have a synergistic effect to remove biofilm. For eliminating or removing biofilm, fungicide, to kill bacteria, must penetrate the EPSs and contact the microbial cells.[5]
- Some extrinsic substances that induce detachment of the matrix of the EPS-detachment-promoting agents can react with EPS with slow diffusion of the agent into the biofilm. Detachment promoting agents attack the EPSs to implement a control strategy for biofilm removal.[5,26]
- Enzymes are proteins with catalytic activity, which can facilitate chemical reaction with the extra cellular matrix of a biofilm.[5,27,28]

c. Biological treatment

Bacteriophages are viruses that can infect bacteria and it can be effectively used to control biofilm formation. The effect of the bacteriophages depends on the susceptibility of the biofilm cell to the bacteriophages and the availability of the receptor site. Bacteriophages are capable of targeting bacteria within biofilms directly instead of attacking EPSs.[5,29,30]

3. Prevention of dispersion

The last step in the biofilm developmental cycle produces new biofilm and causes human infection by free bacteria since dispersed bacteria may carry different phenotypic properties and transcriptional profiles compared to both biofilms and planktonic bacteria, leading to a significantly increased virulence. Thus, it is necessary to control the biofilm dispersion to prevent serious bacterial infection.

21.7 CONCLUSION

Biofilms are formed in food industry as part of survival strategies in adverse environmental conditions and constitute serious health and economic issues. It can harbor and protect various spoilage and pathogenic microorganisms resulting in foodborne infections, spoilage of food, equipment damage, product contamination, and energy losses. It is resistant to various types of antimicrobial agents used as disinfectants.

Control and prevention of biofilm depends on the use of cost effective chemical methods using sodium hydroxide or sodium hypochlorite solutions, physical methods using hot water steam ozone and mechanical methods. However, these methods are not effective in complete inactivation of biofilm on food industry structures. Combination technology is a promising new approach, which involves combination using two or more different control methods to control biofilm. Therefore, more approaches like use of enzymatic disruption (using detergents containing proteases, glysosidases, or DNase), steel surface modification (by coating with silver, copper, or zinc nanoparticles, or by using the novel antibiofilm polymers with lysozyme or bacteriocins), or biosurfactants such as lichenysin (added to industrial detergents) can be used to control biofilms. The biofilm control and prevention has been increasingly difficult due to emergence of multidrug-resistant microorganisms. Hence, development of novel strategies that are more efficient, cost effective eco-friendly methods using novel antibiofilm compounds are warranted.

KEYWORDS

- **biofilm**
- **steps in biofilm formation**
- **quorum sensing**
- **biofilm control**
- **chemical method**
- **physical method**
- **biological method**

REFERENCES

1. Carpentier, B.; Cerf, O. Biofilm and their Consequences with Particular Reference to Hygiene in the Food Industry. *J. Appl. Bacteriol.* **1993,** *75*, 499–511.
2. Wimpenny, J. W. T.; Kinniment, S. L.; Scourfield, M. A. The Physiology and Biochemistry of Biofilm. In *Microbial Biofilms: Formation and Control*; Denyer, S. P., Gorman, S. P., Sussman, M., Eds; Blackwell Scientific: London, 1993; p 51–94.
3. Costerton, J. W.; Irvin, R. T. The Bacterial Glycocalyx in Nature and Disease. *Ann. Rev. Microbiol.* **1981,** 83, 299–324.
4. Sutherland, I. W. Microbial Exopolysaccharides. Their Role in Microbial Adhesion in Aqueous Systems. *Crit. Rev. Microbiol.* **1983,** *10* (2).
5. Zhao, X.; Zhao, F.; Wang, J.; Zhong, N. Biofilm Formation and Control Strategies of Food Borne Pathogens: Food Safety Perspectives. *RSC Adv.* **2017,** *7*, 36670–36683.
6. Trachoo, N. Biofilms and the Food Industry. *Songklanakarin J. Sci. Technol.* **2003,** *25*, 807–815.
7. Chmielewski, R. A. N.; Frank, J. F. Biofilm Formation and Control in Food Processing Facilities. *Compr. Rev. Food Sci. Food Saf.* **2003,** *2*, 22–32.
8. Kumar, C. G.; Anand S. K. Significance of Microbial Biofilms in Food Industry: a Review. *Int. J. Food Microbiol.* **1998,** *42*, 9–27.
9. Mittelman, M. W.. Structure and Functional Characteristics of Bacterial Biofilms in Fluid Processing Operations. *J. Dairy Sci.* **1998,** *81*, 2760–2764.
10. Richards, R. Efficacy of Sanitizers on Listeria Monocytogenes Biofilms. *Food Austr.* **1999,** *51* (12), 624–625.
11. Sinde, E.; Carballo, J. Attachment of Salmonella spp. and Listeria Monocytogenes to Stainless Steel, Rubber and Polytetrafluorethylene: the Influence of Free Energy and the Effect of Commercial Sanitizers. *Food Microbiol.* **2000,** *17*, 439–447.
12. Hammer, B. K.; Bassler, B. L. Quorum Sensing Controls Biofilm Formation in Vibrio Cholera. *Mol. Microbiol.* **2003,** *50*, 101–114.
13. Sauer, K.; Cullen, M. C.; Rickard, A. H.; Zeef, L. A. H.; Davies, D. G.; Gilbert, P. *J. Bacteriol.* **2004,** *186*, 7312–7326.
14. Boulange-Peterman, L. Processes of Bioadhesion on Stainless Steel Surfaces and Cleanability: a Review with Special Reference to the Food Industry. *Biofouling* **1996,** *10* (4), 275–300.
15. Frank, J.; Chmielewski, R. Influence of Surface Finish on the Cleanability on Stainless Steel. *J. Food Prot.* **2001,** *68* (8), 1178–1182.
16. Janczarek, M. Environmental Signals and Regulatory Pathways That Influence Exopolysaccharide Production in Rhizobia. *Int. J. Mol. Sci.* **2011,** *12*, 7898–7933.
17. O'Neill, E.; Pozzi, C.; Houston, P.; Humphreys, H.; Robinson, D. A.; Loughman, A.; Foster; T. J.; O'Gara, J. P. *J. Bacteriol.* **2008,** *190*, 3835–3850.
18. Vergara-Irigaray, M.; Valle, J.; Merino, N.; Latasa, C.; García, B.; Ruiz, D. L. M. I.; Solano, C.; Toledo-Arana, A.; Penadés, J. R.; Lasa, I. *Infect. Immun.* **2009,** *77*, 3978–3991.
19. Bremer, P. J.; Fillery, S.; Mcquillan, A. J. *Int. J. Food Microbiol.* **2006,** *106*, 254–262.
20. Chorianopoulos, N. G.; Giaouris, E. D.; Skandamis, P. N.; Haroutounian, S. A.; Nychas, G. J. E. *J. Appl. Microbiol.* **2008,** *104*, 1586–1596.
21. Zhao, X. H.; Wang, L.; Chu, J.; Li, Y. M.; Li, Y. Y.; Xu, Z. B.; Li, L.; Shirtliff, M. E.; He, X. W.; Liu, Y.; Wang, J. H.; Yang, L. S. *Food Sci. Biotechnol.* **2010,** *19*, 1655–1659.

22. Bakkiyaraj, D.; Sivasankar, C.; Pandian, S. K. *Bioorg. Med. Chem. Lett.* **2012,** *22*, 3089–3094.
23. Farhat, M.; Trouilhé, M. C.; Briand, E.; Molettadenat, M.; Robine, E.; Frère, J. *J. Appl. Microbiol.* **2010,** *108*, 1073–1082.
24. Zhao, X. H.; Wang, J.; Forghani, F.; Park, J. H.; Park, M. S.; Seo, K. H.; Oh, D. H. *J. Microbiol. Biotechnol.* **2013,** *23*, 1708–1716.
25. Karosi, T.; Sziklai, I.; Csomor, P. Low-Frequency Ultrasound for Biofilm Disruption in Chronic Rhinosinusitis with Nasal Polyposis: in vitro Pilot Study. *Laryngoscope* **2013,** *123*, 17–23.
26. Xavier, J. B.; Picioreanu, C.; Rani, S. A.; van Loosdrecht, M. C.; Stewart, P. S. *Microbiology* **2005,** *151*, 3817–3832.
27. Zhao, X. H.; Wang, L.; Chu, J.; Li, Y. Y.; Li, Y. M.; Xu, Z. B.; Li, L.; Shirtliff, M. E.; He, X. W.; Liu, Y.; Wang, J. H.; Yang, L. S. *Food Sci. Biotechnol.* **2010,** *19*, 1191–1197.
28. Xu, Z. B.; Li, L.; Zhao, X. H.; Chu, J.; Li, B.; Shi, L.; Su, J. Y.; Shirtliff, M. E. *Afr. J. Biotechnol.* **2011,** *5*, 1869–1873.
29. Fu, W.; Forster, T.; Mayer, O.; Curtin, J. J.; Lehman, S. M.; Donlan, R. M.; *Antimicrob. Agents Chemother.* **2010,** *54*, 397–404.
30. Carson, L.; Gorman, S. P.; Gilmore, B. F.; Donelli, G.; Bayston, R.; Costerton, W. B.; Shirtliff, M. E. *FEMS Immunol. Med. Microbiol.* **2010,** *59*, 447–455.
31. Galié, S.; García-Gutiérrez, C.; Elisa Miguélez, M.; Villar, C. J.; Lombó, F. Biofilms in the Food Industry: Health Aspects and Control Methods. *Front. Microbiol*. **2018,** *9*, 898.
32. Valle, J.; Toledo-Arana, A.; Berasain, C.; Ghigo, J. M.; Amorena, B.; Penadées, J. R.; Lasa, I. SarA and not sigma B is Essential for Biofilm Development by Staphylococcus Aureus. *Mol. Microbiol.* **2003,** *48*, 1075–1087.
33. Vasseur, P.; Vallet-Gely, I.; Soscia, C.; Genin, S.; Filloux, A. The Pel Genes of the Pseudomonas Aeruginosa PAK Strain are Involved at Early and Late Stages of Biofilm Formation. *Microbiology* **2005,** *151*, 985–997.

Index

C

M

N

O

P